Non-Thermal Processing Technologies for the Grain Industry

Non-Thermal Processing Technologies for the Grain Industry

Edited by
M. Selvamuthukumaran

CRC Press
Taylor & Francis Group
Boca Raton London New York

CRC Press is an imprint of the
Taylor & Francis Group, an **informa** business

First edition published 2021
by CRC Press
6000 Broken Sound Parkway NW, Suite 300, Boca Raton, FL 33487-2742

and by CRC Press
2 Park Square, Milton Park, Abingdon, Oxon, OX14 4RN

Library of Congress Cataloging-in-Publication Data

Names: Selvamuthukumaran, M., editor.
Title: Non-thermal processing technologies for the grain industry / edited
  by M. Selvamuthukumaran.
Description: First edition. | Boca Raton : CRC Press, 2021. | Includes
  bibliographical references and index.
Identifiers: LCCN 2021012525 (print) | LCCN 2021012526 (ebook) | ISBN
  9780367608576 (hardback) | ISBN 9780367625160 (paperback) | ISBN
  9781003109501 (ebook)
Subjects: LCSH: Cereal products--Preservation. | Cereal products industry.
  | Grain trade.
Classification: LCC TP434 .N66 2021 (print) | LCC TP434 (ebook) | DDC
  664/.7--dc23
LC record available at https://lccn.loc.gov/2021012525
LC ebook record available at https://lccn.loc.gov/2021012526

ISBN: 978-0-367-60857-6 (hbk)
ISBN: 978-0-367-62516-0 (pbk)
ISBN: 978-1-003-10950-1 (ebk)

Typeset in Kepler Std
by Deanta Global Publishing Services, Chennai, India

Contents

Preface

Food will rapidly spoil due to the growth of microorganisms, which further renders it unsafe for human consumption. The traditional methods of food preservation, which involve drying, canning, salting, curing, and chemical preservation, can significantly affect the food quality by diminishing the nutrients during heat processing, alter the texture of the products, and leave chemical residues in the final processed products, which in turn has a great impact on consumers' safety and health. To combat this problem, nowadays various non-thermal food processing techniques can be employed in grain processing industries to enhance the consumer demand for delivering wholesome food products to the market. The available non-thermal processing techniques to preserve such food include pulsed electric field processing, high-pressure processing, ultrasonic processing, cold plasma processing, oscillating magnetic field processing, etc.

The book will address various recent advanced non-thermal food processing techniques employed during grain processing. It describes the effect of the application of electric pulses on quality aspects of various grain-based food beverages with respect to their specific acceptable characteristics. It can be a helpful resource for the cereal processing industry in regard to extracting antioxidants from natural sources by using high-pressure technology. It deals with the impact of several non-thermal processing techniques on sensory attributes of treated grain-based food products like color, taste, flavor, and overall acceptability. It recommends special packaging requirements for retaining aroma, flavor, and nutrients for non-thermally processed grain-based food products. It also explains the application of ozone for grain with its advantages, disadvantages, process operations, mechanisms for microbial inactivation, etc. In a nutshell, this book will overall benefit food scientists, food process engineers, academicians, students, and food industry professionals by providing in-depth knowledge about non-thermal processing of grain-based foods for quality retention and also for optimal consumer acceptability.

I would like to express my sincere thanks to all the contributors; without their continuous support, this book would not have seen daylight. We would like also to express our gratitude toward Mr. Steve Zollo and all other CRC Press people, who have made every continuous cooperative effort to make this book a great standard publication on a global level.

M. Selvamuthukumaran

About the Editor

M. Selvamuthukumaran is Associate Professor and Head of the Department of Food Technology at the Hindustan Institute of Technology and Science, Chennai, India. He was a visiting Professor at Haramaya University, School of Food Science and Postharvest Technology, Institute of Technology, Dire Dawa, Ethiopia. He earned his Ph.D. in Food Science from the Defence Food Research Laboratory affiliated with the University of Mysore, India. His core area of research is the processing of underutilized fruits for the development of antioxidant-rich functional food products. He has transferred several technologies to Indian firms as an outcome of his research work. He has received several awards and citations for his research work, published several international papers and book chapters in the area of antioxidants and functional foods, and guided several national and international postgraduate students in the area of food science and technology.

Contributors

Nese Basaran-Akgul
Department of Microbiology and Immunology
School of Medicine
Indiana University
Terre Haute, IN

Pervin Basaran
Department of Food Engineering
Istanbul Technical University
Istanbul, Turkey

Gökçen Baykuş
Department of Food Engineering
Izmir Institute of Technology
Gülbahçe, Turkey

Dilara Devecioglu
Department of Food Engineering
Istanbul Technical University
Istanbul, Turkey

Dilara Nur Dikmetas
Department of Food Engineering
Istanbul Technical University
Istanbul, Turkey

Esra Dogu-Baykut
Department of Gastronomy and Culinary Arts
Istanbul Medeniyet University
Istanbul, Turkey

Zehra Gulsunoglu-Konuskan
Department of Nutrition and Dietetics
Istanbul Aydin University
Istanbul, Turkey

Funda Karbancioglu-Guler
Department of Food Engineering
Istanbul Technical University
Istanbul, Turkey

Meral Kilic-Akyilmaz
Department of Food Engineering
Istanbul Technical University
Istanbul, Turkey

Celale Kirkin
Department of Food Engineering
Istanbul Technical University
Istanbul, Turkey

Kasiviswanathan Muthukumarappa
Department of Agricultural and
    Biosystems Engineering
South Dakota State University
Brookings, SD

Aysegul Mutlu-Ingok
Department of Food Processing
Akcakoca Vocational School
Duzce University
Duzce, Turkey

Vikas Nanda
Department of Food Engineering and
    Technology
Saint Longowal Institute of Engineering and
    Technology
Longowal, India

Kirty Pant
Department of Food Engineering and
    Technology
Saint Longowal Institute of Engineering and
    Technology
Longowal, India

M. Selvamuthukumaran
Department of Food Technology
Hindustan Institute of Technology and
    Science
Chennai, India

Yadigar Seyfi
Department of Food Engineering
Izmir Institute of Technology
Gülbahçe, Turkey

Gabriela John Swam
Department of Nutrition, Food Science, and
    Packaging
San Jose State University
San Jose, CA

Zeynep Tacer-Caba
Department of Gastronomy and Culinary Arts
Bahcesehir University
Istanbul, Turkey

Mamta Thakur
Department of Food Technology and
    Nutrition
School of Agriculture
Lovely Professional University
Jalandhar, India

Sultan Arslan-Tontul
Department of Food Engineering
Selcuk University
Konya, Turkey

Sevcan Unluturk
Department of Food Engineering
Izmir Institute of Technology
Gülbahçe, Turkey

Chapter 1

Recent Advances in Non-Thermal Food Processing Technology for Grain Industries

Yadigar Seyfi, Gökçen Baykuş and Sevcan Unluturk

CONTENTS

1.1 HIGH HYDROSTATIC PRESSURE PROCESSING (HHP)

Research into the effects of high pressures on food dates back to 1899. Bert Hite was the first to design and construct a high-pressure unit for the pasteurization of milk and other food products (Hite 1899). On the other hand, industrial-level usage of HHP devices dates back to the 1990s. The first industrial HHP equipment was installed in 1991 in Japan (Knorr et al. 2011). Some jams, salad dressings, sauces, fruit yogurts and juices are some of the food products processed with HHP. HHP-treated foods have been marketed in Europe and the USA since 1996 (Ortega-Rivas 2012).

High hydrostatic pressure processing (HHP), also known as ultrahigh-pressure processing or pascalization, is a novel non-thermal technology for the food industry. Simply, HHP is a process that is applied to prevent microbial multiplication and improve food quality. The applied pressure may range from 100 to 1000 MPa (whereas the normal atmospheric pressure is equal to 0.1 MPa) (Arora and Chauhan 2019). When heat is applied to foodstuffs, protein and starch are denatured and enzymes and microorganisms become inactivated. The same effect occurs when foods are exposed to extremely high pressures. However, unlike heat treatment, the food's sensory properties are not affected by HHP treatment and microorganisms are inactivated without causing significant changes to the nutritional properties of foods (Ortega-Rivas 2012).

1.1.1 Description of the Process

Hydrostatic pressure is generated in closed systems by heating or mechanical volume reduction. Basically, the HHP process relies on two principles: the isostatic rule and Le Chatelier's principle. According to the isostatic rule, pressure transmission is instantaneous and uniform throughout the food. Le Chatelier's principle states that when a system at equilibrium is disturbed, it returns to equilibrium by countering the disturbance (Hogan, Kelly and Sun 2005).

In an HHP process, the food product to be treated is placed in a pressure vessel capable of supplying the required pressure; the product is submerged in a liquid, which acts as the pressure-transmitting medium. Other than water, liquids containing castor oil, silicone oil, sodium benzoate, ethanol or glycol are also used as the pressure-transmitting medium. HHP equipment mainly consists of a pressure vessel, two end closures, a yoke (a structure used for restraining end closures while under high axial pressure), a pressure-creating device, instrumentation and controls (pressure, temperature, flow and level measurements) (Arora and Chauhan 2019; Naik et al. 2013).

This technology has been adapted from the chemical industry, where pressure is utilized to increase the reaction yield. It has been shown that pressures higher than 400 MPa damage intermolecular and intramolecular bonds (Ortega-Rivas 2012). HHP processing results in some changes to the food structure including protein and enzyme denaturation, starch modification, etc. HHP also results in the inactivation of undesired microorganisms (pathogenic and spoilage) in the food. This is an important consideration for any preservation method (Hogan, Kelly and Sun 2005). Although spores can withstand pressure up to 1,200 MPa, HHP has the ability to inhibit many vegetative cell types (Zhang et al. 2011).

1.1.2 Current Commercial Status of HHP Processing

Industrial HHP application is a batch or semi-continuous process. The choice of equipment depends on the type of food product to be processed. Solid food products or foods with large solid particles can only be processed in batch mode. Liquids, slurries and other pumpable products additionally have the option of semi-continuous production (Hogan, Kelly and Sun 2005).

In today's world, there are many countries that produce lab-scale, pilot-scale or commercial-scale HHP units. These are the USA, Japan, China, Spain, France, Sweden, Poland, Germany, the Netherlands and the UK (Arora and Chauhan 2019).

1.1.3 HHP Processing of Grains

1.1.3.1 Effects of HHP on Microbial Quality of Grains

Food preservation techniques have two main goals: the eradication of pathogenic microorganisms to ensure food safety and the inactivation of spoilage microorganisms to provide food stability and increase the shelf life of the foods (Hogan, Kelly and Sun 2005). It has been found that the effect of high pressure on microorganisms is similar to that of high temperature (Ortega-Rivas 2012).

A number of factors such as the type of microorganism, temperature, magnitude, duration of the HHP treatment and composition of the food affect the microbial inactivation provided by HHP. By considering these factors, appropriate pressure treatment should be applied to assure the inactivation of spoilage, pathogenic and vegetative cells of microorganisms present in foods (Arora and Chauhan 2019).

The effect of high pressure on microbial inactivation is primarily due to cell membrane permeability (Hogan, Kelly and Sun 2005; Ortega-Rivas 2012). A number of studies have shown that high pressure affects all types of vegetative cells rather than spores. For example, *Fusarium* is an important strain of phytopathogenic fungi causing economic losses. Several *Fusarium* species can infect small grains such as wheat, barley, oat and maize (Ferrigo, Raioala and Causin 2016). Fungal contamination can spread rapidly during storage if conditions are suitable. Mycotoxins produced by fungi and their contamination cause both product losses and health hazards (Schmidt, Zannini and Arendt 2018). Several strategies have been developed to prevent mycotoxin accumulation in cereals. For instance, Schmidt et al. (2018) investigated the effect of 13 different treatments on commercial hard winter wheat grain. They found that the application of HHP at 300 MPa for 10 min completely inhibited the growth of *F. culmorum* and prevented its development over the six weeks of storage. In addition, HHP application caused the ergosterol level to remain at the level of detection of 0.5 mg/kg.

1.1.3.2 Effects of HHP on Grain Quality, Components and the Germination and Quality of Sprouted Grains

HHP treatment has been applied to grains and grain products to improve their structural, physicochemical, rheological and textural properties, to enhance their chemical composition by increasing γ-Aminobutyric acid (GABA), arabinoxylan, mineral and vitamin content and to investigate their health benefits (*in vitro* digestibility). In most of the HHP processing studies, wheat varieties and wheat starch, rice varieties and rice starch, corn and corn starch, sorghum, beer and rice wine were used. The physicochemical qualities of these products have been extensively studied before and after HHP processing. These studies on grain quality and process parameters are briefly summarized in Table 1.1.

1.1.4 Regulations

HHP has been applied for decades in the food industry to products ranging from fruit juices to grains. HHP can be used to eliminate or reduce microorganisms, increase the shelf life of the food product, provide reconditioning and work out some other technical problems. There are several regulations issued by governments and authorities. In the United Kingdom (UK), the Pressure Systems Safety Regulations 2000 (PSSR) were published by Health and Safety Executive in 2000. This regulation contains guidelines for safe designs and use of pressure systems (HSE 2014). Furthermore, there is legislation about pressure equipment called the Pressure Equipment (Safety) Regulations 2016 (Google 2016).

In Canada, Novel Foods under Division 28 of the Food and Drug Regulations contain HHP-treated foods. In December 2016, Health Canada took out HHP from the novel food processing methods category due to enough knowledge and data availability (Google n.d.). Also, Safe Food for Canadians Regulations (SFCR) came into force in January 2019 to require food businesses to have evidence showing that the control measures they use are effective at controlling hazards.

1.1.5 Research Needs and Challenges

Knorr (2018) indicated that incomplete R&D and uncertainty about pressure- and temperature-resistant indicator microorganisms are the main challenges of the application of HHP processing in foods.

TABLE 1.1   EFFECTS OF HHP ON GRAIN QUALITY

Materials	Grain Quality	Process Parameters	References
Brown rice (*Oryza sativa* L.)	*In vitro* bioaccessibility of the mineral elements, amino acids, antioxidants and starch	0.1, 500 MPa – 10 min	Xia et al. 2017
	Textural properties (hardness, gumminess, springiness)	200, 600 MPa – 0, 20 min	Yu et al. 2015
	Functionality and quality of wholegrain germinated brown rice	50, 350 MPa – 20 min	Xia et al. 2018
	Chemical composition, microstructure, chemical reactions during the storage	0.1, 450 MPa –10 min	Xia and Li 2018
Rice grains and starch	Changes in bioactive compounds (GABA, arabinoxylan, ɣ-oryzanol)	0.1, 100 MPa – 24 h	Deng et al. 2015
	Structural and physicochemical properties of starch	200, 600 MPa – 30 min	Du et al. 2019
	Quality characteristics of parboiled rice (color, gelatinization temperature, etc.)	100, 600 MPa – 30 min	Xu et al. 2019
	Physicochemical properties and *in vitro* digestibility of waxy rice starch	10, 600 MPa – 20 min	Zeng et al. 2018
	Physicochemical properties and microbial load	300 MPa – 30 min	Lee et al. 2019
	Structural properties	0.1, 600 MPa – 1, 20 min	Wang et al. 2020b
	Structural properties (swelling behavior)	0.1, 600 MPa – 1, 20 min	Wang et al. 2020a
	Chemical composition of germinated rough rice (vitamin E and phenolic acids)	0.1, 100 MPa – 24 h	Kim et al. 2017
	Chemical composition (extractable phenols, total flavonoids and anthocyanins, antioxidant activity) and *in vitro* digestibility of cooked black rice	200, 500 MPa –15 min	Meng et al. 2019
	Enhancement of functional components	30 MPa – 24, 48 h	Kim et al. 2015a
Corn starch	Structural properties (supramolecular structure)	100, 600 MPa – 30 min	Yang et al. 2016
	Physicochemical and structural properties during retrogradation	300, 600 MPa –15 min	Li et al. 2016
	Retrogradation kinetics	400 MPa – 15 min	Kim et al. 2015b
	Structural properties	< 40 MPa – 35 min	Deladino et al. 2017

(*Continued*)

TABLE 1.1 (CONTINUED)   EFFECTS OF HHP ON GRAIN QUALITY

Materials	Grain Quality	Process Parameters	References
Wheat grains and starch	Structural, physicochemical properties, *in vitro* digestibility	300, 600 MPa –30 min	Hu et al. 2017
	Some wheat gluten gel properties (water-holding capacity, non-freezable water, storage modulus)	100, 400 MPa – 10 min	Wang et al. 2018
Buckwheat grains and starch	Physical, optical and thermal properties of buckwheat films	600 MPa –20 min	Kim et al. 2018
	Physicochemical and structural properties of common buckwheat starch and *in vitro* digestibility	120, 600 MPa – 20 min	Liu et al. 2016c
	Physicochemical, textural properties of tartary buckwheat starch and *in vitro* digestibility	120, 600 MPa – 20 min	Liu et al. 2016b
	Chemical composition, structural properties, protein digestibility	600 MPa –30 min	Deng et al. 2015
Maize starch	Microstructural properties	400 MPa – different times	Schneider Teixeira et al. 2018
	Supramolecular structure analysis	600 MPa – 5 min	Li et al. 2019a
Barley flour	Functional, rheological, thermal and structural properties of barley β-glucan concentrate dough	500 MPa – 15 min	Ahmed, Thomas and Arfat 2016
	Physicochemical properties	200, 600 MPa – 20 min	Ueno et al. 2019
Sorghum starch	Physicochemical properties and *in vitro* digestibility	120, 600 MPa –20 min	Liu et al. 2016a
Rice wine	Enological, sensory properties	200, 600 MPPa – 10, 20 min	Yang et al. 2019
Beer	Physicochemical, sensory properties, microbial load	Variable parameters	Santos et al. 2017

1.2 PULSED ELECTRIC FIELD (PEF)

The first application of PEF in food material was reported in 1961 by Heinz Doevenspeck (Misra et al. 2017). The lethal effect of PEF on microorganisms and the cell repair mechanisms of cells was first reported by Sales and Hamilton in 1967. In this work, the factors that affected the performance of PEF were determined to be the electric field treatment, treatment time, pulse width and pulse number (Knorr et al. 2011). On the other hand, short pulses without arcing were also employed to create secondary effects. Based on these effects, the "ELCRACK" and "ELSTERIL" processes were developed by Krupp Maschinentechnik GmbH, a company Germany (Sitzmann and Münch 1988). Flaumenbaum (1968) first reported the application of the electric field for the electro-plasmolysis of apple mash.

The pulsed electric field (PEF) is a new non-thermal preservation method that provides microbial inactivation and preserves food quality by applying a medium- to high-intensity external electric field with a short duration (microseconds) (Toepfl, Heinz and Knorr 2005; Martin-Belloso and Soliva-Fortuny 2011). It induces the formation of pores causing structural changes in the food matrix and a rapid breakdown of the cell membrane. This phenomenon is called electroporation (Toepfl, Heinz and Knorr 2005; Martin-Belloso and Soliva-Fortuny 2011). PEF has been developed to obtain high-quality, microbially safe and stable food products.

1.2.1 Description of the Process

PEF equipment consists of three major parts: a high-voltage pulse generator, a suitable fluid handling system, and the necessary monitoring and control devices (Singh, Kaur and Thangalakshmi 2019). PEF processing induces structural changes and a rapid breakdown of the cell membrane (electroporation) through the application of short, very high-voltage pulses (Ortega-Rivas 2012). Generally, an electric field strength between 10–80 kV/cm is required to inactivate pathogenic microorganisms, to change the structure of enzymes, and to enhance mass transport in some processes such as dehydration and drying (Singh, Kaur and Thangalakshmi 2019; Ortega-Rivas 2012). The microbial inactivation mechanism of PEF relies on the electroporation of cells, which means the formation of pores in the cell and organelles (Zimmermann 1974). The number and size of pores depend on the electric field strength and the processing time. When intensive PEF treatment is administered, both the number and size of pores increase, causing irreversible permeabilization or mechanical disruption of the cell and eventually inducing cell death (Ortega-Rivas 2012).

The effects of PEF treatment on microorganisms have not been fully explored; however, the electromechanical model (Zimmermann et al. 1974) and other statistical models (Abidor et al. 1979; Weaver and Chizmadzhev 1996; Joshi et al. 2002) have been proposed to explain the electroporation process. When a cell is exposed to an external electric field, the transmembrane potential increases, pore formation begins, the number and size of the pores increase and finally, if the electroporation is irreversible, the cells cannot repair themselves and the molecules start to leach out of the cell membrane (Toepfl, Heinz and Knorr 2005).

1.2.2 Current Commercial Status of PEF Processing

PEF systems can be used for the pasteurization of liquid foods like milk and dairy products, fruit juices and beverages, and can be employed in winemaking, sugar production and the microstructural modification of raw vegetables, fish and meat. Also, it can be utilized to increase fruit juice yield (Ortega-Rivas 2012).

Even though there are many studies in this area, only a few industrial-scale instruments are available on the market (Misra et al. 2017). The first commercial PEF system, built by Diversified Technologies, Inc. (DTI), was installed in the Department of Food Science and Technology and Ohio State University (USA). This system can operate with liquid foods at flow rates ranging from 500 to 2,000 L/h (Misra et al. 2017).

The first FDA-approved, PEF-processed juices on the US market were introduced by Genesis Juice Corporation in August 2005. DTI has developed a 25 kW pilot-scale PEF system with a capacity of 100–500 liters per hour to process juices or other products (Kempkes 2016). PEF systems with a capacity of up to 5,000 l/h are now available and used for the preservation of freshly squeezed juice in Europe (Irving 2012; Toepfl 2012). Five companies (DTI, DIL/ELEA, PurePulse, Scandinova and Energy Pulse Systems) are currently known to produce commercial PEF systems focusing on a variety of applications (Kempkes 2016).

1.2.3 PEF Processing of Grains

1.2.3.1 Effect of PEF on Microbial Quality of Grains

PEF technology can provide a 5 log reduction in the number of pathogenic microorganisms but has a limited effect on microbial spores (Morris, Brody and Wicker 2007). There are some microbial factors such as the type, concentration and growth stage of microorganisms as well as media factors such as the pH, antimicrobials, ionic compounds, medium ionic strength and electrical conductivity that affect the PEF processing of foodstuffs. Additionally, the process parameters (pulse wave and width, electrical field intensity, temperature, time) are also important (Morris, Brody and Wicker 2007).

Evrendilek et al. (2019) designed bench-scale PEF equipment for the surface disinfection of eight seeds (winter wheat, winter barley, tomato, cabbage, parsley, onion, lettuce and garden rocket seeds). In order to determine the survival ratio (as a percentage), germination rate and germination energy, the seeds were treated with PEF by applying 12 kV cm^{-1} with a total energy of 240 or 960 J. They obtained a significant reduction in total aerobic mesophilic bacteria and total mold and yeast counts for winter wheat and barley, parsley, onion, lettuce, tomato and garden rocket with the PEF treatments. A significant increase in the germination ratio for winter wheat and barley, lettuce and tomato was reported. In addition, better root development and seedlings were also found for winter barley. Thus, they concluded that PEF treatment could be a promising method for surface disinfection of seeds but requires a scale-up.

It is known that molds and their mycotoxins formed during post-harvest storage can cause serious health hazards and economic losses. The major contamination sources for storage silos are *Aspergillus flavus* and *Aspergillus parasiticus* and their toxins. Rwamurangwa and Mugisha (2017) treated maize crops by applying PEF. They concluded that PEF treatment is a promising method for inhibiting the growth of harmful microorganisms on crops and avoiding aflatoxin elevation.

1.2.3.2 Effects of PEF on Grain Quality and Components

There are a number of studies about the usage of PEF on the quality of cereals. For example, Hong et al. (2020) used PEF treatment to examine the rheological properties of wheat starch and the quality of noodles made from PEF-treated wheat starch. The effects of PEF treatment on the granular morphology, molecular weight, semi-crystalline structure, thermal properties and digestibility of PEF-treated waxy rice starch were investigated by Zeng et al. (2016). Similarly, Li et al. (2019b) studied the effect of PEF on the structural properties and digestibility of wheat starch. Likewise, Lohani and Muthukumaparran (2016) investigated the release of bound phenolics in PEF-treated sorghum flour and found increased levels of phenolic antioxidants. Qian et al. (2014) examined the inactivation of the enzyme lipase in brown rice subjected to PEF processing. Additionally, there are numerous studies investigating the effect of PEF on the α-amylase activity of barley (Zhang et al. 2019), the physicochemical properties of corn starch (Han et al. 2020), the phenolic compounds of brewer's spent grain (Martín-García et al. 2020), the structure of wheat starch (Hong et al. 2020) and the physicochemical properties and thermal and pasting properties of oat flour (Duque et al. 2020).

1.2.3.3 Effects of PEF on Germination and Quality of Sprouted Grains

Seed germination includes events that begin with water uptake by the dry seed and end with the extension of the embryonic axis. This process can take 24–36 hours, depending on the temperature, followed by root elongation and growth of the new pant. Dymek et al. (2012) studied the metabolic responses of germinating barley seeds upon the application of pulsed electric fields (PEF). Malting barley seeds were soaked in aerated water for 24 h and PEF-treated at voltages varying from 0 to 480 V with electric field intensity ranging from 0 to 1200 V/cm, respectively. They found a significant reduction of radicle elongation when 1,200 V/cm was applied. Furthermore, PEF-treated seeds exhibited less α-amylase activity than the controls. Thus, it was concluded that PEF treatment decreased α-amylase concentration in seeds.

In addition to this study, Ahmed et al. (2020) designed a system to explore the effect of PEF treatment on the germination of wheat seeds and the nutritional profile of wheat juice. They applied 25 and 50 pulses to wheat kernels with electric field intensities varying from 2 to 6 kV/cm. The results showed that PEF treatment of 50 pulses at 6kV/cm increased water uptake, the germination rate of seeds and the growth parameters of seedlings. Also, a significant increase in the total phenolic contents, DPPH (2,2-diphenyl-1-picrylhydrazyl), chlorophylls, carotenoids, soluble proteins, minerals and amino acids in PEF-treated seeds' plantlets juice was reported.

1.2.4 Regulations

PEF technology is still under investigation. In the European Union, there is no special legislation on food processed with PEF. The use of this technique is regulated by the Novel Food Regulation (EU) 2015/2283, but the implementation of PEF in production does not automatically mean the food becomes "novel" (Nowosad et al. 2021). Regulations are encouraged in the EU on the use of new processing technologies (Regulation [EU] 2015/2283). Canada, New Zealand, Australia, China and Brazil also have regulations concerning novel foods but the definition of "a novel food" may differ (Magnuson et al. 2013). The term "pasteurization" was redefined in September 2004 by the USDA National Advisory Committee on Microbiological Criteria for Foods (NACMCF). This allows new methods such as PEF to be used (NACMCF 2004) for pasteurization purposes. PEF has been used for the commercial pasteurization of juices in accordance with FDA's juice HACCP regulations (21 C.F.R. 120). The juice processors also have to implement sanitation and good manufacturing practices (GMP) during the production of juice and juice products. According to the FDA, a standard of 5 log reduction of the most resistant pathogen must be met by the processes of production (FDA 2000).

1.2.5 Research Needs and Challenges

Knorr (2018) studied the combination of PEF processing with high temperature to improve the reduction of the microbial load of the product, suggesting that this technology can be used alone or in combination with other methods to make products more energy-efficient and environmentally friendly. Also, the electrode design and materials used in PEF systems still need to be investigated. PEF can be applied for pasteurization and used for other purposes such as the enhancement of drying, freezing or extraction. In addition, PEF technology can be used to develop some functional foods that contain easily absorbed ions of elements that are required for the proper functioning of the human body.

1.3 OZONE TECHNOLOGY

Ozone was discovered by Schonbein in 1839 and was used for water treatment in the early 1990s (Zhu 2018). In the United States, ozone was classified as GRAS (generally recognized as safe) in

1997. In 2001, ozone was officially approved by the FDA for use in the food industry (Mielcke and Ried 2004; Vaz-Velho et al. 2006). Ozone has other commercial uses, such as disinfection of bottled water and swimming pools, prevention of contamination in cooling towers and treatment of wastewater (Rice 1996; Duguet 2004; Guzel-Seydim, Greene and Seydim 2004).

Microorganisms such as bacteria, molds, viruses, protozoa and insects cause quality losses in food products and result in economic and environmental losses for the food industry. Many chemicals like organic acids (acetic acid, propionic acid, lactic acid, sorbic acid, parapens, etc.), plant-derived antimicrobials (carvacrol, thymol, oleuropein, etc.), chitin/chitosan, antimicrobial enzymes (lysozyme, glucose oxidase, lactoperoxidase, etc.), nisin, lactoferrin, reuterin, ozone and electrolyzed water are commonly used in the food industry to eliminate insects, microorganisms and their metabolites like mycotoxins (Søltoft-Jensen and Hansen 2005). Both gaseous and aqueous forms of ozone are used as antimicrobial and disinfectant agents in the food industry (Zhu 2018). Most of the atmospheric ozone is concentrated in a layer called the stratosphere, about 15 to 30 km above the earth's surface. Ozone molecules containing three oxygen atoms are constantly formed and destroyed in the stratosphere (Yousef, Vurma and Rodriguez-Romo 2011). It has a bluish color in high concentrations, but is colorless at low concentrations and room temperature, and has a pungent odor detectable by humans at concentration levels as low as 0.02 mg/L (Yousef, Vurma and Rodriguez-Romo 2011; Zhu 2018). While ozone is partially soluble in water, it is much more soluble in inert non-polar solvents such as carbon tetrachloride or fluorocarbons. It is a powerful oxidant that can achieve disinfection with less contact time and concentration than all weak disinfectants such as chlorine or chlorine dioxide. In addition, ozone is more stable in gaseous form than in the aqueous phase. The solubility of ozone has an inverse relationship with the ambient temperature (Kumar and Sabikhi 2019; Yousef, Vurma and Rodriguez-Romo 2011).

1.3.1 Description of the Process

Ozone generation begins with the splitting of a diatomic oxygen molecule. The resulting free radical oxygen reacts with other diatomic oxygen to form the triatomic ozone molecule. Three methods are generally used to initiate free radical oxygen generation to form ozone, including electrical discharge (corona discharge), electrochemical (cold plasma) and radiation methods (Kumar and Sabikhi 2019; Zhu 2018).

Corona discharge is an electrical discharge that occurs between two electrodes operating at high voltage (about 3 kV and higher) producing ozone from oxygen in the air. In the electrochemical method, the ozone is produced at the anode in an electronegative anions solution. In the radiation method, when the oxygen is exposed to UV light at wavelengths between 140–190 nm, the oxygen breaks into oxygen atoms and these atoms merge with the other oxygen molecules in the ambient air to generate ozone molecules (Kumar and Sabikhi 2019). Electrical discharge systems are commonly used for large-scale ozone production whereas UV-based production systems are used for small quantities (Yousef, Vurma and Rodriguez-Romo 2011).

The germicidal effects of ozone on gram-positive and gram-negative bacteria as well as spores and vegetative cells have been investigated (Fetner and Ingols 1956; Foegeding 1985; Ishizaki, Shinriki and Matsuyama 1986; Restaino, Frampton, Hemphill and Palnikar 1995). An ozone molecule is a strong oxidant. The oxidizing power of ozone decomposes the sulfhydryl groups and amino acids of enzymes, peptides and proteins and the unsaturated fatty acids of the cell membrane, thereby causing the death of the cell (Gonçalves 2009; Kumar and Sabikhi 2019). Ozone decomposes into hydroxyl radicals in the medium where it is applied. Therefore, it is known to have direct or indirect oxidation effects. Direct oxidation reactions of ozone are typical first-order, high-redox-potential reactions. In the indirect oxidation reactions of ozone, the ozone molecule decomposes to form free radicals (OR), which react rapidly to oxidize organic and inorganic compounds. Ozone can also act by fixing the entire molecule to double-bonded atoms, producing two simple molecules with different properties and molecular properties (ozonolysis) (Gonçalves 2009).

1.3.2 Current Commercial Status of Ozone Processing

The use of ozone is becoming an attractive disinfection method to replace chlorine, steam or hot water types of conventional sanitation techniques. This is because it is more effective against resistant viruses and spores compared to chlorine and other disinfectants. Since the use of ozone does not require high temperatures, it offers energy savings. In addition, ozone is produced on-site, thus saving on transportation and storage costs.

As a result, ozone is gaining importance in the food processing industry as a safe, cost-effective and chemical-free method. The patented Oxygreen® process is a new process developed for commercial use in cereal grains (Violleau, Pernot and Surel 2012).

1.3.3 Ozone Processing of Grains

1.3.3.1 Effects of Ozone on Microbial Quality of Grains

Ozone technology is mainly used for microbial inactivation and insect extermination. Some particular fungi such as *Fusarium* sp. and their toxins such as DON (deoxynivalenol), ZEN (zearalenone) and OTA (ochratoxin A) cause quality and economic loss in cereal grains. Sunisha and Pandiarajan (2020) used ozone in different concentrations (500, 1,000 and 1,500 ppm) to determine its effectiveness as a fumigant in stored paddy. They reported a high mortality rate of *Rhyzopertha Dominica* adults in paddy grains stored in a household by an increased ozone concentration. They concluded that ozone could be used as a fumigant but higher concentration levels were required to obtain more reliable mortality rates. Additionally, milling studies of ozone-treated grains indicated that the milling qualities improved upon ozone treatment.

Wang et al. (2016) evaluated the effectiveness of ozonation for the removal of DON in wheat under different conditions such as moisture content, ozone concentration, exposure time and raw material form. They showed that DON reduction significantly improved with increased ozone concentration and exposure time at a high moisture content. Therefore, ozonation was offered to be an effective and rapid method to reduce DON in wheat, especially in whole wheat flour.

Similarly, Piemontese et al. (2018) investigated DON reduction and microbial inactivation in gaseous ozone-treated ($55g\ O_3\ h^{-1}$ for 6 h) wheat grains. They reported that ozonation had a significant ($p < 0.05$) effect on the reduction of the total count (CFU/g) of bacteria, fungi and yeasts in wheat grains. Furthermore, the effect of ozone treatment on microbial inactivation and the removal of mycotoxins in wheat bran, buckwheat flour, rice, ground corn and whole maize flour was thoroughly investigated by Liu et al. (2020), Savi et al. (2020), Krstovic et al. (2020) and Santos Alexandre et al. (2019), respectively.

1.3.3.2 Effects of Ozone on Grain Quality and Components

Ozone has been increasingly applied in the field of grain processing. Studies have examined the effect of ozone application on grain quality in two different parts: the degradation of pesticide residue and the quality of the grain itself. Pesticides are used in the grain during storage to prevent losses due to insect damage. Savi et al. (2015) investigated the removal of residues of organophosphate and pyrethroid pesticides in wheat grains by ozone gas (60 μmol/mol up to 180 min). They concluded that ozone treatment could be used for the degradation of pesticide residues. Savi et al. (2016) also applied ozone to wheat grains to investigate the degradation of residual bifenthrin and pirimiphos-methyl insecticides. They used ozone at a concentration of 60 μmol/mol and obtained a significant reduction in bifenthrin and pirimiphos-methyl residues after 180 and 30 min exposure time, respectively.

Piechowiak, Józefczyk and Balawejder (2018) investigated the effect of the ozonation process on enzymatic activity (proteolytic, amylolytic, lipolytic and lipoxygenase activity) in wheat flours. Process parameters were selected as 5, 15 and 30 ppm ozone concentration and 5, 15 and 30 min exposure times. They observed that the total activity of amylases, proteases and lipases decreased with the increase in the ozone concentration, whereas the activity of lipoxygenase increased significantly. Yadav et al. (2020) used ozone treatment to analyze the physiological response, yield and quality of different wheat cultivars. Gozé et al. (2016) investigated the effect of ozonization on the starch modification of different wheat grain cultivars. Additionally, they evaluated specific effects of gaseous ozone on molecular properties of proteins of different wheat grain cultivars and their consequences on the bread-making quality of the resulting flours (Gozé et al. 2017). Likewise, Yi et al. (2020) examined maize yield in ozone-treated maize. Finally, Obadi et al. (2018) explored the fatty acid concentration of wholegrain flour subjected to ozone gas.

1.3.3.3 Effects of Ozone on Germination and Quality of Sprouted Grains

The germination rate of seeds to increase the grain yield is improved by the adoption of novel technologies. However, the seeds are contaminated from many sources and must be sterilized before the germination stage (Bari et al. 2011). Sterilization of seed is of great importance for the prevention of seed-borne microbial, fungal and viral diseases. Chemical treatments widely used for the sterilization of seeds can cause a loss of physiological quality. Hence, the non-chemical treatments for seed sterilization have gained great interest lately (Pandiselvam et al. 2019). Ozone has been found to increase the seed germination rate, but it has been reported that excess ozone causes some unwanted effects. Various effects of ozone treatment on plant growth and seed germination were investigated by Abeli et al. (2016), Landesmann et al. (2013), Mohammada et al. (2019), Sudhakar et al. (2011) and Vazquez-Ybarra et al. (2015). All of these studies suggested that ozone treatment could be an appropriate option for treating seeds to enhance germination. It has been demonstrated that the activity of ozone could be increased in order to achieve the desired effect by controlling internal and external factors including pH, flow rate, ozone concentration, temperature and the presence of organic matter. However, there were some cases where the ozone treatment affected the chemical components of seeds due to the modifications of starch, lipids and proteins contained in organic matter. The seed germination and viability were also affected due to excessive ozone exposure of the grains.

1.3.4 Regulations

Ozone is a powerful oxidant. Care should be taken if it is used in the form of gas. Toxicity symptoms, such as sharp irritation in the nose and throat, can be seen immediately at a 0.1 ppm dose. Exposure to 0.1 to 0.5 ppm of ozone for 3–6 hours may result in vision loss. Higher levels of ozone (5–10 ppm) can cause an increase in pulse and edema of the lungs. An ozone level of 50 ppm or more is potentially fatal (Kumar and Sabikhi 2019).

The current permissible ozone exposure level time-weighted average (PEL-TWA) in the workplace environment is 0.1 ppm (0.2 mg/m^3) in the United States. The National Institute for Occupational Safety and Hazards (NIOSH) also recommended this level and the US Occupational Safety and Health Administration (OSHA) adopted it. According to OSHA, the "short-term exposure limit (STEL) is 0.3 ppm (0.6 mg/m3) for an exposure of less than 15 minutes and four times per day". The concentration of ozone in the air which is immediately dangerous to life or health (IDLH) is 5 ppm. The FDA approved ozone application for water treatment. Additionally, there is no labeling requirement for ozone-treated products (Yousef, Vurma and Rodriguez-Romo 2011).

1.3.5 Research Needs and Challenges

Most of the studies in the field of ozone treatment showed that more research on ozone is needed to determine its positive effects on the microbial inactivation of grains. Ozone application provides promising solutions for problems such as mycotoxin and pesticide residues but these problems have not actually been fully resolved yet. *In vivo* and *in vitro* toxicological tests may reveal the effects of ozone on human health. Developing safe procedures, improvements and advances in ozone generation and application frameworks will make the procedure even more applicable in the future. Additionally, for complete sterilization of seeds, it is necessary to examine the dissolved ozone concentration and the soaking time of the ozone microbubble water. Although some studies show that ozone does not change the quality profile, it is important to study the effect of ozone on the composition of seeds. Technological approaches combining ozone with other physical methods that can be used to increase the sterilization efficiency of ozone on the seed surface are among future research topics (Pandiselvam et al. 2020).

1.4 RADIATION

Radiation is a form of energy that travels through space in invisible waves (Ortega-Rivas 2012). The electromagnetic spectrum has two major divisions including non-ionizing radiation and ionizing radiation (Ortega-Rivas 2012). Ionizing radiation causes the formation of ions because it has enough energy to remove tightly bound electrons from atoms. Far unltraviolet (UV) light, X-rays and gamma rays are considered to be forms of ionizing radiation. On the other hand, nonionizing radiation does not have enough energy to remove electrons; it just moves atoms in a molecule or causes them to vibrate (Ortega-Rivas 2012). Near UV light, visible light, infrared, microwave and radio waves are regarded as nonionizing radiation.

This section reviews the basic principles and applications of ionizing and nonionizing radiation technologies (especially pulsed UV light) in the grain industry.

1.4.1 Ionizing Radiation (IR)

Ionizing radiation, also known as "irradiation", can destroy harmful bacteria, viruses or insects that may be present in food (Ahn and Lee 2006). The first attempt to irradiate foodstuffs dates back to 1905. J. Appleby and A. J. Banks (UK) got the first patent on irradiation in foodstuff (Shayanfar and Pillai 2019). The first use of X-rays for the elimination of *trichinosis* in raw pork was achieved by B. Schwartz (US) in 1921 (Shayanfar and Pillai 2019). Irradiation of cereals and flours began to be used in the 1950s (Los, Ziuzina and Bourke 2018). In 1958, the FDA approved food irradiation as a food additive (not a physical process) and the USSR (Union of Soviet Socialist

Republics) approved the use of food irradiation on potato and grain in 1958/1959 (Pillai and Shayanfar 2019). In the 1960s, the US FDA approved the use of 0.5 kGy radiation for the prevention of pest contamination in wheat and flour. Since then, the technology has been applied for the preservation and decontamination of various crops (Los, Ziuzina and Bourke 2018).

Compared with other food processing techniques, food irradiation has been studied relatively well. It is rapid, efficient, safe and environmentally friendly, and thus minimizes postharvest losses and increases the shelf life of products (Morris, Brody and Wicker 2007; Schmidt, Zannini and Arendt 2018, 2019). The food irradiation process uses three types of ionizing radiation sources, such as electron beam generators, X-ray accelerators, Cobalt-60 and Cesium-137 gamma sources. Cobalt-60 and Cesium-137 emit ionizing radiation in the form of intense gamma rays. An X-ray accelerator uses an electron beam to target electrons on a metal plate. Some of the energy is absorbed and the rest is converted into X-rays. Electron beam generators produce e-beams accelerated to 99% of the speed light and generate energies of up to 10 MeV. X-ray accelerators, Cobalt-60 and Cesium-137 gamma sources have good penetration ability, while electron-beam generators have low penetrability. However, the impacts of electromagnetic ionizing radiations and electrons are the same in food irradiation (Farkas 2011). Electron beams are the most cost-efficient form of irradiation, but they can only penetrate food to a limited depth. X-rays, on the other hand, are expensive but have good penetration ability that is suitable for bulk operations. Gamma rays are relatively inexpensive and highly penetrative, making them a cost-effective alternative for food irradiation (Ortega-Rivas, 2012). A gamma ray is the form of electromagnetic radiation with the shortest wavelength and highest energy.

1.4.1.1 Description of the Process
Gamma rays arise from the disintegration of radioactive atomic nuclei and the decay of certain subatomic particles. The energetic electromagnetic radiation strips electrons from food molecules, converting them to electrically charged particles or ions. Gamma rays are incapable of inducing radioactivity (Morris et al. 2007).

Irradiation equipment using gamma rays consists of a high-energy isotope source which is shielded within a pool of water below the process area. Thus, the radiation source is kept inside a shielded concrete radiation chamber. During use, the radiation source is raised into an active position. The packaged food is loaded onto conveyors and transported through the radiation field in a circular path (Ortega-Rivas 2012). The product transport mechanism is used to bring the product inside the irradiation chamber, pass it around the source in a predetermined pattern and remove it from the chamber once the irradiation is complete (Kunstadt n.d.)

The primary effect of food irradiation is produced by energetic electrons. When ionizing radiation is passing through an item, it may result in ionization (removal of electrons), dissociation (loss of hydrogen atoms) and excitation (raising the energy of molecule to a higher energy level) types of chemical reactions in irradiated foods (Unluturk 2016). The main product of the ionization reaction is reactive free radicals produced as a result of the primary effect. The secondary effects will occur when the free radicals undergo reactions such as recombination, electron capture or dimerization (Unluturk 2016; Alper 1977). Thus, microbial inactivation by irradiation occurs directly or indirectly. Irradiation directly affects the deoxyribonucleic acid (DNA) of microorganisms causing base damage, DNA strand breakage and cross-linking (Los, Ziuzina and Bourke 2018; Morris, Brody and Wicker 2007). The presence of oxygen and water also have a profound effect on the radiolytic process. Ionizing radiation acts indirectly through the radiolysis of water, thus creating reactive chemical species that can damage nucleic acids, proteins and lipids (Jan et al. 2017; Los, Ziuzina and Bourke 2018; Morris, Brody and Wicker 2007). All of these direct and indirect effects result in cell death.

1.4.1.2 Current Commercial Status of Ionizing Radiation Processing

The use of irradiation has been approved for different types of foods in over 55 countries world-wide (Farkas 2011). The list of irradiated products is limited to spices, herbs, seasonings, some fresh or dried fruits and vegetables, seafood, meat and meat products, poultry and egg products (Unluturk 2016). Furuta and Todorki (2009) published a study on the current state of irradiation in the world. According to this study, in the United States, ionizing radiation is mostly used in the disinfection of grains, fruits, spices, dry vegetables, meat and seafood, that is, irradiating approximately 116,400 tons of food materials in total. In Asia and the Oceania region, 4,582 tons of grains and fruits were irradiated in 2005. In other regions, especially Ukrainians are reported to be irradiating close to 50,000 tons of wheat and 28,000 tonnes of barley in their Odessa facility each year. In South Africa, a large number of food items including mangoes, strawberries, chicken wieners and some vegetables are irradiated and sold. Japan irradiates about 20,000 tons of potatoes each year to prevent sprouting (Kunsdatd n.d.). The quantity of irradiated grain constituted 20% of total irradiated foods in the world in 2005.

1.4.1.3 Effects of Irradiation on Microbial Quality of Grains

Irradiation was used for microbial inactivation or inhibition of microbial toxins. El-Dine et al. (2015) treated wheat (*Triticum aestivum* L.), maize (*Zea mays*) and polished rice (*Oryza sativa* L.) with gamma irradiation at radiation doses of 4 kGy and 8 kGy to inhibit AFB1 (aflatoxin B1) produced mainly by the *Aspergillus* species. They found that the percentages of reduction of aflatoxin B1 at 4 KGy are 15.54%, 22.25% and 27.46% for maize, wheat and rice respectively, whereas the 8 KGy radiation dose removed about 60.26% of the toxin in maize, 64.68% in rice and 69.29% in wheat samples. It was indicated that a higher dose is needed to remove the toxin until it reaches the legal limit (5 ppb) and this technique can be used for toxin reduction without affecting the nutritional value of grains. Similarly, Markov et al. (2015) used gamma radiation to reduce the aflatoxin (B_1) level in maize samples. They treated maize samples with radiation doses ranging from 1 to 10 kGy. It was concluded that the 5 and 10 kGy gamma radiations decreased the AFB1 level significantly, reaching a maximum of 94.5%.

1.4.1.4 Effects of Irradiation on Grain Quality and Components

The irradiation process can be used for several purposes. For instance, Bashir et al. (2017) used gamma radiation to examine the starch granules and physicochemical properties of brown rice. Also, El-Karamany et al. (2015) examined the effect of gamma radiation on the rheological and baking properties of wheat grains. All other studies about the effects of irradiation on the quality of grains and their components are summarized in Table 1.2.

1.4.1.5 Regulations

There are excessive international legislative limitations on irradiation. The US Food and Drug Administration (FDA) is the agency responsible for the regulation of radiation sources in the United States. Because, as defined by the FDA, radiation is a "food additive" as opposed to a food process it is therefore evaluated according to food additive regulations (Morris, Brody and Wicker 2007). Every food approved for irradiation has specific guidelines in terms of minimum and maximum dosage set as safe by the FDA. Packaging materials containing food processed

TABLE 1.2   THE EFFECTS OF IRRADIATION ON GRAIN QUALITY

Material	Method	Method Parameters	Application Purpose	References
Brown rice	γ-rays	5–20 kGy	Determine physicochemical, antioxidant and thermal properties	Kumar et al. 2017
Barley, wheat, oat	UV	0.48 J/cm^2	The influence of UV irradiation on grains originating from genotypes with different sensitivity to oxidative stress	Kurdziel et al. 2018
Wheat flour	γ-rays	0.5–10kGy	Physicochemical, thermal and functional properties	Bashir et al. 2017
Corn starch	γ-rays	Up to 15kGy	Physicochemical properties	Teixeira et al. 2018
Buckwheat, oat starches	γ-rays	5–20kGy	Physicochemical, rheological and thermal properties	Dar et al. 2018
Oat grains	γ-rays	5–20kGy	Physicochemical, structural, thermal and antioxidant properties	Mukhtar et al. 2017
Wheat flour	γ-rays	0–3.5kGy	Physicochemical properties	El-Karamany 2015
Oat	γ-rays	0–10kGy	The antioxidant and antiproliferative potential of β-glucan	Masoodi et al. 2015

by irradiation should also be subject to approval. The United States Department of Agriculture (USDA) amends these rules for use with meat, poultry and fresh fruit.

The WHO (World Health Organization), the FAO (Food and Agriculture Organization of the United Nations) and the IAEA (International Atomic Energy Agency) formed a Joint Expert Committee on Food Irradiation (IJEFCI), which reviewed the data generated between the years 1970 to 1982 and concluded that the maximum recommended dose for foods must be 15 kGy with the average dose not exceeding 10 kGy (Ortega-Rivas 2012). This dose level does not show toxicological hazard, nutritional loss or microbial problems. In addition to the approval of using irradiation on fruits, the regulation also included the type and sources of irradiation; gamma rays and electron beam sources cannot exceed 10 MeV and X-ray sources cannot exceed 5 MeV. Also, the internationally recognized logo called the radura and statements like "treated with irradiation" or "treated by irradiation" must be displayed on the labels of the food items. Currently, over 100 items in more than 40 countries allow the use of irradiation (Jan et al. 2017).

1.4.2 Nonionizing Radiation

Short-wave ultraviolet light (UV-C) and pulsed light (PL) are nonionizing radiation technologies that use continuous UV-C light at 254 nm or short pulses of intense white light in the range of 100 to 1,100 nm. They are capable of decreasing the microbial populations of both vegetative cells and spores (Maftei, Ramos-Villarroel, Ramos-Villarroel, Martín-Belloso and Soliva-Fortuny 2014). The bactericidal effect of continuous UV-C light was first discovered in 1928. However, the use of PL for the inactivation of microorganisms took place in Japan in the late 1970s, with a reported patent by Hiramoto (1984). Later, extensive work was carried out to sterilize pharmaceutical products. Its use has been approved by the FDA (1996) for application "in the production, processing and handling of food" (Unni and Chauhan 2019; Palmieri and Cacae 2005). Since then, pulsed light has been used as a sterilizer and sanitizer in the food industry (Palmieri and Cacae 2005).

There are few studies regarding the sterilization of grain surfaces using UV-C irradiation (Murata, Ishiguro and Miyazato 1994; Hidaka and Kubota 2006; Koutchma 2009). In this section, only PL treatment and its impact on grains will be discussed because PL has gained increasing attention due to the short treatment time required to achieve the desired microbial inactivation with it (Ferrario and Guerrero 2016). Additionally, it has higher penetration depth and emission power compared to continuous UV light (Krishnamurthy, Demirci and Irudayaraj 2007).

PL treatment with a short time and high-power pulses (pulse rate is 1–20 pulses per second and pulse width is 300 ns–1 ms) of a spectrum of UV light was applied to foods (Unni and Chauhan 2019; Los, Ziuzina and Bourke 2018; Demirci and Krishnamurthy 2011). Pulsed UV light has been employed for the decontamination of air, the inactivation of vegetative cells and spores in a variety of food products, the surface treatment of bakery products, the disinfection of drinking water and liquid foods, etc. (Ortega-Rivas 2012; Ferrario, Alzamora and Guerrero 2015).

1.4.2.1 Description of the Process

A typical pulsed UV light system consists of a power supplier, capacitor, switches to obtain pulses, UV treatment chamber, UV lamp, sample holder and cooling duct (Unni and Chauhan 2019). The operational principles of pulsed light first involve the conversion of electrical power (in the form of AC) to high-power DC, which is then used to charge a capacitor. The capacitor stores electrical energy for the flash lamp. After the capacitor has been charged to a certain voltage, a high-voltage switch releases the charge in the capacitor into a flash lamp filled with an inert gas such as xenon or krypton. The lamp converts the DC pulses to light pulses of the same duration and frequency as those of the received DC pulses (Demirci and Krishnamurthy 2011).

Pulsed light inactivates microorganisms by a combination of photothermal and photochemical effects (Jan et al. 2017). The photochemical effect of UV light is due to the absorption of the light energy by highly conjugated carbon double bonds in proteins and nucleic acids, which disrupt the DNA and RNA structures of microorganisms (Jan et al. 2017; Demirci and Krishnamurthy 2011). Pulsed light contains a wide spectrum of white light from UV to infrared wavelengths (100–1,100 nm), in which 45% lies in the visible range, 30% in the infrared range and 25% in the UV range. The UV component of pulsed light has a photochemical effect, but the effect is mostly photothermal as most of the energy is in the visual spectrum. The surface of the food is exposed to a large amount of energy and the vegetative cells are destroyed due to the increase in the surface temperature (Ortega-Rivas 2012). Jan et al. (2017) stated that if the fluence exceeds $0.5 \, J/cm^2$, the inactivation occurs due to the heating effect.

1.4.2.2 Research Needs and Challenges

There are a considerable number of works in the field of pulsed light microbial disinfection. The use of pulsed light for microbial inactivation in food preservation tends to gain high acceptance. Pulsed light provides a high energy dose in a shorter time in situations requiring rapid disinfection. Lamp and sample heating, the penetration depth of light in the materials and the excess ozone generated by the UV component of the pulsed light are the major challenges in the use of pulsed light (Kundwal, Lani and Tamuri 2015). It has been stated that pulsed light (PL) has not been taken up on a large scale by the food industry despite being approved by FDA in 1996 (Rowan 2019). Guidelines are required for making PL experimental data and exposure conditions compatible.

1.5 NON-THERMAL PLASMA (COLD PLASMA) PROCESSING

In recent years, consumers who have adopted a healthy lifestyle have been demanding convenient and healthy foods with a safe and long shelf life. Therefore, innovative food processing methods that do not affect the nutritional, functional and sensory properties of foods, e.g. non-thermal plasma, are needed to improve microbial food safety and quality. Non-thermal plasma processing (NTP), also known as cold plasma, is a green processing technology for food manufacturing and was recently recognized as one of the emerging techniques for disinfection and modification of food properties (Arora and Chauhan 2019).

The basic concepts of plasma physics and plasma engineering were first introduced in the 17th century. More on the history of plasma research can be found in the book by Roth (1995). The term "plasma" was first introduced by Irving Langmuir in 1928. An increase in the energy input of matter causes a phase change. A phase change is a physical process in which a substance goes from one phase to another, e.g. from solid to liquid to gas; on the other hand, a further increase in energy beyond a certain level in a gaseous state delivers an excited stage of molecules, called plasma. Plasma has been described as the fourth state of matter which refers to a partially or completely ionized gas containing positive and negative ions, free electrons and neutral molecules. Since 1970, plasma technology has been extensively used for etching processes necessary for the fabrication of microelectronic circuits (Manos and Flamm 1989). With the development of technology, plasma started to be applied in medicine, water treatment and food preservation.

This section briefly describes the process principles of plasma technology and presents the latest developments and applications for it in the grain industry.

1.5.1 Description of the Process

The state of matter changes when energy is supplied to it; solids become liquid and liquids become gases. If more energy is supplied to a gas, it becomes partially or fully ionized gas and goes into the energy-rich plasma state, the fourth state of matter.

Plasma is obtained by exciting gaseous molecules and atoms using different sources of energy. Generally, plasma is electrically energized by means of different kinds of electrical discharges. The electrical discharges create an electric field between two electrodes and the gas that is found between these electrodes is excited to generate plasma. In addition, thermal and magnetic fields and radio or microwave frequencies can be used to increase the kinetic energy of the electrons. Non-thermal (cold) plasma is generated by means of different electrical discharges. Various kinds of discharges that supply energy to the gas environment may be used for this purpose. The most widely used discharges are corona, dielectric barrier, radiofrequency plasma, microwave, gliding arc discharges and plasma jet (Misra et al 2017; Mir, Manzoor and Mir 2016). Plasma is obtained from various types of gases. While simple gases such as air or nitrogen may be used, the mixture of noble gases like helium, argon or neon can find an application area in generating plasma.

Two kinds of plasma are known and the classification is based on their generation condition. These are thermal (equilibrium) and non-thermal (quasi-equilibrium and/or nonequilibrium) plasma. If the heat is applied to achieve ionized gas, the plasma is called thermal plasma. The temperature in order to achieve this equilibrium is very high (about 20,000 K) and obtained under high-pressure conditions. In this type of plasma, all chemical species, electrons and ions exist in thermal equilibrium (Misra et al. 2016). Nonequilibrium (cold) plasma is attained by using electrical discharges in gases. The electrical discharges are associated with some chemical processes within the plasma such as ionization and recombination processes. Cold plasma is produced at atmospheric pressure or vacuum pressure and requires less energy in its generation compared to thermal plasma. In non-thermal plasma, electrical energy is transferred to the plasma components instead of being used for heating the entire gas, resulting in energetic electrons and gas species (Hati et al. 2018).

1.5.2 Current Commercial Status of NTP Processing

Commercial developments in atmospheric plasma technology have been mostly focused on specific needs in the food industry, such as the processing of bottle caps to aid in labeling with food-grade inks (without organic solvents) or the modification of packaging polymers (Keener and Misra 2016). Cold plasma has drawn some attention from researchers and food processors because it offers some advantages over HHP, PEF, UV and ozone treatments. The lab-scale experiments showed that it can be a suitable method for microbial inactivation in fresh produce (Misra et al. 2014a, 2014b; Schnabel et al. 2014) and the surface disinfection of food packaging materials (Misra et al. 2016; Mir, Manzoor and Mir 2016). In addition, treatment of cheese and milk with atmospheric pressure plasma, DBD plasma and atmospheric corona discharge offered promising results (Song et al. 2009; Segat et al. 2015). Plasma has great potential for use in the fruit and vegetable processing industry (Mir et al. 2016). Spice products have also been successfully treated with plasma. Hertwig et al. (2015) investigated the efficacy of plasma treatment on selected herbs and spice. While experimental data are promising for the use of non-thermal plasma technology in food processing industries, further studies are needed to fully understand the safety-related issues before this technology can be commercialized.

1.5.3 NTP Processing of Grains

1.5.3.1 Effects of NTP on Microbial Quality of Grains

Microorganisms can affect the safety, quality and properties of the grains. Some molds producing harmful mycotoxins such as *Aspergillus* spp. pose serious health risks for humans (Dasan et al. 2017; Los et al. 2018). In addition, molds and mycotoxins cause great losses in cereal grains during storage. Other than the use of pesticides, some conventional methods such as drying, mechanical debranning, chlorine and hypochlorite, irradiation, ozone, etc. have been applied to control the microbial spoilage of cereals. However, these methods resulted in unsatisfactory efficacy and losses of the nutritional quality and technological properties of cereals and generated harmful environmental impacts (Los et al. 2018). Due to these limitations on current methods used for microbial inactivation in cereals, it appears that there is a great demand for new technologies. Cold atmospheric pressure plasma technology can be a non-thermal alternative to these methods. It can also be applied to inactivate bacterial spores.

Cold plasma consists of various reactive species that play a role in microbial inactivation such as UV photons, free radicals, free electrons, neutral or excited atoms and molecules (Scholtz et al. 2015). These reactive species can inactivate microorganisms both individually and synergistically. The inactivation mechanisms and kinetics of non-thermal plasma on microorganisms have been reviewed by Liao et al. (2017) and Pignata et al. (2017). The underlying mechanisms for plasma-mediated inactivation by cold plasma are still not fully understood but there are some biological and physical scenarios. The biological scenario includes DNA damage, lipid peroxidation, protein modulation and inducing apoptosis. The physical one includes electrostatic disruption and electroporation contributing to the inactivation of microorganisms (Liao et al. 2017).

Applications of cold plasma in cereal grains have been reviewed by Los et al. (2018). Scholtz et al. (2019) and Randeniya and Groot (2015) reviewed the effects of NTP on various grains. These reviews demonstrated that the inactivation efficacy of cold plasma depends on the grain property, treatment parameters and plasma generation system. Especially, the moisture content of the grain and the microbiological load (inoculation level) are found to be important. A higher moisture content increases the gas humidity and this causes the quenching of the plasma species and results in weaker plasma. In addition, a larger microbiological load leads to a shielding effect and decreases plasma efficiency. Also, the uneven shape of the grains is another challenge in grain disinfection. The geometry and structures of the grain surfaces are suitable to shelter microorganisms. Therefore, the microorganisms harbored on the surface due to the irregular shape of the grains is a concern for the grain industry (Butscher et al. 2016; Dasan et al. 2017). Treatment parameters are also important for the microbiological disinfection of grains. In fact, it is impossible to compare the exact treatment parameters due to the variety of devices used in plasma technology. However, greater inactivation is often observed at higher voltages and longer treatment times (Charoux 2020). In the grain industry, not only the inactivation of vegetative microorganisms but also the inactivation of spores is a difficulty and spore inactivation requires a longer treatment time.

1.5.3.2 Effects of NTP on Grain Quality and Components

In the cereal industry, cold plasma can not only be used for decontamination but also to improve the nutritional value and growth of seeds. For example, brown rice that has a great nutritional value is less desirable due to the different eating sensation. Plasma processing was used to improve the cooking properties of brown rice (Mir, Manzoor and Mir 2016). Textural properties of brown rice were determined by applying cold plasma at low pressure before cooking (Sarangapani et al. 2015; Chen et al. 2014). It was concluded that plasma treatment caused an increase in the surface etching of brown rice, allowing rapid water uptake and reducing the cooking time. In addition, the textural properties varied significantly.

The application of cold plasma can change some of the technological properties of grains due to the reactive species environment. These changes may be desirable and improve the grain quality. On the other hand, some unwanted alterations can be also observed, so the disinfection of grains without quality loss is a challenge in cold plasma treatment.

Nowadays, the functionality of products is gaining importance for the consumer. Therefore, the loss of bioactive substances during the processing of grains is a major concern too. Studies have shown that cold plasma treatment has no significant effect on phenolic compounds and antioxidant activity. For example, it was found that the number of phenolic components in oat, black seed and wheat seed did not change significantly after cold plasma treatment (Sera et al. 2010; Charoux et al. 2020; Tolouie et al. 2018). However, Saberi et al. (2018) observed an increase in the carotenoid and anthocyanin content of wheat samples after cold plasma treatment. They concluded that plasma treatment conditions and the food matrix could affect the results. Additionally, protein content is another parameter in evaluating grain quality. He et al. (2016) revealed that the effect of cold plasma treatment on protein content was different for different varieties.

1.5.3.3 Effects of NTP on Germination and Quality of Sprouted Grains

Treatment of grains with NTP significantly affects germination and the initial growth of plants. Grains undergo physical distortion after cold plasma treatment. The plasma induces etching of the seed surface and improves the activation of the seed. Plasma-derived reactive oxygen species erode the seed coat, increasing the boll permeability for water, oxygen and nutrition. Therefore, the dormancy is broken more easily and activation of the seed becomes faster. Consequently, the growth rate of the plasma-treated seeds is faster than that of untreated seed samples. Recent studies showed that cold plasma treatment could improve the seed germination in arabidopsis (*Arabidopsis thaliana*), wheat (*Triticum* spp.), tomato seed (*Lycopersicon esculentum*), radish seed (*Raphanus sativus*), oat (*Avena sativa*) and hemp seed (*Cannabis sativa* L.) (Cui et al. 2019; Jiang et al. 2014; Kucerova et al. 2019; Magurenau et al. 2018; Matra 2016; Sera et al. 2010). However, the seed germination and growth rate were strongly dependent on the seed type and cultivars (Sera et al. 2017). In contrast, some studies reported that NTP treatment might cause an adverse effect in some plant species during seed germination (Scholtz et al. 2019).

1.5.4 Regulations

Each country has its own regulations for the approval of new technologies. In the United States, new technologies have to receive approval from three federal agencies: the Environmental Protection Agency (EPA), the Food and Drug Administration (FDA), and the United States Department of Agriculture (USDA).

The FDA considers cold plasma technology a "food additive" like other new technologies. FDA defines a "food additive" as any substance that affects the characteristics of any food including any substance intended for use in producing, packaging, manufacturing, any source of radiation, etc. Atmospheric cold plasma technology used for the treatment of food and food packages can be also regulated under the US EPA as a pesticide. Additionally, any use of atmospheric cold plasma in meat, poultry or eggs must get approval from the US Department of Agriculture's Food Safety Inspection Service (USDA-FSIS) (Bourke et al. 2018).

In Europe, a novel food is approved only if it does not present a risk to public health, is not nutritionally disadvantageous when compared to similar food and is not misleading to consumers. In the first step, a scientific assessment of the new technology is required prior to authorization to ensure its safety. This is performed by the European Food Safety Authority (EFSA). The EFSA collects revealed scientific information to develop an expert report about the advantages and disadvantages of the novel food's applications. Then, a novel application can receive authorization. Currently, uncertainty remains in the EC regulatory approval process for cold plasma technologies due to the lack of definition within the evaluation criteria (e.g. "risk to public health", "nutritionally disadvantageous" and "not misleading to the consumer") (Bourke et al. 2018).

1.5.5 Research Needs and Challenges

Regulatory approval, designing the plasma source, and process control are the three major challenges for the use of atmospheric plasma in food manufacturing (Keener and Misra 2016). Additionally, the use of cold plasma in the grain industry has some limitations because the efficiency of this method depends on the specific properties of the grains and their surface. More studies are required to evaluate the effects of cold plasma processing on grain quality. Cytotoxicity and genotoxicity studies are needed to confirm the lack of toxicity of the chemical species generated during NTP treatments. Finally, more research can be done on scaling up so that the applicability of plasma processing in food plants may be possible in the future.

1.6 ULTRASOUND

Ultrasound is sound waves with frequencies of 20 kHz or more that exceed the hearing limit of the human ear. The history of ultrasound traces back to the discovery of the piezoelectric properties of quartz by the Curie brothers, Jacques and Pierre in 1880. Then, Barnaby and Thornyeroft conducted a study about cavitation in 1895. Ultrasound is actually one of the topics of acoustics. Although the previous studies were about high acoustic frequencies in the 19th century, the history of the ultrasound began in 1915 with Paul Langevin. During World War I, Langevin designed an ultrasound transducer to detect submarines. Following that, Wood and Lomis (1927) conducted a study about using high-intensity ultrasonic waves to form emulsions. After that, with the rapid improvement in technology, new transducer materials were produced and made the manufacturing of commercial ultrasonic systems possible. Since 1970, ultrasound technology has gone further and now is considered an emerging technology covering a wide range of applications in the industrial, environmental and medical areas.

Ultrasound is described as a novel processing technique and used in various areas of food processing. Ultrasound waves are categorized as power ultrasound (16–100 kHz), high-frequency ultrasound (0.1–1 MHz) and diagnostic ultrasound (1–10 MHz). Low-intensity or high-frequency (100 kHz at < 1 W/cm^2) ultrasound is used as a non-destructive technique to assess the properties of foods, such as composition, structure and physical state, whereas high-intensity or low-frequency ultrasound (20–100 kHz, 10–1000 W/cm^2) is employed in various food processing operations such as homogenization, extraction, degassing, de-foaming, emulsification, crystallization and cleaning, as well as the inactivation of microorganisms in foods (Piyasena, Mohareb and McKellar 2003).

1.6.1 Description of the Process

In the food industry, high-power (or high-intensity) ultrasound is applied due to its capability of physical, chemical and mechanical disruption. Ultrasound is generated by a system that consists of three basic parts: an electrical power generator, a transducer and a coupler. Transducers convert electrical or mechanical energy to sound energy. There are three types of transducer that are mainly used in generating ultrasound: liquid-driven, piezoelectric and magnetostrictive. The coupler is responsible for emitting the ultrasound waves from the transducer into the medium (Bhargava et al. 2021). When high-power ultrasound propagates in a liquid, alternating compression and expansion regions are created in the medium. These regions of pressure change cause cavitation and the formation of bubbles. These microbubbles first expand during the expansion cycle and then collapse violently in the succeeding compression cycles, creating shock waves. These shock waves are regions of very high temperatures up to 5,000 K and pressures of up to 50 MPa with high shearing effects and a localized sterilization effect (Piyasena et al. 2003). Microbiological inactivation and etching of the surfaces of the food are mainly caused by cavitation. Also, bubble formation ends up releasing free radicals and has a role in microbial cell death. The cavitation threshold of a medium is determined by a number of factors including dissolved gas, hydrostatic pressure, the specific heat of the liquid and the gas in the bubble, the tensile strength of the liquid and temperature (Rahman 1999).

1.6.2 Current Commercial Status of Ultrasound Processing

Recently, Bhargava et al. (2021) reviewed the various applications of ultrasound in different unit operations and specific food divisions (meat, fruits and vegetables, cereals, dairy, etc.) along with the advantages and drawbacks of the technology.

Ultrasonic probes and ultrasonic water baths are the two most common types of equipment used in the food industry. Ultrasonic water baths are widely used for cleaning and sanitation in food and beverage processing. They are inexpensive and easy to use. Also, different capacities of this equipment are available ranging from 10 to 2,500 L (Sao Jose et al. 2014). The ultrasonic probe or horn system is more powerful than the ultrasonic water bath since it introduces vibrations directly into the sample. These types of systems are used for the sonication of small volumes of samples (Chemat et al. 2011).

Industrial ultrasonic equipment is essential for industrial-level ultrasonic liquid processing. A typical large flow chamber, developed by Hielscher (n.d.), provides 16 kW for flow rates > 10m³/h. Special sono-reactors (from 30 to 1,000 L) have been developed by REUS (n.d.) to perform industrial trials or to scale up laboratory experiments. Although a number of attempts have been made to manufacture ultrasound equipment for use in food processing, more effort is required to scale up ultrasound equipment for industrial use.

1.6.3 Ultrasound Processing of Grains

1.6.3.1 Effects of Ultrasound on Microbial Quality of Grains

Post-harvest preservation of grains is becoming an important issue due to limited agricultural areas and rapid growth in the global population. Therefore, a lot of research is being done to reduce post-harvest grain losses. Microbial spoilage due to filamentous fungi and bacteria is one of the main reasons for losses of grains. Ultrasound is one of the methods that can be used in the microbial post-harvest decontamination of grains. Generally, high-intensity ultrasound combined with other techniques such as sanitizing chemicals or heat is used for microbial decontamination (Seymour et al. 2002). However, there are a few studies investigating the individual application of ultrasound in the grain industry (Butz and Tauscher 2002; Scoten and Beuchat 2002; Chemat et al. 2011). Although the results of these studies contradict each other, they concluded that ultrasound alone is not a sufficient method for successful decontamination. In addition, this technology has been found to be difficult to apply to cereal grains because the treatment must be carried out in a liquid medium. It is also important to evenly distribute the cavitation of the liquid around the grains for better results. Therefore, the usage of ultrasound for grain disinfection causes a further challenge for industrial application (Schmidt et al. 2018).

1.6.3.2 Effects of Ultrasound on Germination and Quality of Sprouted Grains

Breaking seed dormancy to start germination, high germination percentage and germination speed are important factors for the early production of valuable grains. In recent years, ultrasonic waves have been used to break dormancy and to improve the germination characteristics of grains. In ultrasound application, water is used as the liquid medium in which ultrasonic waves lead to cavitation, causing a mechanical pressure on the grains. This pressure results in cell wall fluidity, creating micropores and micro-cracks. Consequently, grains become more permeable to water and oxygen entry, which are essential for germination (Nazari and Eteghadipor 2017).

Goussous et al. (2010) performed an experiment on chickpea (*Cicer arietinum*) and wheat (*Triticumaestivum*) to investigate the effect of sonication on germination percentages. As a result, the germination percentages of chickpea and wheat increased by 36% and 2% respectively when compared to control samples. Similarly, sonication treatment improved the germination of wheat (*Triticumaestivum*) and lentil seeds by 4% and 6% respectively (Aladjanjiyan 2003). In another study, it was found that treatment with ultrasound increased the germination speed of peas by 93% (Chiu and Sung 2013). Likewise, Goussous et al. (2010) reported that the germination speed of chickpea and wheat increased up to 20% and 36%, respectively. In conclusion, laboratory-scale observation has shown that early germination of grains is possible by ultrasound treatment.

1.6.4 Regulations

Rasco (2017) reviewed the laws and regulations for novel food processing technologies. According to Rasco, the Food and Drug Administration (FDA) regulates about 80% of the food in the United States. The FDA determines the regulatory status of substances added to food, including any food ingredient or other substances that can become a component of the food from a food contact surface or package. Companies are legally responsible for the safety of food processes that could create unintentional food additives. Therefore, any new technology that might have an impact on product safety would have to be considered. During ultrasound processing, there is localized heating, so that the leaching of materials from packaging might be a factor to consider. In ultrasound processing operation, a Hazard Analysis and Critical Control Point (HACCP) system is used to identify and detect the risks of the process. In ultrasound treatment, parameters such as the temperature and physical properties of the product, the frequency and power of the ultrasound and the nature of the probe are key points that need to be evaluated as critical control points (Chemat and Hoarau 2004).

1.6.5 Research Needs and Challenges

Ultrasound is an emerging technology that may have the potential to improve grain germination and to reduce post-harvest losses due to microbiological issues. Currently, it is evaluated as environmentally friendly and green technology. However, the use of ultrasound alone in the food industry for microbial inactivation is currently unfeasible. Additionally, ultrasound, when applied at high intensities, generates heat, which has an adverse effect on the organoleptic properties and nutritional quality of grains. In addition, the requirement of liquid medium usage in grain decontamination might be a challenge for the grain industry, whereas the efficacy of ultrasound treatment is enhanced when integrated with other techniques for the preservation of food. The microbial inactivation efficiency of ultrasound and its inactivation mechanisms in combination with other processing technologies should be addressed. There is still room for future research in order to manufacture equipment on an industrial scale.

BIBLIOGRAPHY

21C.F.R.120. "Title 21 – Food and Drugs Chapter I – Food and Drug Administration Department of Health and Human Services Subchapter B – Food for Human Consumption (Continued)." https://www.accessdata.fda.gov/scripts/cdrh/cfdocs/cfcfr/CFRSearch.cfm?fr=179.21.

Abeli, T., D. B. Guasconi, A. Mondoni, D. Dondi, A. Bentivoglio, A. Buttafava, P. Cristofanelli, et al. "Acute and Chronic Ozone Exposure Temporarily Affects Seed Germination in Alpine Plants." *Plant Biosystems – An International Journal Dealing with all Aspects of Plant Biology* 151(2) (2017): 304–15.

Abidor, I. G., V. B. Arakelyan, L. V. Chernomordik, Yu A. Chizmadzhev, V. F. Pastushenko, and M. P. Tarasevich. "Electric Breakdown of Bilayer Lipid Membranes: I. The Main Experimental Facts and Their Qualitative Discussion." *Journal of Electroanalytical Chemistry and Interfacial Electrochemistry* 104 (1979): 37–52.

Ahmed, Jasim, Linu Thomas, and Yasir Ali Arfat. "Effects of High Hydrostatic Pressure on Functional, Thermal, Rheological and Structural Properties of B-D-Glucan Concentrate Dough." *LWT* 70 (2016): 63–70.

Ahmed, Zahoor, Muhammad Faisal Manzoor, Nazir Ahmad, Xin-An Zeng, Zia ud Din, Ume Roobab, Abdul Qayum, et al. "Impact of Pulsed Electric Field Treatments on the Growth Parameters of Wheat Seeds and Nutritional Properties of Their Wheat Plantlets Juice." *Food Science and Nutrition* 8(5) (2020): 2490–500.

Ahn, Doug U., and E. J. Lee. "Mechanisms and Prevention of Quality Changes in Meat by Irradiation." In: *Food Irradiation Research and Technology*, edited by C. H. Sommers and X. Fan, 127–42. Ames, IA: Blackwell Publishing, 2006.

Aladjadjiyan, Anna, and Teodora Ylieva. "Influence of Stationary Magnetic Field on the Early Stages of the Development of Tobacco Seeds (*Nicotiana tabacum* L.)." *Journal of Central European Agriculture (jcea @agr.hr)* 4(2) (2003): 132.

Alper, Tikvah. "The Role of Membrane Damage in Radiation-Induced Cell Death." In: *Membrane Toxicity*, edited by Morton W. Miller and Adil E. Shamoo, 139–65. Boston, MA: Springer US, 1977.

Aron Maftei, N., A. Y. Ramos-Villarroel, A. I. Nicolau, O. Martín-Belloso, and R. Soliva-Fortuny. "Pulsed Light Inactivation of Naturally Occurring Moulds on Wheat Grain." [In eng]. *Journal of the Science of Food and Agriculture* 94(4) (March 2014): 721–26.

Arora, S. K., and O. P. Chauhan. "High-Pressure Processing Principles and Engineering Aspects." In: *Non-Thermal Processing of Foods*, edited by O. P. Chauhan. Boca Raton, FL: CRC Press/Taylor & Francis, 2019.

Bari, Latiful, Katsuyoshi Enomoto, Daisuke Nei, and Shinichi Kawamoto. "Development of Effective Seed Decontamination Technology to Inactivate Pathogens on Mung Bean Seeds and Its Practical Application in Japan." *Japan Agricultural Research Quarterly: JARQ* 45(2) (2011): 153–61.

Bashir, Khalid, Tanya L. Swer, Kumar S. Prakash, and Manjeet Aggarwal. "Physico-Chemical and Functional Properties of Gamma Irradiated Whole Wheat Flour and Starch." *LWT – Food Science and Technology* 76 (2017): 131–39.

Bhargava, Nitya, Rahul S. Mor, Kshitiz Kumar, and Vijay Singh Sharanagat. "Advances in Application Ultrasound in Food Processing: A Review." *Ultrasonics Sonochemistry* 70 (2021): 105293.

Bourke, P., D. Ziuzina, D. Boehm, P. J. Cullen, and K. Keener. "The Potential of Cold Plasma for Safe and Sustainable Food Production." *Trends in Biotechnology* 36(6) (2018): 615–26.

Butscher, Denis, Daniel Zimmermann, Markus Schuppler, and Philipp Rudolf von Rohr. "Plasma Inactivation of Bacterial Endospores on Wheat Grains and Polymeric Model Substrates in a Dielectric Barrier Discharge." *Food Control* 60 (2016): 636–45.

Butz, Peter, and Bernhard Tauscher. "Emerging Technologies: Chemical Aspects." *Food Research International* 35(2–3) (2002): 279–84.

Charoux, Clémentine M. G., Louis Free, Laura M. Hinds, Rajani K. Vijayaraghavan, Stephen Daniels, Colm P. O'Donnell, and Brijesh K. Tiwari. "Effect of Non-Thermal Plasma Technology on Microbial Inactivation and Total Phenolic Content of a Model Liquid Food System and Black Pepper Grains." *LWT* 118 (2020): 108716.

Chemat, Farid, and Nicolas Hoarau. "Hazard Analysis and Critical Control Point (Haccp) for an Ultrasound Food Processing Operation." *Ultrasonics Sonochemistry* 11(3) (2004): 257–60.

Chemat, Farid, Huma Zill e, and Muhammed Kamran Khan. "Applications of Ultrasound in Food Technology: Processing, Preservation and Extraction." *Ultrasonics Sonochemistry* 18(4) (2011): 813–35.

Chen, Hua Han, Hung Chia Chang, Yu Kuo Chen, Chien Lun Hung, Su Yi Lin, and Yi Sheng Chen. "An Improved Process for High Nutrition of Germinated Brown Rice Production: Low-Pressure Plasma." *Food Chemistry* 191 (2016): 120–27.

Chiu, Kai, and Jih Sung. "Use of Ultrasonication to Enhance Pea Seed Germination and Microbial Quality of Pea Sprouts." *International Journal of Food Science and Technology* 49(7) (2013): 1699–1706.

Cui, Dongjie, Yue Yin, Jiaqi Wang, Zhiwei Wang, Hongbin Ding, Ruonan Ma, and Zhen Jiao. "Research on the Physio-Biochemical Mechanism of Non-Thermal Plasma-Regulated Seed Germination and Early Seedling Development in Arabidopsis." *Frontiers in Plant Science* 10 (2019): 1322.

Dar, M. Z., K. Deepika, K. Jan, T. L. Swer, P. Kumar, R. Verma, K. Verma, et al. "Modification of Structure and Physicochemical Properties of Buckwheat and Oat Starch by Γ-Irradiation." [In eng]. *International Journal of Biological Macromolecules* 108 (Mar 2018): 1348–56.

Dasan, Beyhan, Ismail Boyacı, and Mehmet Mutlu. "Nonthermal Plasma Treatment of *Aspergillus* spp. Spores on Hazelnuts in an Atmospheric Pressure Fluidized Bed Plasma System: Impact of Process Parameters and Surveillance of the Residual Viability of Spores." *Journal of Food Engineering* 196 (2017): 139–49.

Deladino, L., A. Schneider Teixeira, F. J. Plou, A. S. Navarro, and A. D. Molina-García. "Effect of High Hydrostatic Pressure, Alkaline and Combined Treatments on Corn Starch Granules Metal Binding: Structure, Swelling Behavior and Thermal Properties Assessment." *Food and Bioproducts Processing* 102 (2017): 241–49.

Demirci, Ali, and Kathiravan Krishnamurthy. "Pulsed Ultraviolet Light." In: *Nonthermal Processing Technologies for Food*, edited by Howard Q. Zhang, Gustavo V. Barbosa-Cánovas, V. M. Balasubramaniam, C. Patrick Dunne, Daniel F. Farkas and James T. C. Yuan. Ames, IA: Blackwell Publishing Ltd., 2011.

Deng, Yun, Olga Padilla-Zakour, Yanyun Zhao, and Shishi Tao. "Influences of High Hydrostatic Pressure, Microwave Heating, and Boiling on Chemical Compositions, Antinutritional Factors, Fatty Acids, In Vitro Protein Digestibility, and Microstructure of Buckwheat." *Food and Bioprocess Technology* 8(11) (2015): 2235–45.

Du, Jingjing, Zhikai Yang, Xiaonan Xu, Xiaona Wang, and Xianfeng Du. "Effects of Tea Polyphenols on the Structural and Physicochemical Properties of High-Hydrostatic-Pressure-Gelatinized Rice Starch." *Food Hydrocolloids* 91 (2019): 256–62.

Duguet, J. P. "Basic Concepts of Industrial Engineering for the Design of New Ozonation Processes." *Ozone News* 32 (2004): 15–19.

Duque, M. Sheba Mae, Sze Ying Leong, Dominic Agyei, Jaspreet Singh, Nigel Larsen, and Indrawati Oey. "Understanding the Impact of Pulsed Electric Fields Treatment on the Thermal and Pasting Properties of Raw and Thermally Processed Oat Flours." *Food Research International* 129 (2020): 108839.

Dymek, Katarzyna, Petr Dejmek, Valentina Panarese, António A. Vicente, Lars Wadsö, Christine Finnie, and Federico Gómez Galindo. "Effect of Pulsed Electric Field on the Germination of Barley Seeds." *LWT – Food Science and Technology* 47(1) (2012): 161–66.

El-Karamany, A. "Effect of Irradiation on Rheological and Baking Propertes of Flour Wheat Grain." *Journal of Food and Dairy Sciences* 6(12) (2015): 753–69.

Evode, Rwamurangwa. "Approach on High Electric Fields for Remedying the Micro-Organisms Attack on Crops." *Agricultural Research & Technology: Open Access Journal* 10(3) (2017): 555786. doi:10.19080/ARTOAJ.2017.10.555786.

Evrendilek, Gulsun A., Berna Karatas, Sibel Uzuner, and Igor Tanasov. "Design and Effectiveness of Pulsed Electric Fields towards Seed Disinfection." *Journal of the Science of Food and Agriculture* 99(7) (2019): 3475–80.

Farkas, József, and Csilla Mohácsi-Farkas. "History and Future of Food Irradiation." *Trends in Food Science and Technology* 22(2) (2011): 121–26.

FDA. "Title 21 – Food and Drugs Chapter I. Food and Drug Administration Department of Health and Human Services Subchapter B – Food for Human Consumption (Continued). Part 179. – Irradiation in the Production, Processing and Handling of Food." Washington, DC: Office of the Federal Register, U.S. Government Printing Office, 1996.

FDA. "Kinetics of Microbial Inactivation for Alternative Food Processing Technologies." U.S. Department of Health and Human Services, 2000. https://www.fda.gov/files/food/published/Kinetics-of-Microbial-Inactivation-for-Alternative-Food-Processing-Technologies.pdf.

Ferrario, Mariana, Stella Alzamora, and Sandra Guerrero. "Study of the Inactivation of Spoilage Microorganisms in Apple Juice by Pulsed Light and Ultrasound." *Food Microbiology* 46 (2015): 635–42.

Ferrigo, D., A. Raiola, and R. Causin. "Fusarium Toxins in Cereals: Occurrence, Legislation, Factors Promoting the Appearance and Their Management." [In eng]. *Molecules* 21(5) (May 13 2016): 627.

Fetner, R. H., and R. S. Ingols. "A Comparison of the Bactericidal Activity of Ozone and Chlorine against Escherichia coli at 1°." *Journal of General Microbiology* 15(2) (1956): 381–85.

Flamm, Daniel L., and G. Kenneth Herb. "1 – Withdrawn: Plasma Etching Technology – An Overview." In: *Plasma Etching*, edited by Dennis M. Manos and Daniel L. Flamm, 1–89. Boston, MA: Academic Press, 1989.

Flaumenbaum, B. L. "Anwendung der Elektroplasmolyse bei der Herstellung von FruchtsäFten." *Flüssiges* 35 (1968): 19–22.

Foegeding, P. M. "Ozone Inactivation of Bacillus and Clostridium Spore Pópulations and the Importance of the Spore Coat to Resistance." *Food Microbiology* 2(2) (1985): 123–34.

Gonçalves, Alex Augusto. "Ozone – An Emerging Technology for the Seafood Industry." *Brazilian Archives of Biology and Technology* 52(6) (2009): 1527.

Google. "The Pressure Equipment (Safety) Regulations." 2016. https://www.legislation.gov.uk/uksi/2005/schedule/2/made.

Google. "Foods Treated with High Pressure Processing (HPP)." Accessed November 25, 2020. https://www.inspection.gc.ca/preventive-controls/high-pressure-processing/eng/1498504011398504256677.

Gozé, Perrine, Larbi Rhazi, Lyès Lakhal, Philippe Jacolot, André Pauss, and Thierry Aussenac. "Effects of Ozone Treatment on the Molecular Properties of Wheat Grain Proteins." *Journal of Cereal Science* 75 (2017): 243–51.

Gozé, Perrine, Larbi Rhazi, André Pauss, and Thierry Aussenac. "Starch Characterization after Ozone Treatment of Wheat Grains." *Journal of Cereal Science* 70 (2016): 207–13.

Guzel-Seydim, Zeynep B., Annel K. Greene, and A. C. Seydim. "Use of Ozone in the Food Industry." *LWT – Food Science and Technology* 37(4) (2004): 453–60.

Han, Zhong, Yu Han, Jun Wang, Zhongyi Liu, Roman Buckow, and Junhu Cheng. "Effects of Pulsed Electric Field Treatment on the Preparation and Physicochemical Properties of Porous Corn Starch Derived from Enzymolysis." *Journal of Food Processing and Preservation* 44(3) (2020): e14353.

Hati, Subrota, Maulik Patel, and Deepika Yadav. "Food Bioprocessing by Non-Thermal Plasma Technology." *Current Opinion in Food Science* 19 (2018): 85–91.

He, Zhuoliang, Yongwei Wang, Jun Wang, and Chen Tang. "Influence of Cold Plasma Treatment on Ear Characters and Group Quality of Wheat." In: *2016 ASABE Annual International Meeting*, 1. St. Joseph, MI: ASABE, 2016.

Hertwig, Christian, Kai Reineke, Jörg Ehlbeck, Dietrich Knorr, and Oliver Schlüter. "Decontamination of Whole Black Pepper Using Different Cold Atmospheric Pressure Plasma Applications." *Food Control* 55 (2015): 221–29.

Hidaka, Y., and K. Kubota. "Study on the Sterilization of Grain Surface Using UV Radiation: Development and Evaluation of UV Irradiation Equipment." *Japan Agricultural Research Quarterly: JARQ* 40(2) (2006): 157–61.

Hiramoto, Tatsumi. *Method of Sterilization*. Tokyo, Japan, 1984.

Hite, B. H. *The Effect of Pressure in the Preservation of Milk*, 15–35. Morgantown, WV: West Virginia Agricultural Experimental Station Bulletin, 1899.

Hogan, E., A. L. Kelly, and D.-W. Sun. "High Pressure Processing of Foods: An Overview." In: *Emerging Technologies for Food Processing*, edited by Da-Wen Sun, 3–27. CA: Elsevier Ltd., 2005.

Hong, Jing, Di An, Xin-An Zeng, Zhong Han, Xueling Zheng, Mengjie Cai, Ke Bian, and Rana Muhammad Aadil. "Behaviors of Large a-Type and Small B-Type Wheat Starch Granules Esterified by Conventional and Pulsed Electric Fields Assisted Methods." *International Journal of Biological Macromolecules* 155 (2020): 516–23.

Hong, Jing, Chaopeng Li, Di An, Chong Liu, Limin Li, Zhong Han, Xin-An Zeng, Xueling Zheng, and Mengjie Cai. "Differences in the Rheological Properties of Esterified Total, a-Type, and B-Type Wheat Starches and Their Effects on the Quality of Noodles." *Journal of Food Processing and Preservation* 44(3) (2020): e14342.

HSE. "Safety of Pressure Systems Pressure Systems Safety Regulations 2000." (2014). https://www.hse.gov.uk/pubns/priced/l122.pdf.

Hu, Xiao-Pei, Bao Zhang, Zheng-Yu Jin, Xue-Ming Xu, and Han-Qing Chen. "Effect of High Hydrostatic Pressure and Retrogradation Treatments on Structural and Physicochemical Properties of Waxy Wheat Starch." *Food Chemistry* 232 (2017): 560–65.

Irving, D. "We Zijn Nu Al Aan Het Opschalen." *VMT Voedingsmiddelentechnologie* 16 (2012): 11–13.

Ishizaki, K., N. Shinriki, and H. Matsuyama. "Inactivation of Bacillus Spores by Gaseous Ozone." [In eng]. *Journal of Applied Bacteriology* 60(1) (Jan 1986): 67–72.

Jan, Awsi, Monika Sood, Sajad Sofi, and Tsering Norzom. "Non-Thermal Processing in Food Applications: A Review." *International Journal of Food Science and Nutrition* 2 (2017): 171–80.

Joshi, R. P., Q. Hu, K. H. Schoenbach, and H. P. Hjalmarson. "Improved Energy Model for Membrane Electroporation in Biological Cells Subjected to Electrical Pulses." [In eng]. *Physical Review. Part E: Statistical, Nonlinear and Soft Matter Physics* 65(4 Pt 1) (2002): 041920.

Keener, K. M., and N. N. Misra. "Chapter 14 – Future of Cold Plasma in Food Processing." In: *Cold Plasma in Food and Agriculture*, edited by N. N. Misra, Oliver Schlüter and P. J. Cullen, 343–60. San Diego, CA: Academic Press, 2016.

Kempkes, Michael. *Industrial Pef Systems*. Bedford, MA, 2016.

Kim, Min Young, Sang Hoon Lee, Gwi Yeong Jang, Meishan Li, Youn Ri Lee, Junsoo Lee, and Heon Sang Jeong. "Changes of Phenolic-Acids and Vitamin E Profiles on Germinated Rough Rice (*Oryza sativa* L.) Treated by High Hydrostatic Pressure." *Food Chemistry* 217 (2017): 106–11.

Kim, Min Young, Sang Hoon Lee, Gwi Young Jang, Hye Jin Park, Meishan Li, Shinje Kim, Youn Ri Lee, et al. "Effects of High Hydrostatic Pressure Treatment on the Enhancement of Functional Components of Germinated Rough Rice (*Oryza sativa* L.)." *Food Chemistry* 166 (2015a): 86–92.

Kim, Sang-Kab, Seung-Hyun Choi, Hyun-Wook Choi, Jae-Heung Ko, Wooki Kim, Dae-Ok Kim, Byung-Yong Kim, and Moo-Yeol Baik. "Retrogradation Kinetics of Cross-Linked and Acetylated Corn Starches under High Hydrostatic Pressure." *Food Science and Biotechnology* 24(1) (2015b): 85–90.

Kim, Sujin, So-Young Yang, Ho Hyun Chun, and Kyung Bin Song. "High Hydrostatic Pressure Processing for the Preparation of Buckwheat and Tapioca Starch Films." *Food Hydrocolloids* 81 (2018): 71–76.

Knorr, D., A. Froehling, H. Jaeger, K. Reineke, O. Schlueter, and K. Schoessler. "Emerging Technologies in Food Processing." [In eng]. *Annual Review of Food Science and Technology* 2 (2011): 203–35.

Knorr, D. "Emerging Technologies: Back to the Future." *Trends in Food Science and Technology* 76 (2018): 119–23.

Koutchma, Tatiana, Fornee Larry, and Carmen Moraru. *Ultraviolet Light in Food Technology: Principles and Applications*, 1–49. New York: CRC Press, 2009. doi:10.1281420059519.ch4.

Krishnamurthy, Kathiravan, Ali Demirci, and Joseph Irudayaraj. "Inactivation of Staphylococcus aureus in Milk Using Flow-through Pulsed UV-Light Treatment System." *Journal of Food Science* 72(7) (2007): M233–39.

Krstović, Saša, Jelena Krulj, Sandra Jakšić, Aleksandra Bočarov-Stančić, and Igor Jajić. "Ozone as Decontaminating Agent for Ground Corn Containing Deoxynivalenol, Zearalenone, and Ochratoxin A." *Cereal Chemistry* n/a, no. n/a (2020).

Kučerová, Katarína, Mária Henselová, Ľudmila Slováková, and Karol Hensel. "Effects of Plasma Activated Water on Wheat: Germination, Growth Parameters, Photosynthetic Pigments, Soluble Protein Content, and Antioxidant Enzymes Activity." *Plasma Processes and Polymers* 16(3) (2019): 1800131.

Kumar, C. T. Manoj, and Latha Sabikhi. "Ozone Application in Food Processing." In: *Non-Thermal Processing of Foods*, edited by O. P. Chauhan, 190–206. Boca Raton, FL: CRC Press/Taylor & Francis Group, 2019.

Kumar, Pradeep, Kumar S. Prakash, Kulsum Jan, Tanya L. Swer, Shumaila Jan, Ruchi Verma, Km Deepika, et al. "Effects of Gamma Irradiation on Starch Granule Structure and Physicochemical Properties of Brown Rice Starch." *Journal of Cereal Science* 77 (2017): 194–200.

Kume, Tamikazu, Masakazu Furuta, S. Todoriki, Naoki Uenoyama, and Yasuhiko Kobayashi. "Status of Food Irradiation in the World." *Radiation Physics and Chemistry* 78(3) (2009): 222–26.

Kundwal, Moses, Mohd Lani, and Abd Rahman Tamuri. "Microbial Inactivation Using Pulsed Light- a Review." *Jurnal Teknologi* 77(13) (2015): 99–109.

Kurdziel, Magdalena, Maria Filek, and Maria Łabanowska. "The Impact of Short-Term UV Irradiation on Grains of Sensitive and Tolerant Cereal Genotypes Studied by Epr." *Journal of the Science of Food and Agriculture* 98(7) (2018): 2607–16.

Landesmann, Jennifer, Gundel Pedro, Maria Martinez-Ghersa, and Claudio Ghersa. "Ozone Exposure of a Weed Community Produces Adaptive Changes in Seed Populations of Spergula Arvensis." *PloS ONE* 8(9) (2013): e75820.

Lee, Ji Hae, Koan Sik Woo, Cheorun Jo, Heon Sang Jeong, Seuk Ki Lee, Byong Won Lee, Yu-Young Lee, Byoungkyu Lee, and Hyun-Joo Kim. "Quality Evaluation of Rice Treated by High Hydrostatic Pressure and Atmospheric Pressure Plasma." *Journal of Food Quality* (2019): 4253701.

Li, Guantian, Fan Zhu, Guang Mo, and Yacine Hemar. "Supramolecular Structure of High Hydrostatic Pressure Treated Quinoa and Maize Starches." *Food Hydrocolloids* 92 (2019a): 276–84.

Li, Qian, Qiu-Yan Wu, Wei Jiang, Jian-Ya Qian, Liang Zhang, Mangang Wu, Sheng-Qi Rao, and Chun-Sen Wu. "Effect of Pulsed Electric Field on Structural Properties and Digestibility of Starches with Different Crystalline Type in Solid State." *Carbohydrate Polymers* 207 (2019b): 362–70.

Li, Wenhao, Xiaoling Tian, Peng Wang, Ahmed S. M. Saleh, Qingui Luo, Jianmei Zheng, Shaohui Ouyang, and Guoquan Zhang. "Recrystallization Characteristics of High Hydrostatic Pressure Gelatinized Normal and Waxy Corn Starch." *International Journal of Biological Macromolecules* 83 (2016): 171–77.

Liao, Xinyu, Donghong Liu, Qisen Xiang, Juhee Ahn, Shiguo Chen, Xingqian Ye, and Tian Ding. "Inactivation Mechanisms of Non-Thermal Plasma on Microbes: A Review." *Food Control* 75 (2017): 83–91.

Liu, Chong, Yanyan Zhang, Huan Li, Limin Li, and Xueling Zheng. "Effect of Ozone Treatment on Processing Properties of Wheat Bran and Shelf Life Characteristics of Noodles Fortified with Wheat Bran." *Journal of Food Science and Technology* 57(10) (2020): 3893–902.

Liu, Hang, Huanhuan Fan, Rong Cao, Christopher Blanchard, and Min Wang. "Physicochemical Properties and In Vitro Digestibility of Sorghum Starch Altered by High Hydrostatic Pressure." *International Journal of Biological Macromolecules* 92 (2016a): 753–60.

Liu, Hang, Xudan Guo, Yunlong Li, Hongmei Li, Huanhuan Fan, and Min Wang. "In Vitro Digestibility and Changes in Physicochemical and Textural Properties of Tartary Buckwheat Starch under High Hydrostatic Pressure." *Journal of Food Engineering* 189 (2016b): 64–71.

Liu, Hang, Lijing Wang, Rong Cao, Huanhuan Fan, and Min Wang. "In Vitro Digestibility and Changes in Physicochemical and Structural Properties of Common Buckwheat Starch Affected by High Hydrostatic Pressure." *Carbohydrate Polymers* 144 (2016c): 1–8.

Lohani, U. C., and K. Muthukumarappan. "Application of the Pulsed Electric Field to Release Bound Phenolics in Sorghum Flour and Apple Pomace." *Innovative Food Science and Emerging Technologies* 35 (2016): 29–35.

Los, A., D. Ziuzina, and P. Bourke. "Current and Future Technologies for Microbiological Decontamination of Cereal Grains." [In eng]. *Journal of Food Science* 83(6) (2018): 1484–93.

Magnuson, Bernadene, Ian Munro, Peter Abbot, Nigel Baldwin, Rebeca Lopez-Garcia, Karen Ly, Larry McGirr, Ashley Roberts, and Susan Socolovsky. "Review of the Regulation and Safety Assessment of Food Substances in Various Countries and Jurisdictions." [In eng]. *Food Additives and Contaminants. Part A: Chemistry, Analysis, Control, Exposure and Risk Assessment* 30(7) (2013): 1147–220.

Magureanu, M., R. Sîrbu, D. Dobrin, and M. Gidea. "Stimulation of the Germination and Early Growth of Tomato Seeds by Non-Thermal Plasma." *Plasma Chemistry and Plasma Processing* 38(5) (2018): 989–1001.

Markov, Ksenija, Branka Mihaljević, Ana-Marija Domijan, Jelka Pleadin, Frane Delaš, and Jadranka Frece. "Inactivation of Aflatoxigenic Fungi and the Reduction of Aflatoxin B1 In Vitro and In Situ Using Gamma Irradiation." *Food Control* 54 (2015): 79–85.

Martín-Belloso, Olga, and Robert S. Oliva-Fortuny. "Pulsed Electric Fields Processing Basics." In: *Nonthermal Processing Technologies for Food*, edited by Howard Q. Zhang, Gustavo V. Barbosa-Cánovas, V. M. Balasubramania, C. Patrick Dunne, Daniel F. Farkas and James T. C. Yuan, 157–76. Ames, IA: Blackwell Publishing Ltd., 2011.

Martín-García, Beatriz, Urszula Tylewicz, Vito Verardo, Federica Pasini, Ana María Gómez-Caravaca, Maria Fiorenza Caboni, and Marco Dalla Rosa. "Pulsed Electric Field (Pef) as Pre-Treatment to Improve the Phenolic Compounds Recovery from Brewers' Spent Grains." *Innovative Food Science and Emerging Technologies* 64 (2020): 102402.

Matra, Khanit. "Non-Thermal Plasma for Germination Enhancement of Radish Seeds." *Procedia Computer Science* 86 (2016): 132–35.

Meng, Ling, Wencheng Zhang, Xianhan Zhou, Zeyu Wu, Ailing Hui, Yiwen He, Han Gao, and Pengpeng Chen. "Effect of High Hydrostatic Pressure on the Bioactive Compounds, Antioxidant Activity and In Vitro Digestibility of Cooked Black Rice During Refrigerated Storage." *Journal of Cereal Science* 86 (2019): 54–59.

Mielcke, J., and A. Ried. "Current State of Application of Ozone and UV for Food Processing." In: *Proceedings of the Food Protection International Conference 2004*. Monte da Caparica, Portugal, 2004.

Mir, Shabir, Manzoor Shah, and Mohammad Mir. "Understanding the Role of Plasma Technology in Food Industry." *Food and Bioprocess Technology* 9(5) (2016): 734–50.

Misra, N. N., Mohamed Koubaa, Shahin Roohinejad, Pablo Juliano, Hami Alpas, Rita S. Inácio, Jorge A. Saraiva, and Francisco J. Barba. "Landmarks in the Historical Development of Twenty First Century Food Processing Technologies." *Food Research International* 97 (2017): 318–39.

Misra, N. N., T. Moiseev, S. Patil, S. Pankaj, P. Bourke, J. P. Mosnier, K. Keener, and P. J. Cullen. "Cold Plasma in Modified Atmospheres for Post-Harvest Treatment of Strawberries." *Food and Bioprocess Technology* 7(10) (2014a): 3045–54.

Misra, N. N., S. K. Pankaj, T. Walsh, F. O'Regan, P. Bourke, and P. J. Cullen. "In-Package Nonthermal Plasma Degradation of Pesticides on Fresh Produce." *Journal of Hazardous Materials* 271 (2014b): 33–40.

Misra, N. N., O. Schlüter, and P. J. Cullen. "Chapter 1 – Plasma in Food and Agriculture." In: *Cold Plasma in Food and Agriculture*, edited by N. N. Misra, Oliver Schlüter and P. J. Cullen, 1–16. San Diego, CA: Academic Press, 2016.

Mohamed, Neeven, Rasha El-Dine, Metwally Kotb, and Aida Saber. "Assessing the Possible Effect of Gamma Irradiation on the Reduction of Aflatoxin B1, and on the Moisture Content in Some Cereal Grains." *American Journal of Biomedical Sciences* 7(1) (2015): 33–39.

Mohammad, Zahra, Ahmad Kalbasi-Ashtari, Gerald Riskowski, and Alejandro Castillo. "Reduction of Salmonella and Shiga Toxin-Producing Escherichia coli on Alfalfa Seeds and Sprouts Using an Ozone Generating System." *International Journal of Food Microbiology* 289 (2019): 57–63.

Morris, Caroline, Aaron L. Brody, and Louise Wicker. "Non-Thermal Food Processing/Preservation Technologies: A Review with Packaging Implications." *Packaging Technology and Science* 20(4) (2007): 275–86.

Mukhtar, R., A. Shah, N. Noor, A. Gani, I. A. Wani, B. A. Ashwar, and F. A. Masoodi. "Γ-Irradiation of Oat Grain – Effect on Physico-Chemical, Structural, Thermal, and Antioxidant Properties of Extracted Starch." [In eng]. *International Journal of Biological Macromolecules* 104, (2017): 1313–20.

Murata, S., E. Ishiguro, and M. Miyazato. "Simulation of Survival Curves in Several Molds Irradiated with Gamma and Ultraviolet Rays – A Quantitative Analysis on Tailing Problem." *Science Bulletin of the Faculty of Agriculture Kyushu University* 48 (1994): 151–62.

NACMCF. *Requisite Scientific Parameters for Establishing the Equivalence of Alternative Methods of Pasteurization*, edited by National Advisory Committee on Microbiological Criteria for Foods. Washington, DC: United States Department of Agriculture, 2004.

Naik, L., R. Sharma, Y. S. Rajput, and G. Manju. "Application of High Pressure Processing Technology for Dairy Food Preservation – Future Perspective: A Review." *Journal of Animal Production Advances*, 3(8) (2013): 232–241.

Nazari, Meisam, and Mohammad Eteghadipour. "Impacts of Ultrasonic Waves on Seeds: A Mini-Review." *Agricultural Research and Technology* 6 (2017): 1–5.

Negi, P., and Rastogi, N. K. "Use of Plasma in Food Processing." Chap. 10. In: *Non-Thermal Processing of Foods*, edited by O. P. Chauhan, 173. Boca Raton, FL: CRC Press/Taylor & Francis, 2019.

Nowosad, Karolina, Monika Sujka, Urszula Pankiewicz, and Radosław Kowalski. "The Application of Pef Technology in Food Processing and Human Nutrition." *Journal of Food Science and Technology* 58(2) (2021): 397–411.

Obadi, Mohammed, Ke-Xue Zhu, Wei Peng, Anwar Noman, Khalid Mohammed, and Hui-Ming Zhou. "Characterization of Oil Extracted from Whole Grain Flour Treated with Ozone Gas." *Journal of Cereal Science* 79 (2018): 527–33.

Ortega-Rivas, Enrique. "High-Voltage Pulsed Electric Fields." In: *Non-Thermal Food Engineering Operations*, edited by Gustavo V. Barbosa-Cánovas, 275–300. New York: Springer Science+Business Media, 2012.

———. "Ionizing Radiation: Irradiation." In: *Non-Thermal Food Engineering Operations*, edited by Gustavo V. Barbosa-Cánovas. New York: Springer Science+Business Media, 2012.

———. "Pulsed Light Technology." In: *Non-Thermal Food Engineering Operations*, edited by Gustavo V. Barbosa-Cánovas. New York: Springer Science+Business Media, 2012.

———. "Ultrahigh Hydrostatic Pressure." Chap. 14 In: *Non-Thermal Food Engineering Operations*, edited by Enrique Ortega-Rivas, 301–24. New York: Springer Science+Business Media, 2012.

Palmieri, Luigi, and Domenico Cacace. "High Intensity Pulsed Light Technology." In: *Emerging Technologies for Food Processing*, edited by Da-Wen Sun, 279–304. CA: Elsevier Ltd., 2005.

Pandiselvam, R., V. P. Mayookha, A. Kothakota, L. Sharmila, S. V. Ramesh, C. P. Bharathi, K. Gomathy, and V. Srikanth. "Impact of Ozone Treatment on Seed Germination – A Systematic Review." *Ozone: Science and Engineering* 42(4) (2020): 331–46.

Pandiselvam, R., S. Subhashini, E. P. Banuu Priya, A. Kothakota, S. V. Ramesh, and S. Shahir. "Ozone Based Food Preservation: A Promising Green Technology for Enhanced Food Safety." *Ozone: Science and Engineering* 41(1) (2019): 17–34.

Piechowiak, Tomasz, Radosław Józefczyk, and Maciej Balawejder. "Impact of Ozonation Process of Wheat Flour on the Activity of Selected Enzymes." *Journal of Cereal Science* 84 (2018): 30–37.

Piemontese, Luca, Maria Cristina Messia, Emanuele Marconi, Luisa Falasca, Rosanna Zivoli, Lucia Gambacorta, Giancarlo Perrone, and Michele Solfrizzo. "Effect of Gaseous Ozone Treatments on Don, Microbial Contaminants and Technological Parameters of Wheat and Semolina." *Food Additives and Contaminants. Part A* 35(4) (2018): 760–71.

Pignata, C., D. D'Angelo, E. Fea, and G. Gilli. "A Review on Microbiological Decontamination of Fresh Produce with Nonthermal Plasma." [In eng]. *Journal of Applied Microbiology* 122(6) (2017): 1438–55.

Piyasena, P., E. Mohareb, and R. C. McKellar. "Inactivation of Microbes Using Ultrasound: A Review." *International Journal of Food Microbiology* 87(3) (2003): 207–16.

Qian, Jian-Ya, Yu-Ping Gu, Wei Jiang, and Wei Chen. "Inactivating Effect of Pulsed Electric Field on Lipase in Brown Rice." *Innovative Food Science and Emerging Technologies* 22 (2014): 89–94.

Rahman, M. S. "Light and Sound in Food Preservation." In: *Handbook of Food Preservation*, edited by M. S. Rahman, 673–86. New York: Marcel Dekker, 1999.

Randeniya, Lakshman K., and Gerard J. J. B. de Groot. "Non-Thermal Plasma Treatment of Agricultural Seeds for Stimulation of Germination, Removal of Surface Contamination and Other Benefits: A Review." *Plasma Processes and Polymers* 12(7) (2015): 608–23.

Rasco, Barbara. "Laws and Regulations for Novel Food Processing Technologies." In *Ultrasound: Advances in Food Processing and Preservation*, edited by Daniela Bermudez-Aguirre, 499–524. New York: Academic Press, 2017.

Restaino, L., E. W. Frampton, J. B. Hemphill, and P. Palnikar. "Efficacy of Ozonated Water against Various Food-Related Microorganisms." [In eng]. *Applied and Environment Microbiology* 61(9) (1995): 3471–75.

REUS. "Spherical Sono Reactor." Accessed November 25, 2020. http://www.etsreus.com/en/home-eng/.

Rice, Rip G. "Applications of Ozone for Industrial Wastewater Treatment – A Review." *Ozone: Science and Engineering* 18(6) (1996): 477–515.

Roth, J. R. *Industrial Plasma Engineering. Vol. 1: Principles*. Bristol: Inst. Phys. Publ, 1995.

Rowan, Neil J. "Pulsed Light as an Emerging Technology to Cause Disruption for Food and Adjacent Industries – Quo Vadis?" *Trends in Food Science and Technology* 88 (2019): 316–32.

Saberi, Mahin, Seyed Ali Mohammad Modarres-Sanavy, Rasoul Zare, and H. Ghomi. "Amelioration of Photosynthesis and Quality of Wheat under Non-Thermal Radio Frequency Plasma Treatment." *Scientific Reports* 8(1) (2018): 11655.

Santos, Lígia, Fabiano Oliveira, Elisa Ferreira, and Amauri Rosenthal. "Application and Possible Benefits of High Hydrostatic Pressure or High-Pressure Homogenization on Beer Processing: A Review." *Food Science and Technology International* 23(7) (2017): 108201321771467.

Santos Alexandre, Allana, Nanci Castanha, Naiara Costa, Amanda Santos, Eliana Badiale-Furlong, Pedro Augusto, and Maria Calori-Domingues. "Ozone Technology to Reduce Zearalenone Contamination in Whole Maize Flour: Degradation Kinetics and Impact on Quality." *Journal of the Science of Food and Agriculture* 99(15) (2019): 6814–21.

São José, Jackline Freitas Brilhante de, Nélio José de Andrade, Afonso Mota Ramos, Maria Cristina Dantas Vanetti, Paulo César Stringheta, and José Benício Paes Chaves. "Decontamination by Ultrasound Application in Fresh Fruits and Vegetables." *Food Control* 45 (2014): 36–50.

Sarangapani, Chaitanya, Yamuna Devi, Rohit Thirundas, Uday S. Annapure, and Rajendra R. Deshmukh. "Effect of Low-Pressure Plasma on Physico-Chemical Properties of Parboiled Rice." *LWT – Food Science and Technology* 63(1) (2015): 452–60.

Savi, G. D., K. C. Piacentini, T. Bortolotto, and V. M. Scussel. "Degradation of Bifenthrin and Pirimiphos-Methyl Residues in Stored Wheat Grains (Triticum aestivum L.) by Ozonation." [In eng]. *Food Chemistry* 203 (2016): 246–51.

Savi, Geovana D., Thauan Gomes, Sílvia B. Canever, Ana C. Feltrin, Karim C. Piacentini, Rahisa Scussel, Daysiane Oliveira, et al. "Application of Ozone on Rice Storage: A Mathematical Modeling of the Ozone Spread, Effects in the Decontamination of Filamentous Fungi and Quality Attributes." *Journal of Stored Products Research* 87 (2020): 101605.

Savi, Geovana D., Karim C. Piacentini, and Vildes M. Scussel. "Reduction in Residues of Deltamethrin and Fenitrothion on Stored Wheat Grains by Ozone Gas." *Journal of Stored Products Research* 61 (2015): 65–69.

Schmidt, M., E. Zannini, and E. K. Arendt. "Recent Advances in Physical Post-Harvest Treatments for Shelf-Life Extension of Cereal Crops." [In eng]. *Foods* 7(4) (2018): 45.

Schmidt, Marcus, Emanuele Zannini, and Elke K. Arendt. "Screening of Post-Harvest Decontamination Methods for Cereal Grains and Their Impact on Grain Quality and Technological Performance." *European Food Research and Technology* 245(5) (2019): 1061–74.

Schnabel, Uta, Rijana Niquet, Oliver Schlüter, Holger Gniffke, and Jörg Ehlbeck. "Decontamination and Sensory Properties of Microbiologically Contaminated Fresh Fruits and Vegetables by Microwave Plasma Processed Air (Ppa)." *Journal of Food Processing and Preservation* 39(6) (2014): 12273.

Schneider Teixeira, A., L. Deladino, M. A. García, N. E. Zaritzky, P. D. Sanz, and A. D. Molina-García. "Microstructure Analysis of High Pressure Induced Gelatinization of Maize Starch in the Presence of Hydrocolloids." *Food and Bioproducts Processing* 112 (2018): 119–30.

Scholtz, V., B. Sera, J. Khun, M. Sery, and J. Julák. "Effects of Nonthermal Plasma on Wheat Grains and Products." *Journal of Food Quality* 2019 (2019): 1–10.

Scholtz, Vladimir, Jarmila Pazlarova, Hana Souskova, Josef Khun, and Jaroslav Julak. "Nonthermal Plasma – A Tool for Decontamination and Disinfection." *Biotechnology Advances* 33(6, Part 2) (2015): 1108–19.

Scouten, A. J., and L. R. Beuchat. "Combined Effects of Chemical, Heat and Ultrasound Treatments to Kill Salmonella and Escherichia coli O157:H7 on Alfalfa Seeds." [In eng]. *Journal of Applied Microbiology* 92(4) (2002): 668–74.

Segat, A., N. N. Misra, P. J. Cullen, and N. Innocente. "Atmospheric Pressure Cold Plasma (Acp) Treatment of Whey Protein Isolate Model Solution." *Innovative Food Science and Emerging Technologies* 29 (2015): 247–54.

Sera, B., M. Sery, B. Gavril, and I. Gajdova. "Seed Germination and Early Growth Responses to Seed Pre-Treatment by Non-Thermal Plasma in Hemp Cultivars (*Cannabis sativa* L.)." *Plasma Chemistry and Plasma Processing* 37(1) (2017): 207–21.

Sera, B., P. Špatenka, M. Šerý, N. Vrchotova, and I. Hrušková. "Influence of Plasma Treatment on Wheat and Oat Germination and Early Growth." *IEEE Transactions on Plasma Science* 38(10) (2010): 2963–68.

Seymour, I. J., D. Burfoot, R. L. Smith, L. A. Cox, and A. Lockwood. "Ultrasound Decontamination of Minimally Processed Fruits and Vegetables." *International Journal of Food Science and Technology* 37(5) (2002): 547–57.

Shah, A., F. A. Masoodi, A. Gani, and B. A. Ashwar. "Effect of Γ-Irradiation on Antioxidant and Antiproliferative Properties of Oat B-Glucan." *Radiation Physics and Chemistry* 117 (2015): 120–27.

Shayanfar, Shima, and Suresh Pillai. "Electron Beam Processing of Foods." Chap. 16. In: *Non-Thermal Processing of Foods*, edited by O. P. Chauhan, 315–27. Boca Raton, FL: CRC Press/Taylor & Francis Group, 2019.

Singh, R., B. P. Kaur, and S. Thangalakshmi. "Pulsed Electric Field Processing Principles and Engineering Aspects." In: *Non-Thermal Processing of Foods*, edited by O. P. Chauhan, 89–106. Boca Raton, FL: CRC Press/Taylor & Francis Group, 2019.

Sitzmann, W., and E. W. Muench. "[The Elcrack-Procedure. A New Method for Processing Material of Animal Origin]. [German]. Das Elcrack-Verfahren. Ein Neues Verfahren zur Verarbeitung Tierischer Rohstoffe." *Fleischmehlindustrie* 40 (1988): 22–30.

Sj, Goussous, Nezar Samarah, Ahmad Alqudah, and Mohammmad Othman. "Enhancing Seed Germination of Four Crop Species Using an Ultrasonic Technique." *Experimental Agriculture* 46(2) (2010): 231–42.

Søltoft-Jensen, Jakob, and Flemming Hansen. "New Chemical and Biochemical Hurdles." In: *Emerging Technologies for Food Processing*, edited by Da-Wen Sun, 387–410. CA: Elsevier Ltd., 2005.

Song, H. P., B. Kim, J. H. Choe, S. Jung, S. Y. Moon, W. Choe, and C. Jo. "Evaluation of Atmospheric Pressure Plasma to Improve the Safety of Sliced Cheese and Ham Inoculated by 3-Strain Cocktail Listeria monocytogenes." [In eng]. *Food Microbiology* 26(4) (2009): 432–36.

Sudhakar, N., D. Nagendra-Prasad, N. Mohan, B. Hill, and M. Gunasekaran. "Assessing Influence of Ozone in Tomato Seed Dormancy Alleviation." *American Journal of Plant Sciences* 02(3) (2011): 443–48.

Sunisha, K., and T. Pandiarajan. "Studies on the Effect of Ozone on Mortality of Insects in Stored Paddy: Changes in Milling Qualities Upon Storage." *Journal of Entomology and Zoology Studies* 8(2) (2020): 736–41.

Technology, Hielscher Ultrasound. "Industrial Ultrasonic Devices." Accessed November 25, 2020. https://www.hielscher.com/industry.htm.

Teixeira, Bruna S., Rafael H. L. Garcia, Patricia Y. I. Takinami, and Nelida L. del Mastro. "Comparison of Gamma Radiation Effects on Natural Corn and Potato Starches and Modified Cassava Starch." *Radiation Physics and Chemistry* 142 (2018): 44–49.

Toepfl, Stefan. "Pulsed Electric Field Food Processing -Industrial Equipment Design and Commercial Applications." *Stewart Postharvest Review* 8 (2012): 1–7.

Toepfl, Stefan, Volker Heinz, and Dietrich Knorr. "Overview of Pulsed Electric Field Processing for Food." In: *Emerging Technologies for Food Processing*, edited by Da-Wen Sun, 67–91. San Diego, CA: Elsevier Ltd., 2005.

Tolouie, Haniye, Mohammad Amin Mohammadifar, Hamid Ghomi, Amin Seyed Yaghoubi, and Maryam Hashemi. "The Impact of Atmospheric Cold Plasma Treatment on Inactivation of Lipase and Lipoxygenase of Wheat Germs." *Innovative Food Science and Emerging Technologies* 47 (2018): 346–52.

Ueno, Shigeaki, Shoji Sasao, Hsiuming Liu, Mayumi Hayashi, Toru Shigematsu, Yasuko Kaneko, and Tetsuya Araki. "Effects of High Hydrostatic Pressure on B-Glucan Content, Swelling Power, Starch Damage, and Pasting Properties of High-B-Glucan Barley Flour." *High Pressure Research* 39(3) (2019): 509–24.

Unluturk, Sevcan. "Impact of Irradiation on the Microbial Ecology of Foods: Modeling the Microbial Ecology." Chap. 8. In: *Quantitative Microbiology in Food Processing*, edited by Sant'Ana A., 176–93. Chichester, UK: John Wiley & Sons, 2016.

Unni, L. E., and O. P. Chauhan. "Use of Pulsed Light in Food Processing." In: *Non-Thermal Processing of Foods*, edited by O. P. Chauhan, 173–85. Boca Raton, FL: CRC Press/Taylor & Francis Group, 2019.

Vaz-Velho, M., M. Silva, J. Pessoa, and P. Gibbs. "Inactivation by Ozone of *Listeria innocua* on Salmon-Trout During Cold-Smoke Processing." *Food Control* 17(8) (2006): 609–16.

Vazquez-Ybarra, J. A., C. B. Pena-Valdivia, C. Trejo, A. Villegas-Bastida, S. Benedicto-Valdez, and P. SánchezGarcía. "Promoción Del Crecimiento de Plantas de Lechuga (Lactucasativa L.) con Dosissubletales de Ozonoaplicadas Al Medio de Cultivo." *Revista Fitotecnia Mexicana* 38(4) (2015): 405–13.

Violleau, Frederic, Gabriel Pernot, and O. Surel. "Effect of Oxygreen (R) Wheat Ozonation Process on Bread Dough Quality and Protein Solubility." *Journal of Cereal Science* 55 (2012): 392–96.

Wang, Bingzhi, Fengru Liu, Shuizhong Luo, Peijun Li, Dongdong Mu, Yanyan Zhao, Xiyang Zhong, Shaotong Jiang, and Zhi Zheng. "Effects of High Hydrostatic Pressure on the Properties of Heat-Induced Wheat Gluten Gels." *Food and Bioprocess Technology* 12(2) (2018): 220–27.

Wang, Chao, Yong Xue, Laraib Yousaf, Jinrong Hu, and Qun Shen. "Effects of High Hydrostatic Pressure on Swelling Behavior of Rice Starch." *Journal of Cereal Science* 93 (2020a): 102967.

———. "Effects of High Hydrostatic Pressure on the Ordered Structure Including Double Helices and V-Type Single Helices of Rice Starch." *International Journal of Biological Macromolecules* 144 (2020b): 1034–42.

Wang, Li, Huili Shao, Xiaohu Luo, Ren Wang, Yongfu Li, Yanan Li, Yingpeng Luo, and Zhengxing Chen. "Effect of Ozone Treatment on Deoxynivalenol and Wheat Quality." *PloS ONE* 11(1) (2016): e0147613.

Weaver, James C., and Yu A. Chizmadzhev. "Theory of Electroporation: A Review." *Bioelectrochemistry and Bioenergetics* 41(2) (1996): 135–60.

Wood, R. W., and Alfred L. Loomis. "Xxxviii. The Physical and Biological Effects of High-Frequency Sound-Waves of Great Intensity." *The London, Edinburgh, and Dublin Philosophical Magazine and Journal of Science* 4(22) (1927): 417–36.

Xia, Qiang, and Yunfei Li. "Ultra-High Pressure Effects on Color, Volatile Organic Compounds and Antioxidants of Wholegrain Brown Rice (Oryza sativa L.) during Storage: A Comparative Study with High-Intensity Ultrasound and Germination Pretreatments." *Innovative Food Science and Emerging Technologies* 45 (2018): 390–400.

Xia, Qiang, Liping Wang, and Yunfei Li. "Exploring High Hydrostatic Pressure-Mediated Germination to Enhance Functionality and Quality Attributes of Wholegrain Brown Rice." *Food Chemistry* 249 (2018): 104–10.

Xia, Qiang, Liping Wang, Congcong Xu, Jun Mei, and Yunfei Li. "Effects of Germination and High Hydrostatic Pressure Processing on Mineral Elements, Amino Acids and Antioxidants In Vitro Bioaccessibility, as Well as Starch Digestibility in Brown Rice (*Oryza sativa* L.)." *Food Chemistry* 214 (2017): 533–42.

Xu, Xiaonan, Weilong Yan, Zhikai Yang, Xiaona Wang, Yu Xiao, and Xianfeng Du. "Effect of Ultra-High Pressure on Quality Characteristics of Parboiled Rice." *Journal of Cereal Science* 87 (2019): 117–23.

Yadav, Durgesh Singh, Amit Kumar Mishra, Richa Rai, Nivedita Chaudhary, Arideep Mukherjee, S. B. Agrawal, and Madhoolika Agrawal. "Responses of an Old and a Modern Indian Wheat Cultivar to Future O3 Level: Physiological, Yield and Grain Quality Parameters." *Environmental Pollution* 259 (2020): 113939.

Yang, Yijin, Yongjun Xia, Guangqiang Wang, Leren Tao, Yu Jianshen, and Lianzhong Ai. "Effects of Boiling, Ultra-High Temperature and High Hydrostatic Pressure on Free Amino Acids, Flavor Characteristics and Sensory Profiles in Chinese Rice Wine." *Food Chemistry* 275 (2019): 407–16.

Yang, Zhi, Peter Swedlund, Yacine Hemar, Guang Mo, Yanru Wei, Zhihong Li, and Zhonghua Wu. "Effect of High Hydrostatic Pressure on the Supramolecular Structure of Corn Starch with Different Amylose Contents." *International Journal of Biological Macromolecules* 85 (2016): 604–14.

Yi, Fu-jin, Jia-ao Feng, Yan-jun Wang, and Fei Jiang. "Influence of Surface Ozone on Crop Yield of Maize in China." *Journal of Integrative Agriculture* 19(2) (2020): 578–89.

Yousef, Ahmed E., Mustafa Vurma, and Luis A. Rodriguez-Romo. "Basics of Ozone Sanitization and Food Applications." In: *Nonthermal Processing Technologies for Food*, edited by G. V. Barbosa-Cánovas, H. Q. Zhang, V. M. Balasubramaniam, D. F. Farkas, C. P. Dunne and J. T. C. Yuan, 291–314. Ames, IA: Blackwell Publishing Ltd., 2011.

Yu, Yong, Lingyan Ge, Songming Zhu, Yao Zhan, and Qiuting Zhang. "Effect of Presoaking High Hydrostatic Pressure on the Cooking Properties of Brown Rice." [In eng]. *Journal of Food Science and Technology* 52(12) (2015): 7904–13.

Zeng, Feng, Qun-yu Gao, Zhong Han, Xin-an Zeng, and Shu-juan Yu. "Structural Properties and Digestibility of Pulsed Electric Field Treated Waxy Rice Starch." *Food Chemistry* 194 (2016): 1313–19.

Zeng, Feng, Tao Li, Qunyu Gao, Bin Liu, and Shujuan Yu. "Physicochemical Properties and In Vitro Digestibility of High Hydrostatic Pressure Treated Waxy Rice Starch." *International Journal of Biological Macromolecules* 120(A) (2018): 1030–38.

Zhang, Howard Q., Gustavo V. Barbosa-Cánovas, V. M. Balasubramaniam, C. Patrick Dunne, Daniel F. Farkas, and James T. C. Yuan. *Nonthermal Processing Technologies for Food*. Ames, IA: Blackwell Publishing Ltd., 2011.

Zhang, Liang, Chao-Qun Li, Wei Jiang, Mangang Wu, Sheng-Qi Rao, and Jian-Ya Qian. "Pulsed Electric Field as a Means to Elevate Activity and Expression of A-Amylase in Barley (Hordeum vulgare L.) Malting." *Food and Bioprocess Technology* 12(6) (2019): 1010–20.

Zhu, Fan. "Effect of Ozone Treatment on the Quality of Grain Products." *Food Chemistry* 264 (2018): 358–66.

Zimmermann, U., G. Pilwat, and F. Riemann. "Dielectric Breakdown of Cell Membranes." [In eng]. *Biophysical Journal* 14(11) (1974): 881–99.

Chapter 2

Current Trends in the Use of Pulsed Electric Fields for Quality Retention in Grain-Based Beverages

Pervin Basaran

CONTENTS

2.1 INTRODUCTION

Fermented dairy products have been considered the best carriers for probiotics; nevertheless, nearly 75% of the world's population shows symptoms of lactose intolerance, and on top of that, allergenic casein proteins, heart disease, and the emerging risk of cancer caused by saturated fatty acid content all limit the use of milk-based products (Prado et al., 2008; Gasparin et al., 2010; Davoodi et al., 2013). Even further, the current market for beverages in most western countries has been dominated by soft drinks with added sugars or artificial sweeteners. A potential link between the over-consumption of sugary soft drinks and obesity and other metabolic disorders has provoked the beverage industry to seek various food products with low calories but high bioactivity (Greenwood et al., 2014). An increased consumer demand for convenient food products suited to a fast lifestyle and limitations in the consumption of dairy products and sugary soft drinks have shifted consumers' interests toward naturally functional beverages made from seeds and grains that will serve as probiotic carriers (Gupta and Bajaj, 2017). The high bioactive compound content of grains also makes them suitable for diverse nutritional lifestyles such as veganism, vegetarianism, flexitarian diets and dyslipidaemic individuals (Panghal et al., 2018; Sethi et al., 2016). From an economic perspective, grain-based beverages show an exponential increase in consumption and are the fastest-growing segment of the functional food market. By 2025 functional beverages are expected to account for 40% of overall consumer food demand (Nazir et al., 2019). Some traditional grain-based functional drinks have emerged from their home regions to pass across trans-national commercial borders and become unconventional stable products in developed countries (Soni and Dey, 2014).

Traditional cereal-based beverages (Table 2.1) are drinks developed by the people of a local area using home-grown materials and ancestral techniques transferred from generation to generation as a part of immaterial cultural heritage and folk gastronomy (Nemo and Bacha, 2020). In many countries, these indigenous beverages contribute to the local population's gastronomic identity and are a fundamental part of daily cuisine for cultural or ceremonial occasions (Evans et al., 2015). In Africa lactic acid bacteria (LAB)–fermented beverages are regarded as a nutritious, energizing food ideal for those doing hard manual work. Traditional drinks are usually used not only to hydrate and refresh or for their alcoholic content, but also for the enjoyment gained from certain flavors or other characteristics. Oshikundu, a beverage produced from millet and sorghum, is consumed as a part of the traditional initiation of girls into womanhood and is a symbol of hospitality (Misihairabgwi and Cheikhyoussef, 2017). Omoladu, a sorghum-millet beer, is of sociocultural importance in Namibia, being prepared during weddings or to welcome newborn babies. In Turkey nursing mothers largely consume boza to enhance their milk production. In the case of Haria, produced from rice in India, the beverage has been used for its ethnomedicinal properties (protection from gastrointestinal ailments such as indigestion, acid formation, flatulence, vomiting, anorexia, and diarrhea) (Ghosh et al., 2014). The consumption of fermented cereal beverages with probiotics is known to have antimicrobial activity against diarrheal diseases (Sharma et al., 2017). The common processing steps for cereal-based beverages involve combinations of (i) soaking, (ii) addition of the source of enzymes (e.g., sprouting), (iii) cooking raw materials to gelatinize the starch, (iv) milling (i.e., breaking down and grinding to reduce the size), (v) extraction in water, (vi) spontaneous or inoculum-added fermentation, (vii) separating the liquid phase, (viii) homogenization to obtain colloidal suspension or emulsion and (ix) pasteurization to improve microbial stability. This can be in different orders depending on the beverage (Sethi et al., 2016). Based on the production process, grain-based beverages are grouped into three: alcoholic, lactic acid and non-fermented products (Table 2.1).

Conventionally, pasteurization and UHT treatment are applied to preserve the microbial stabilities and suspension in beverages. Food processors continually seek alternative or complementary manufacturing strategies for a longer shelf life of traditional cereal beverages while maintaining the desired flavors and quality, which is the major offer of emerging non-thermal processing technologies such as pulsed electric field (PEF) and its derivatives (e.g., HIPEF) (Silva et al., 2020). PEF presents some superior features over the conventional thermal processing of cereal beverages that include shortened processing times, accelerated heat and mass transfer, emulsion stability, better control of Maillard reactions, enhancement of final sensory quality and functionality, extended shelf life and enzyme deactivation (Hernandez-Hernandez, 2019).

PEF treatment, also referred to as electroporation or electropermeabilization, is the process where high-voltage bursts of electrical energy are applied to food placed between two electrodes. A PEF operation unit consists of a voltage pulse generator (nanoseconds to several milliseconds), a fluid handling system, a static or continuous treatment chamber suitable for voltage changes (100–300 V/cm to 20–80 kV/cm), and a monitoring/controlling system

TABLE 2.1   SOME EXAMPLES OF CEREAL-BASED BEVERAGES WITH TYPICAL PRODUCTION REGIONS AND UNIQUE LOCAL NAMES

Cereal	Microbial Inocula	Name by Country	Source
Alcoholic Beverages			
Sorghum and millet	Yeast, LAB	Sorghum beer – West Africa; opaque beer – Zimbabwe; Chibuku – South and East Africa	Dicko et al., 2006
Rice or wheat	*Rhizopus* sp., *Mucor* sp., *Aspergillus* sp., *S. cerevisiae* *B. subtilis* and LAB	Takju – Korea	Park and Lee, 2002
Yellow rice	*Aspergillus, Rhizopus, Bacillus, Lactobacillus, Wickerhamomyces* and *Saccharomycopsis*	Shaoxing – China	Xu et al., 2018
Barley, rye and wheat	N/A	Sourish Shchi – Russia	Dankovtsev et al., 2002
Rice, maize or sorghum	N/A	Tchoukoutou, jnard – Africa	Nout, 2009
Maize	N/A	Sekete – Nigeria	Sanni, 1989
Sorghum and millet	N/A	Omalodu, epwaka, otombo, okatokele – Namibia	Misihairabgwi and Cheikhyoussef, 2017
Maize and millet	N/A	Kweete – Uganda	Mwesigye and Okurut, 1995
Finger millet	N/A	Ajon – Uganda	Mwesigye and Okurut, 1995
Rice	*Bifidobacterium sp.*, LAB, yeast and molds	Haria, rice beer – central and eastern India	Ghosh et al., 2014
Brown wheat	N/A	HosGoas – Namibia	Misihairabgwi and Cheikhyoussef, 2017
Maize	*Saccharomyces, Leuconostoc citreum, Weisella cibaria, Chryseobacterium bovis* and *Bacillus safensis*	Teshuino – Mexico	Lacerda Ramos and Schwan Freitas, 2017
Rye and barley or oat	Yeast	Taar – Estonia	Soukand et al., 2015
Barley or millet	N/A	Chyang, kodo ko jaanr – India, Nepal, Bhutan, China	Ray et al., 2016; Kanwar et al., 2011
Sorghum	N/A	Pito, pirukutu – Ghana, Nigeria, Africa	Aboagye et al., 2020
Sorghum	N/A	Impeke, urwarwa, kanyanga – Burundi	Aloys and Angeline, 2009

(Continued)

TABLE 2.1 (CONTINUED)   SOME EXAMPLES OF CEREAL-BASED BEVERAGES WITH TYPICAL PRODUCTION REGIONS AND UNIQUE LOCAL NAMES

Cereal	Microbial Inocula	Name by Country	Source
Guinea, corn, millet	N/A	Burukutu – Ghana	Aboagye et al., 2020
Corn and sorghum	*Lactobacillus* spp.	Gowe – Benin	Adinsi et al., 2014
Maize	*Lactobacillus paracasei, S. cerevisiae, Pichia kluyveri*	Maize beverage – Brazil	Menezes et al., 2018
Tiger nut	N/A	Horchata – Spain	Rosella-Soto, 2018
Millet, maize and sorghum	*Lactobacillus* and *Lactococcus*	Malwa, bantu, pompe, kaffir – Africa	Muyanja et al., 2010
Rice	*Lb. pentosus, Lb. plantarum* subsp. *plantarum, Lb. sakei* subsp. *sakei*	Omegisool – Korea	Oh and Jung, 2015
Non-Alcoholic Beverages			
Cereal	Microbial Source	Name by Country	Source
Sorghum, millet	*Lb. plantarum*	Obiolor – Nigeria	Ajiboye et al., 2014
Maize	*Lb. plantarum, Weissella confuse, Streptococcus sp, Bacillus sp. S. cerevisiae, Pichia fermentans, Candida*	Calugi – Brasil	Miguel et al., 2012
Maize or sorghum	*Lb. rhamnosus* and *Lb acidophilus*	Amehewu, mageu – South Africa	Taylor and Duodu, 2019
Pearl millet and sorghum	*Lb. plantarum, L. lactis ssp. lactis, Lb. delbrueckii ssp. delbrueckii, Lb. fermentum, Lb. pentosus, Lb. curvatus*	Oshikundu – Namibia	Misihairabgwi and Cheikhyoussef, 2017
Maize	N/A	Maxau – Namibia	Misihairabgwi and Cheikhyoussef, 2017
Sorghum and pearl millet	*Lb. plantarum, Lb. fermentum* and *L. lactis*	Kununzaki – Nigeria; hulumur – Sudan; motoho oamabele – South Africa	Taylor and Duodu, 2019
Oat	LAB	Kaera kiesa, kaerapiim, kile – Estonia	Soukand et al., 2015
Maize	LAB	Agua-agria – Mexico	Taylor, 2016
Sorghum	LAB	Motoho oa mabela – South Africa	Taylor, 2016
Rice	*S. thermophilus, L. acidophilus, L. bulgaricus, B. bifidum*	Ricera – India	Panghal et al., 2018
Sorghum and millet	N/A	Urwaga, mwenge, munkoyo, bantu beer, kafir beer – Africa	Rooney and Waniska, 2000

(Continued)

TABLE 2.1 (CONTINUED)   SOME EXAMPLES OF CEREAL-BASED BEVERAGES WITH TYPICAL PRODUCTION REGIONS AND UNIQUE LOCAL NAMES

Cereal	Microbial Inocula	Name by Country	Source
Rice	*Leuc. mesenteroides, S. faecalis, Lb. plantarum, L. brevis, L. leichmannii, L. fermenti, L. cellobiosus, P. acidilactici*	Uji, ben-saalga – Africa	Nout, 2009
Wheat, malt	N/A	Sobia – Saudi Arabia	Gassem, 2002
Sorghum and/or millet	*Lb. plantarum, L. paracasei, L. fermentum, L. brevis, L. delbrueckii, Streptococcus thermophilus*	Bushera – Uganda	Mukisa et al., 2012
Millet or barley	*S. cerevisiae, Saccharomycopsis fibuligera, Candida glabrata, Lb. plantarum, Lb. bifermentans,* etc.	Chhang – India	Asrani et al., 2019
Millet	*S. cerevisiae, Saccharomyces fibuligera, Enterococcus faecium, Lb. casei, Pediococcus pentosaceus*	Sura – India	Asrani et al., 2019
Barley	Yeast and lactic acid bacteria	Kacchi – India	Asrani et al., 2019
Barley	*Alternaria, Curvularia, Bacillus*	Yu – India	Asrani et al., 2019
Maize, sorghum, and finger millet	*Lactobacillus, Streptococcus,* yeast	Togwa – West Africa, Tanzania, China, Japan	Kitabatake et al., 2003
Maize or sorghum	*Lactococcus lactis* subsp. *lactis*	Mahewu – South Africa, Zimbabwe	Mugochi et al., 2001
Rye	LAB	Kvass – Eastern Europe, Russia	Jargin, 2009
Rice	*Aspergillus*	Amazake – Japan	Matzugo et al., 2018
Maize	LAB, yeast, fungi	Pozol – Mexico, Guatemala	Perez-Almendariz et al., 2020
Maize	LAB and yeast	Tepache – Mexico	Armendariz and Cardoso-Ugarte, 2020
Rice or maze	*L. fermentum, L. delbrueckii, L. reuteri,* yeast, *Aspergillus* and *Penicillium*	Ogi – Africa	Stanton et al., 2001
Soya and millet	*Pediococcus acidilactici, Weisella confusa, L. fermentum, L. reuteri, L. salivarius, L. paraplantarum*	Fura – India	Behera and Panda, 2020

(Continued)

TABLE 2.1 (CONTINUED)   SOME EXAMPLES OF CEREAL-BASED BEVERAGES WITH TYPICAL PRODUCTION REGIONS AND UNIQUE LOCAL NAMES

Cereal	Microbial Inocula	Name by Country	Source
Barley	*Saccharomyces, Candida, Pichia, Saccharomycopsis Saccharomycopsis, LAB, Mucor, Rhizopus*	Marcha – India	Sha et al., 2016
Corn or rice	*Saccharomyces*	Chicha – Latin America, Brazil, Argentina	Puerari et al., 2015
Millet	*Lb. plantarum, Lb. fermentum* and *L. lactis*	Kununzaki – Nigeria	Adesulu-Dahunsi et al., 2020
Maize, wheat, oat, sorghum	*Lactobacillus* and *Weisella*	Atole agrico – Mexico, Central America	Vivas et al., 1987
Oat	LAB, *Bifidobacterium animalis* ssp. *lactis*	Vellie – Finland	Taylor, 2016
Oat	*Lactobacillus plantarum*	Proviva™ – Sweden	Prado et al., 2007
Oat, maize, rice, alfalfa, barley, linseed, mung beans, rye, wheat, millet, buckwheat	*Lb. acidophilus, Lb. delbrueckii* and *S. cerevisiae*	Wholegrain Probiotic Liquid™ – Australia, USA	Salmeron et al., 2017
Barley, oat, rye, millet, maize, wheat	*S. cerevisiae, Leuconostoc mesenteroides* and *Lb. confuses*	Vefa Boza™ – Turkey	Genc et al., 2002
Non-Fermented Beverages			
Maize	No fermentation	Tejuino – Mexico	Perez-Almendariz and Cardose-Ugarte, 2020
Soy milk	No fermentation	Asia	Adebayo-Tayo et al., 2020
Maize and cacao	Nixtamalized with ashes	Tejate – Mexico, southwest USA	Goze-Amore, 2015
Basil seeds	No fermentation	Falooda – Bahrain	Alalwan et al., 2017

(Alexandre et al., 2019). The effectiveness of PEF application in beverages depends on several process parameters including electric field intensity, treatment duration, explicit energy applied, pulse shape, pulse width, frequency, temperature, treatment mode (batch, continuous), configuration of treatment chamber (collinear, coaxial and parallel), physicochemical characteristics of the treated food (composition, water activity, antimicrobials, ionic compounds [strength], conductivity, pH, solid content, food matrix, suspensions, temperature, orientation in the electric field) and characteristics of the treated microbial cells (size, shape, cell wall composition, membrane characteristics) (Amiali et al., 2012; Barba et al., 2015; Vorobiev and Lebovka, 2021). The use of PEF has been reported in almost all stages of beverage processing, from the handling of raw material to the processing or storage of semi-finished and finished products before packaging. PEF application in grain beverage production would support several aspects of the final quality. The antimicrobial effect, extraction influence, improvement of the fermentation progress, modification of the food matrix, modification of food components, preservation of micronutrients and removal of anti-nutritional factors are among the major objectives of PEF application for cereal beverages.

2.2 ANTIMICROBIAL EFFECT OF PEF TREATMENT IN BEVERAGES

Fermentation remains the best choice for improving the nutritional, sensory and shelf-life properties of cereal-based beverages (Taylor, 2016). Raw cereals carry low levels of organoleptic-active compounds and give flat and unpleasant odors and flavors, and depending on the beverage, natural microflora or added starter cultures biochemically transform the biochemical content of the cereals (Zhou et al., 1998). There are two major types of fermented cereal beverages: alcoholic beverages and beverages based on lactic acid fermentation (Table 2.1). Grains with high starch content are a good carbon source for microorganisms. Starch is saccharified by amylases which have been synthesized during the grain's malting process. Although barley-based beer is the oldest and most widely known cereal-based alcoholic beverage, wheat, corn, sorghum, rice and millet are also used in traditional alcoholic beverages worldwide (Table 2.1). In spontaneous cereal-based fermentation, LAB (*Bifidobacteria*, *Lactobacillus*, *Pediococcus*, *Lactococcus*, *Leuconostoc*, *Streptococcus* spp., etc.), *Enterobacter* spp., yeasts (*Candida*, *Debaryomyces*, *Rhodotorula*, *Endomycopsis*, *Hansenula*, *Pichia*, *Saccharomyces* and *Trichosporon* spp.), *Bacillus* ssp. and filamentous fungi (*Amylomyces*, *Aspergillus*, *Mucor* and *Rhizopus* spp.) are mainly involved (Nemo and Bacha, 2020; Miguel et al., 2012; Menezes et al., 2018; Panghal et al., 2018; Santos et al., 2014). LAB, either in singular form or in combinations, predominantly contributes to the creation of a great variety of sour cereal beverages (Table 2.1). Fermentation by LAB prolongs shelf life and enhances sensory properties, in addition to improving protein digestibility and the bioavailability of minerals and other micronutrients (such as vitamins, exopolysaccharides, enzymes, antioxidant, antimicrobial compounds, probiotics and cholesterol-lowering substances) (Taylor 2016, Luana et al., 2014; Peyer et al., 2016a,b; Adebayo-Tayo, 2020; Lee and Kim, 2019; Novik and Savich, 2020). The microorganisms responsible for spontaneous fermentation originate from the cereals, the fermentation containers, and the surrounding environment; thus, the quality of the final beverage product varies enormously. The composition of the initial microbial populations results in different physicochemical properties and nutritional attributes as non-standardized products (Onyango et al., 2004; Phiri et al., 2020). In this sense, PEF presents great potential for the standardization of the initial microflora of raw-material preparation for fermentation and the improvement of the final product quality.

In manufacturing and storage, microbial safety is the main challenge for health and economic reasons. Pathogenic microorganisms such as *Escherichia coli*, *Bacillus cereus*, *Staphylococcus aureus*, *Salmonella*, *Clostridium perfringens* and *Shigella* spp. might survive and multiply in beverages to reach doses that would pose a serious threat to consumers (Aboagye et al., 2020). *Coliforms* and *Enterococci* are indicators of fecal contamination or unhygienic processing conditions. *Enterobacteriaceae* are important food safety issues related to cereal fermentation and cause unpleasant sensory properties, such as a sharp manure-like odor (Westling et al., 2016). Although more than 150 indigenous mold species can be found on grains, the filamentous fungi (*Aspergillus* spp., *Penicillium* spp. and *Fusarium* spp.) are the main safety concern due to their production of mycotoxins, which can be carcinogenic, mutagenic, genotoxic, teratogenic, neurotoxic and estrogenic (Buszewska-Forajta, 2020). *Fusarium* infections are the most problematic contamination in beer production. *Fusarium* spp. produce small polypeptides with high hydrophobicity (hydrophobins) that originate nucleation centers and the growth of bubbles (Shokribousjein et al., 2011). As a result, malt infected with *Fusarium* causes premature yeast flocculation during fermentation and an increased beverage staling (Oliveira et al., 2014).

Although it is not possible to prevent the introduction of microbial contaminants into beverage-processing facilities, it is crucial to minimize their presence in the final product (Akins-Lewenthal, 2012). Since the use of electric fields to kill microorganisms was first reported in the early 1960s, an enormous number of PEF applications have been reported in the last few decades. Microbial inactivation in liquids generally requires a strong electric field generated by very high voltage ranging from 20 kV/cm to 80 kV/cm (Huang et al., 2014). PEF can reduce the microbial load without elevating the temperature of the food; therefore, it has great potential to pasteurize fermented or non-fermented cereal beverages. The efficacy of PEF inactivation of microorganisms depends on microbial parameters (microbial type, growth phase, size and shape of microbes, cell wall characteristics and concentration) and food parameters such as pH, composition, water activity, ionic compounds and conductivity (Raso et al., 2014; Timmerman et al., 2019). PEF has been shown to be capable of inactivating a large number of serious pathogenic bacteria such as *E. coli*, *Listeria* spp., *Staphylococcus aureus*, *Salmonella* ssp. and *Pseudomonas fluorescens* in various food products (Cregenzan-Alberti et al., 2015; Sanz-Puig, 2016). Nevertheless, there are few reports about the application of PEF in grain-based beverages. Specifically, Li and Zhang (2004) reported that a soy drink treated with PEF at 40 kV/cm showed inactivation of *E. coli* more than 5 log in a duration of less than 1 min. The PEF treatment also effectively deactivated the natural flora of the soy drink and prolonged its microbial shelf life under refrigeration conditions, while the color and viscosity had not changed significantly (Li and Zhang, 2004). Selma et al. (2003) investigated the effectiveness of PEF to inactivate *E. aerogenes* and *E. coli* in the traditional Spanish drink horchata and observed 1.1 and 2.6 log reductions, respectively. In a later study, Selma et al. (2006) observed an insignificant reduction of *Listeria* in the same beverage and determined by these results that *Listeria* has an enhanced self-repairing mechanism for membranes as compared to other species. Morales et al. (2010) reported that treatment at 35 kV/cm for 1400 µs reduced *L. innocua* 5 log in a fruit juice–soy milk beverage, which resulted in a shelf life of nearly two months, similar to that gained by thermal sterilization. PEF-pretreated brewer's spent grain (BSG) extracts showed antimicrobial activity against *Salmonella typhimurium* and *L. monocytogenes* (Kumari et al., 2019), where the individual effect of PEF has not been reported. Generally, Gram-negative bacteria are more sensitive to PEF than Gram-positive bacteria and yeasts, and vegetative cells are more sensitive than bacterial spores and fungal ascospores in food products (Barbosa-Canovas et al., 1999). Studies show that PEF has a limited lethal effect on spores. For instance, Wang et al. (2020) applied PEF treatment (300 kV/cm, 30°C) for the inactivation of *B. subtilis* spores, showing only a 0.6 log reduction. As pH changes, the sensitivity of a microorganism also varies. Precisely, Gram-positive bacteria are more PEF-resistant in neutral pH than in acidic conditions, whereas Gram-negative cultures are more resistant in acidic conditions than in neutral conditions (Garcia et al., 2007). Because of their varying susceptibility, upon PEF treatment microbial cells exhibit three possible conditions: dead, undamaged or sublethally injured (Wang et al., 2018). PEF-induced sublethal damage causes a significant delay in microbial ability to grow but results in spoilage in the product during storage (Wang et al., 2018). Overall, at low electric field intensities, the cell membrane damage is reversible, and the metabolic activity and membrane integrity are reduced without any loss of viability (Saldana et al., 2009). The protective effect of microorganisms against PEF treatment could be partially related to the presence of proteins in the medium (Jaeger et al., 2009). Bacteria and yeast are vulnerable to PEF treatment at various degrees when suspended in alcohol content such as beer or wine. It was demonstrated that 4 ms of PEF treatment at 20 kV/cm was sufficient to inactivate all microorganisms (5 log reduction) present in red wine (Delsart et al., 2016). The susceptibility of both *L. brevis* and *Zygosaccharomyces baili* increased when they were suspended in alcoholic beverages (Beveridge et al., 2004). PEF treatment above the values of 13 kV/cm inactivated *L. plantarum* in beer, the bacteria were more susceptible to the PEF treatment at a higher (3.5%) ethanol concentration compared to 0.5% (Walkling-Ribeiro et al., 2011; Ulmer et al., 2002). Ethanol fermentation by *S. cerevisiae* is a common application for the production of grain-based beverages. To retain the quality properties, such as flavor stability, of the final product during storage, fermentation needs to be terminated by thermal pasteurization. A 2.2 log reduction was reported when *S. cerevisiae* ascospores in beer when exposed at 45 kV/cm (Milani, et al., 2015). PEF also poses a post–fermentation process option by protecting the beverage from further acidification. Martinez et al. (2016) and Martinez et al. (2018) investigated the use of PEF for the stimulation of *S. cerevisiae* yeast autolysis and the subsequent release of mannoprotein for a wide array of temperatures, pH values and ethanol concentrations during storage. Dimopoulos et al. (2018) reported the autolysis effect of PEF treatment (5–20 kV/cm, 1–2,000 pulses, 15-μs pulse width) on *S. cerevisiae* (8 log CFU/mL in distilled water); PEF accelerated the progress of autolysis cell disintegration up to 78% compared to untreated cells.

2.3 THE EXTRACTION EFFECT OF PEF IN CEREAL PRODUCTS

When a low-intensity electric field (lower than 20 kV/cm) is applied to living plant cells, an electric potential passes through the membrane and disintegrates the internal membranous structures, inducing reversible electroporation. Especially as the transmembrane potential across the membrane reaches a critical value of approximately 1 V, the biological membrane of the plant cell is disrupted and pores are formed in the weak areas, which results in a loss of semi-permeability, promoting the discharge of intracellular metabolites from plant material to the surrounding solution (Barba et al., 2015; Toepfl et al., 2014). There have been reports where PEF has found practical applications like the pretreatment step preceding oil and sugar extraction in which PEF improves the rate of mass transfer and can substitute and support conventional disintegration methods like grinding to increase the final yield. PEF pretreatment has been shown as an emerging technology to improve sunflower oil yield up to 2.3% (Puertolas et al., 2016). PEF has been successfully used as a pretreatment to improve the extraction of sugar from sugar beets, saving up to 50% of the downstream energy investment in the process (Sack et al., 2010). PEF pretreatment was applied for winemaking before the macerating stage, and pore formation in the grape cell membranes increased permeability, which enhanced mass transport and the extraction of color attributes, sucrose and other secondary metabolites in wine (Brianceau et al., 2015). While PEF intensities ranging from 0.5 to 2 kV/cm are applied for fresh plant materials, it was found necessary to apply higher intensities, e.g., 20 kV/cm, for drier materials (Boussetta et al., 2014). Barbosa-Pereira et al. (2018) used PEF pretreatment to recover polyphenols from coffee silver skin and cocoa bean shells, and results showed that extraction was significantly increased compared with conventional heat applications. The improved extraction of plants' biologically active compounds using PEF pretreatment can also be due to an enzymatic process that triggers *in vivo* damage to the membranous structure in cells. In another study, the recovery of isoflavonoids from soybeans was increased by 20% while the yield of germ oil and phytosterols from corn was improved by 88.4% and 32% respectively when a mild application of PEF (0.6 kV/cm) was used as a pretreatment (Guderjan et al., 2005). PEF technology has been said to have potential as a greener (less amount of energy) and relatively cheaper technique for the recovery of nutritionally valuable compounds with high quality and purity compared to other innovative extraction methods (Galanakis et al., 2015; Rosello-Soto et al., 2015). PEF (5.12 kV/cm and a number of pulse 40 n) was effective pretreatment for extracting compounds from cinnamon (Pashazade et al., 2020). PEF processing as a pretreatment can induce osmotic yield and enhance mass transfer out of the cells while leaving the food matrix relatively unchanged as compared to thermal processing (Barba et al., 2015; Baier et al., 2015). For instance, the effect of PEF on mass transfer kinetics (water loss, solid gain and water activity) and quality characteristics (color, antioxidant capacity and total phenolic compounds) during the dehydration of goji berries was investigated (Dermesonlouoglou, 2018). The results demonstrated that combined PEF and air drying (60°C) compared to conventional air drying alone led to the total processing time decrease of 33% and retained better color and higher antioxidant capacity and total phenolic content (Dermesonlouoglou, 2018). Baier et al. (2015) investigated PEF as a pretreatment for pea rehydration and compared the results with those of the conventional thermal process. PEF improved mass transport while preserving more than 99% of nutritionally valuable proteins in whole peas, without affecting quality (Baier et al., 2015). In a recent study, PEF pretreatment at 2.8 kV/cm with 3,000 pulses followed by aqueous extraction at 55°C significantly increased yields ($p < 0.05$) of carbohydrates, proteins and starches and reduced sugar in extracts from dark brewer's spent grains (BSG), a valuable industrial beer byproduct (Kumari et al., 2019).

Gentle raw material preparation (pretreatment) techniques that do not cause losses in nutritional and organoleptic attributes should be considered in the procedures to improve the solid–water extraction yield of cereal components. Among the parameters that affect cell electropermeabilization are the physicochemical characteristics (pH and conductivity) and state (suspension, solid, semi-solid) of the treated food matrix (Vorobiev and Lebovka, 2021). PEF is also characterized by its low energy consumption, and its continuous operability with shorter processing times offers a novel processing concept as a pretreatment before conventional water extraction in the cereal beverage industry.

2.4 PROGRESSIVE EFFECT OF PEF ON FERMENTATION

When PEF is applied at sublethal low electric field intensities, microbial activities (cell number, fermentation rate and production of some metabolites) are stimulated rather than inactivated (Mattar et al., 2014, 2015). Since the majority of cereal-based beverages are fermented products, PEF has a potential application to initiate the fermentation process during industrial production. Studies show that the stimulation increased the fermentation rate, mainly in the initial adaptation phase. Applying 285 V/cm field intensity to the *Hanseniaspora* sp. yeast cell suspension shortened the lag phase by approximately two-thirds (Al Daccache et al., 2020). An increase by 85% in the amount of yeast biomass was also recorded after PEF treatment at 285 V/cm, and PEF treatment did not induce irreversible deformations of the yeasts. Electron microscopic observations show no irreversible deformations of the yeast cells (Al Daccache et al., 2020). Mattar et al. (2015) showed that PEF (6 kV/cm) before fermentation increased the fructose consumption of *S. cerevisiae* up to four times at the end of the lag phase, and a 20 h decrease of the overall fermentation time was achieved compared to the control (without electric treatment). LAB fermentation is one of the main production options for cereal-based products. Recent studies have shown that PEF treatment of LAB affects growth and fermentation characteristics. *L. acidophilus* and *L. delbrueckii* treated by PEF (1 kV/cm) reached the logarithmic phase one hour earlier than the control, and both cultures showed a longer exponential phase compared to untreated cells (Najim and Aryana, 2013). The growth of *L. acidophilus* increased by 4.5%–21% at PEF 7.5 kV/cm as compared to the control cells (Lye et al., 2012). Although the underlining mechanisms are not explained yet, PEF-assisted fermentation offers practical applications to improve inoculum performance in cereal beverages.

2.5 MODIFICATION OF MATRIX IN GRAINS BY PEF

Beverages contain many constituents that partition into different phases within the product, e.g., aqueous, interfacial, gas, soluble solids and insoluble particles, etc. At low or ambient temperature, PEF treatment is applied to enhance the suspension by changing the structural, physicochemical and functional properties of macromolecules (starch, proteins, phenolic compounds, etc.) that improve the quality of the final product (Duque et al., 2020a). For instance, PEF treatment has been shown to lower granular crystallinity in starches. When the effect of PEF treatment on raw and thermally processed oat flours was examined, it was found that both sample types consistently showed differences in their particle size distribution and lower granular crystallinity in their starches (Giteru et al. 2018; Duque et al. 2020b).

Grain starch supplies most of the sugars from which the alcohol is derived in most of the world's beers. Germinated grains (malt) are more nutritious than raw barley, so much so that digestibility, bioavailable vitamins, minerals, amino acids, proteins and other phytochemicals are improved (Aboulfazli et al., 2016). Controlled application of physical energy forms such as PEF stimulates seed germination and is expected to be one of the emerging applications of PEF in beverage production (Dymek et al., 2012). No damage has been reported in the overall rate of metabolic activity of barley upon PEF treatment, while an increase in the germination power of seeds was reported. In fact, almost 100% of PEF-exposed barley seeds germinated, whereas the control showed only 76% germination (Dymek et al., 2012). A low-intensity electric field temporarily or permanently causes the loss of semi-permeability of biological membranes. As a result of electroporation, cellular compounds are released, which lowers the turgor pressure in cells and softens the grain texture. Even further, the dipole–dipole interactions introduced by PEF improve water absorption and the delignification of lignocellulosic biomass in grains (Rifna et al., 2019).

2.6 THE EFFECT OF PEF TREATMENT ON GRAIN INGREDIENTS

Cereals are complex mixtures of fibers, starch, sugars, proteins, lipids, vitamins, aroma compounds, pigments, antioxidants, minerals and other organic compounds. Many nutritional compounds in cereals are thermally sensitive and vulnerable to changes during conventional thermal food-processing methods. Another advantage of this technique is that no excess water is needed to achieve uniform cooking; consequently, nutrients and bioactive compounds are not lost significantly. In this sense, PEF offers a greener and milder processing strategy to modify cereal properties that are important for maintaining or enhancing beverage quality and consumer perception. A limited number of studies related to the effect of PEF on food with grain content has been reported; here some are discussed.

2.6.1 Carbohydrates

Grains in general are rich in carbohydrates ranging from 65% to 75% depending on the source (Poutanen et al., 2008). According to the degree of polymerization, carbohydrates in grains are classified as mono- and disaccharide sugars, oligosaccharides, starch and cell wall polysaccharides (cellulose, arabinoxylan and β-D-glucan) (Andersson et al., 2013). β-D-glucan in grains are characterized by their properties of solubility or enhanced ability to bind water, therefore the presence of β-glucans in beverages may fix high viscosity even at low concentration (Luana et al., 2014; Sullivan et al., 2013). The β-glucan content in oat is approximately 2.3%–8.5% (Flander et al., 2007), and recently, Duque et al. (2020a) reported a significant (up to 94%) increase in β-glucan content in the PEF-treated (2 kV/cm) oat suspension, indicating the ability of PEF treatment to enhance the release of soluble β-glucan from the aleurone layer of the oat bran.

Cereal grains are thermally processed in the production of essentially all beverages, and water plays a fundamental role in thermal processing as the medium for starch gelatinization. The effect of PEF treatment on the physical and functional properties of starch isolates from various sources (wheat, potato, oat, pea, waxy rice, oca, tapioca, corn and potato) has been studied, and even more studies are expected to be published (Duque et al., 2020a; Maniglia et al., 2020). PEF processing affected the physical, thermal and pasting properties of the starches whether in flour or isolated from the grain. PEF pretreatment resulted in a considerably higher amount of total starch in the case of dark BSG extracts from barley, however, no effect was observed for light BSG samples (Kumari et al., 2019). PEF treatment changed proportions of rapidly digestible starch, slowly digestible starch and resistant starch in the samples of wheat, potato and pea at different extents. For wheat starch, rapidly digestible starch and resistant starch increased significantly, whereas for pea starch, rapidly digestible starch increased and resistant starch did not alter significantly compared to the control samples (Li et al., 2019). No obvious alteration (shape, size or appearance) in starch granules was observed after PEF treatment of wheat and pea (Li et al., 2019), whereas Zeng et al. (2016) reported the degradation and reduction of the molecular weight of corn starch. Some studies have reported that PEF treatment led to a greater impact on aggregation, formation of groves and channels, granular rearrangement and destruction of the molecular structure in the crystalline and amorphous regions of starch, cellular/structural alterations and depolymerization of starch chains depending on the plant species (Duque et al., 2020a; Hong et al., 2015; Han et al., 2009). Duque et al. (2020b) investigated the effects of PEF on the functional thermal and pasting properties of native and gelatinized oat starches. Both raw oat flour that retains the native structure of starch and other macromolecules and thermally processed oat flour as typically conducted in the industry were exposed to PEF treatment. Both PEF-treated oat flours exhibited improvement in the stability of the paste without affecting the processing time (Duque et al., 2020b). PEF treatment of both raw and thermally processed oat flours appeared to lead to an overall decrease in viscosity, and such results can be attributed to the occurrence of starch depolymerization (Han et al., 2009). In other words, various phenomena such as starch granule swelling, fracture, rupture or partial gelatinization in oats caused by the initial cooking steps were intensified by PEF treatment. PEF-treated raw oat flour samples particularly exhibited a lower setback viscosity compared to their corresponding control, which reflects the retrogradation tendency of amylose in starch paste (Rafiq et al., 2016; Wang et al., 2015). Duque et al. (2020a) also concluded that PEF treatment affects the ability of amylopectin external glucan chains to form double helices for re-association and crystal formation (Duque et al., 2020a). From PEF treatment–induced rearrangement or destruction of starch molecular structure, Duque et al. (2020a) concluded that the thermal properties of food material with high starch content (such as grains) can be modified (Duque et al., 2020b). Due to structural changes in grain carbohydrates, PEF treatment results in partial softening, which is a useful result for pretreatment of cereals during the industrial production of cereal beverages. The conformational changes to α-amylase and α-amylase/pectin complexes were assessed recently by Jin et al. (2020). After PEF treatment at 20kV/cm, the content of β-sheet increased and the enzyme lost 80% of its activity. The size of α-amylase/pectin complex sizes increased while they demonstrated various shapes (branches or circles). It has been suggested that the controlled application of PEF at lower than 30 kV/cm can be a promising way to remove one or more reactive compounds from the Maillard reaction (Janositzet et al., 2011).

2.6.2 Proteins

Cereal beverages are very suitable products for a high-protein nutrition lifestyle. Especially cereal bran proteins are regarded as having high nutritional value, which can be attributed to higher and balanced amounts of essential amino acids (Apprich et al., 2014). Among all grains, soy beverages provide similar protein content to cow's milk and are generally considered to be complete for an adult (Jeske et al., 2018). Even further, due to general consumer acceptance of organoleptic properties and health-related bioactive components, soy milk–based drinks are considered one of the primary alternatives to dairy-based probiotic beverages worldwide (Bansal et al., 2016; Costa et al., 2017). The impact of PEF treatment on protein yield varies depending on the source. For instance, proteins were preserved when PEF was used for pretreatment for canola oil extraction (Zhang et al., 2017a). A study by Parniakov et al. (2015) reported improved recovery of proteins (twofold) with PEF pretreatment (13.3 kV/cm) before the aqueous extraction at 50°C compared to untreated papaya seeds. PEF treatment leading to the disorganization of primary and higher protein structure, resulting in the release of free reducing amino acids and small peptides, has been reported (Ganeva et al., 2020). Kumari et al. (2019) reported that PEF pretreatment significantly increased the extraction yields of proteins for both light and dark BSG samples as compared to untreated samples. PEF-based extraction at 40 kV/cm caused an increase (7.5%) in total free amino acids; valine, isoleucine, and leucine were the main amino acids responsible for this increase, representing an increase of 113%, 100%, and 66% in dark BSG, respectively (Kumari et al., 2019).

Earlier studies have demonstrated that the interactions between an enzyme and its substrate (protein or carbohydrates) are greatly affected by PEF treatment at various levels depending on the food composition and electric field duration. PEF may cause a change in the charge distribution of the protein surface and destroy the noncovalent bonds, such as hydrophobic interactions, electrostatic interactions and hydrogen bonds, and may cause protein unfolding and aggregation (Giteru et al., 2018; Zhang et al. 2017). Zhao et al. (2009a) suggested that protein molecules are polarized at low PEF strength and generate free radicals; afterward, proteins unfold, the hydrophobic amino acids are exposed to water and eventually unfolding proteins can aggregate. When the proteins are exposed to an external electric field, the proton migrates from the NH_3^b group to the COO group and results in an additional ion–dipole interaction (Mattison et al., 1998). Li et al. (2007) studied the effects of PEF up to 40 kV/cm on the physicochemical properties of soybean protein isolates. They reported a significant increase in surface free sulfhydryls and hydrophobicity, which resulted in induced dissociation, denaturation and reaggregation of soy proteins with increasing PEF strength. After PEF treatment, both raw and thermally processed oat flours revealed recognizable alteration of protein secondary structure (Duque et al. 2020a). The native structure is critical to protein functionality such as solubility, emulsification, gelation, water- and fat-holding capacities and foaming ability, therefore, effects of PEF need to be explored further for any given cereal beverage. In beverages, proteins and polysaccharides form complexes as a result of electrostatic interactions that in return affect the stability and texture of the drinks (Jin et al., 2020). PEF has been reported to change the charge distribution of starch and pectin (Giteru et al., 2018). Jin et al. (2020) investigated the effects of PEF on the complexes of the α-amylase enzyme and pectin, and pectin solubilization was found to be temperature dependent. It is possible that the structural modification of proteins by PEF treatment can influence their relevance for the function of an ingredient in a food product. Controlled PEF application can be used to affect the structure and function of cereal beverage proteins to improve sensory quality, product appearance and palatability (Granato et al., 2012).

2.6.3 Phenolic Compounds

Earlier studies have demonstrated that PEF can improve the retrieval, recovery and bioaccessibility of phenolic compounds and retain the high antioxidant properties of fruit juices (blackberry, tomato, green tea infusions, grapes, apples, etc.) compared to those thermally treated (Galanakis et al., 2015; Zhao et al., 2009b; Ribas-Agusti et al., 2019). Among the quality parameters of cereal-based products are the quantification of total phenolics and flavonoids. The form and the content of phenolic compounds may vary depending on the species, cultivar, growing location and environmental conditions; for instance, phenolics in wheat and rye bran range from 3 to 6mg/g (Wang et al., 2014). Phenolics are not distributed evenly in grains in that the outmost layers of the grains possess high phenolic content that can lead to poor consumer acceptance due to bitterness (Heinio et al., 2011). Studies on cereals have reported that PEF treatments may reduce or increase the phenolic content depending on intensity, duration and the type of cereal tested. For instance, a previous study reported an increase in the phenolic content when PEF treatment was applied as the pretreatment step for sorghum flour (Lohani and Muthukumarappan, 2017). PEF-assisted extraction showed a significant effect on the levels of phenolics in the case of light BSG (Kumari et al., 2019). Martin-Garcia et al. (2020) reported that PEF as a pretreatment at the electric field strength of 2.5 kV/cm frequency for 14.5 s increased free and bound phenolic recovery of BSG by 2.7 and 1.7 times respectively, as compared to samples without PEF treatment. In particular, low- and moderate-intensity PEF have been shown to increase phenolic compounds. The beverages made with oat flakes contained appreciable levels of polyphenols like tocols, phenolic compounds and avenanthramides (Peterson, 2001). The highest concentration was found in fermented beverages, and especially lactic acid fermentation was reported to improve the extraction of phenols (Katina and Poutanen, 2013). In a study of pigmented rice varieties, cooking decreased the total free phenolic compound and anthocyanin content, and after cooking there was an increase in the insoluble-bound phenolics in some samples (Melini et al., 2019). Pericarp removal produces reductions in fiber and certain associated components, causing a decline in the antioxidant capacity and a loss of phenolic compounds. Thermal processing causes the loss of important nutrients and bioactive compounds. Bound phenolics especially are reduced to half after nixtamalization (Morales and Zepeda, 2017), and ferulic acid, the most abundant phenolic compound in corn, might be degraded up to 80% (Mora-Rochin et al., 2010). PEF processing may increase or decrease the assayed phenolic content but not always the actual content in grains. The results obtained from mostly fruit beverages demonstrate that PEF can provide a high potential for the removal of phenolic compounds in whole-grain cereal beverages when it is needed.

2.6.4 Micro-Ingredients

A study by Zhang et al. (2017b) provided evidence that PEF treatment can inhibit the degradation of chlorophyll and carotenoids in spinach. Thermal processing causes substantial losses of carotenoids; on the other hand, non-thermal PEF processing results in no or insignificant losses of carotenoids depending on the chemical structure, but mixed results for the bioaccessibility of these components have been reported (Barba et al., 2017). PEF treatment affects the unsaturated bond of carotenoids, which changes the conformation of bonds from cis to trans (Zhang et al., 2017b).

Vitamins are bioactive compounds and are highly sensitive to degradation during food processing operations. Even further, food processing greatly affects the bioaccessibility of micronutrients and bioactive compounds. Generally, PEF application in food processing maintains the activity of vitamins. The effect of HIPEF (35 kV/cm in bipolar mode, 4-μs pulse width) on soy milk–fruit juice beverages was analyzed to determine in vitro bioaccessibility of vitamin C and phenolic compounds as well as hydrophilic antioxidant activity (Rodríguez-Roque et al., 2015). The results demonstrated that HIPEF at the above-mentioned conditions improved or did not change the bioaccessibility of vitamin C and some phenolic compounds (Rodríguez-Roque et al., 2015). Cereals and pseudocereal-derived products are important sources of folates; however, B-group vitamins are destroyed during cooking (Martinez-Villaluenga et al., 2020). For instance, in traditional fermented sorghum beverages, the thiamin and riboflavin requirements are low (1%) due to vitamin destruction during thermal processing (Taylor and Krugen, 2019). There is no report on the effect of PEF application on mineral content levels in cereal products. The influence of HIPEF on the mineral content of fruit juice and milk beverages was assessed after processing and during chilled storage (Salvia-Trujillo et al., 2017). During storage, mineral concentration in both beverages remained highly stable, irrespective of the HIPEF or thermal processing applied (Salvia-Trujillo et al., 2017). Ahmad Khan et al. (2018) studied the effects of a low and high electric field (2.5 kV/cm and 10 kV/cm) on the level of minerals present in chicken and beef samples. They observed that PEF treatment reduced the concentration of Ca, Na and Mg and increased the concentration of Cr in beef compared to non-treated controls, while chicken samples had higher Cu concentrations than control samples.

Electroporation causes leakage of intercellular content, which would enhance the activity of some enzymes, but in some cases, PEF can inactivate enzymes, which otherwise cause a color change depending on the material properties and treatment conditions (Lopez Gamez, 2020). PEF often has little or a limited effect on enzymes at processing conditions sufficient for microbial inactivation (50–1000 kJ/kg). Only at high specific energy input (e.g., > 1,000 kJ/kg) and ambient or mild temperature conditions does PEF cause significant inactivation of enzymes (Terefe et al., 2010). In regard to enzyme activity, peroxidase and lipoxygenase catalyze some reactions affecting phenolic compounds and polyunsaturated fatty acids, and as a result, rancid off-flavors and off-odor are generated. The peroxidase and lipoxygenase of HIPEF-treated fruit juice–soy milk beverages were moderately inactivated (17.5%–29% and 34%–39%), whereas thermal treatment achieved 100% and 51% inactivation (Fauster et al., 2021). PEF has proven to be moderately efficient in polyphenol oxidase inactivation (Sanchez-Vega et al., 2009). Electrical pulses can affect the three-dimensional structure of an enzyme, which is governed by weak non-covalent forces (e.g., hydrogen bonds and hydrophobic interactions) (Terefe et al., 2014). PEF treatment enhanced the β-glucosidase activity for the bioconversion of isoflavones from glucosidase to aglycones in biotin soy milk (Ewe Wan-Abdullah et al., 2012). An increment of approximately 29% for α-amylase activity was achieved in barley malting at 8 kV (500 Hz) (Zhang et al., 2019). Polyphenol compounds are secondary metabolites found in vacuoles and their activity increases after PEF treatment when polyphenol oxidases are released from plastids or chloroplasts (Nowacka et al., 2019). Lipolytic enzymes cause the rapid release of free fatty acids, which are further subject to oxidation and an increase of rancidity (Kim et al., 2014). Rice bran is a byproduct of milling and a rich source of oil (> 20%), and thermal treatment is applied as a prerequisite step to inactivate endogenous lipase enzymes responsible for the production of off-flavors and rancidity (Wu et al., 2020). Qian et al. (2014) demonstrated that PEF (15 kV/cm) was effective in inactivating lipase (60%) in brown rice grains, and PEF was not harmful to the subsequent milling of brown rice (Qian et al., 2014). Pectin methylesterase activity increases with ionic strength (Toepfl et al., 2014). PEF-induced disintegration of the membrane and the transport of intracellular ions through the cell wall is expected to facilitate pectin demethylesterification in cereal beverages.

Cereals are also a source of natural antinutrients (phytases, oxalate, etc.). PEF presents an option for the elimination of these antinutrients from cereal-based beverages. Phytase, mainly concentrated in the grain's outer layers, is considered an antinutrient since it binds and reduces the bioavailability and absorption of essential divalent ions (magnesium, iron, zinc and calcium) (Gupta et al., 2015). Besides, certain strains of lactic acid bacteria involved in beverage fermentation are also a source of phytases (Gobbetti et al., 2005). In a current study, Liu et al. (2018) demonstrated that PEF treatment of oca tubers at different electric field strengths up to 1.2kV/cm can reduce the levels of oxalate contents by 50%. PEF pretreatment and electroporation is also a suitable method for the targeted removal of protease inhibitor activity or allergic food components in order to enhance the nutritional quality of the seeds as beverage ingredients.

2.7 RECENT ADVANCES AND FUTURE PROSPECTS

Depending on the raw material and processing technology applied, there are numerous nutritional improvements acquired for cereal-based beverages. These include increased protein quantity, quality and digestibility, increased vitamin levels, decreased anti-nutritional factors such as tannins and phytates, improved mineral availability, assembly of prebiotics and probiotic microorganisms in accordance with functional needs and finally, improved deliciousness, viscosity and organoleptic acceptability for a wide range of consumers. Yet, further studies should be conducted to develop pretreatment models with all functional parameters suitable for the pre- and post-processing of high-solid, complex cereal beverages. Sorghum, pearl millet and finger millet have been the staple crop and the main source of dietary energy in dry and semiarid tropic regions of Africa and Asia. The consumption of beverages made from these unconventional cereal varieties can be encouraged for gluten-intolerant consumers (Anunciacao et al., 2017). Production and commercialization of traditional and novel beverages made from these grains need particular attention paid to innovative technologies that can preserve nutritional value. However, some sorghum and millet varieties contain anti-nutritional factors such as trypsin and amylase inhibitors, phytate, phenolics and tannins, which restrict protein and carbohydrate bioavailability (McKevith, 2004). The effects of innovative PEF applications on the levels of anti-nutrient compounds in cereals need to be evaluated for beverage production.

PEF processing has proven its effectiveness for stabilizing various fruit juices or milk and maintaining their freshness and quality attributes; however, there is scarce information about the effect of this technology on cereal beverages (fermented or non-fermented) products. One of the technological challenges for fermented beverages is to ensure that the probiotic microorganisms are viable throughout their shelf life and that sensory changes during shelf life are acceptable to consumers. The main drawback of probiotics-containing beverages is that strains of lactic acid bacteria continue to produce lactic acid through active metabolism and cause undesirable post-acidification and a resulting decrease in pH during storage (Ferdousi et al., 2013). The metabolism of probiotic cultures in foods should be controlled, without negatively affecting their viability and gastrointestinal functional properties (Lanciotti et al., 2007; Bevilacgua et al., 2016; Ferdousi et al., 2013). Further detailed studies are needed to investigate the selective effects of PEF treatment along with attenuation applications for the control of LAB fermentation. Thus there is a need to study the effect of PEF on the microbial stability, physicochemical assets and viscosity-related enzymes of cereal beverages while considering the inactivation of target microorganisms case by case depending on the beverage properties. The uncontrolled fermentation process by diverse microbial communities results in differences in product quality, taste, acceptability and microbial stability, and the quality and stability of these products vary considerably. Inconsistency in the organoleptic properties of cereal-based beverages is considered one major drawback and a major challenge to overcome for the industry. PEF presents an industrial option for standardization of the initial microbial load and variation. Pretreatment and post-treatment of fermented products by PEF can help the standardization of safe and stable final products. The effects of food composition (proteins and other matrix materials) on the resistance of microorganisms to PEF are many-sided; the PEF treatment parameters that were tested in a simulated system should be investigated yet again in actual cereal beverage systems to ensure that the intended level of microbial reduction is achieved in complex cereal beverages. Even further, the usefulness of PEF in treating high-electrical-conductivity food products (such as fermented acidic beverages) is a concern, as the given energy is transformed into heat (Amiali and Ngadi, 2012; Li and Farid, 2016). New microbial threats continue to arise with more aggressive pathogens adapting to the environment and producing mycotoxins. The emergence of microorganisms resistant to conventional preservation techniques (thermal treatments, etc.) increases the necessity of techniques such as PEF and its derivatives to inhibit unwanted microbial growth.

The consumer acceptance of newly developed or modified products is another area of concern for functional beverage development. Many factors (personal needs, cultural beliefs, allergies, intolerances, etc.) affect consumers' attitudes toward any food. Taste and other sensory characteristics of foods are expected to be faultless since they strongly influence food choices, in many cases surpassing health issues; for instance, a consumer survey reported that consumer acceptance of lactic acid–fermented cereal-based beverages was based on the aroma (59%) whereas nutritional aspects were considered less important (only 36%) (Azzurra and Paola, 2009; Yu and Bogue, 2013; Dongmo et al., 2016). From the sensory point of view, consumer acceptance of cereal-based beverages even after PEF treatment remains challenging. For example, among the beverages, one challenging task is to replace sucrose with naturally occurring sweeteners that effectively address concerns of sugar intake for patients with diabetes and provide a healthy choice for consumers. Flavor improvement is therefore a key step for the development and consumption habits of grain-based foods. Different cereals have complex nutrient compositions and when mixed in a certain proportion, they can considerably modify the nutritional and organoleptic properties of the beverages. The main volatile compounds reported in studies on fermented cereal beverages have been associated with carbohydrate (mainly carboxylic acids and aldehydes, ketones, esters) and amino acid (mainly aldehydes and alcohols) metabolism. PEF, along with other processing technologies, triggers chemical responses that are directly related to the sensory quality of final products. PEF pretreatment for improving desirable volatiles, proteins, phenolic compounds, carbohydrates and isothiocyanates within the final product needs to be investigated. In particular, PEF offers the post-fermentation preservation of aroma compounds without further acidification of the beverage. Beverage manufacturers have established quantitative criteria that are used to determine the end of the shelf life of the beverage, whereas consumers judge the quality of a finished beverage product by its visual appearance (clarity, cloudiness, etc.) upon purchase. Further investigations of the quantitative and qualitative analysis of key sensory compounds and nutritional/functional components need to be conducted for each cereal beverage.

PEF as a non-thermal processing technology is a great tool for the upscaling and industrialization of traditionally processed beverages to improve their large-scale production and longer commercial shelf life and enhance their attractiveness and sensory properties with preserved functional properties. To date, the majority of the existing research activities have been obtained at lab scale with small or moderate-sized batch-treatment PEF units. To successfully introduce novel and functional cereal-based beverages processed with PEF into the market, science-based demonstration and authentication of health claims related to beverage content is crucial. Furthermore, an appropriate marketing strategy, including clear transparency accompanied by a clear regulatory framework, needs to be implemented if traditional cereal beverages are to be novelly processed. The promotion of the health and nutritional benefits of functional health foods with local identities and the standardization of their processing technologies (e.g., PEF) can further benefit local beverage manufacturers.

PEF (for extraction, modification, preservation) is among the most significant non-thermal technologies that have the potential to be commercialized in the coming decade or so (Hernandez-Hernandez et al., 2019). The efficiency of PEF for the preservation of high solids and viscous heat-labile liquids such as cereal beverages is still a challenge and further processing strategies need to be investigated. Some studies have demonstrated that alternative treatment-chamber configurations influence the effectiveness of continuous PEF pasteurization. Schottoff et al. (2020) reported the efficiency of continuous PEF treatment for the eradication of *L. innocua* in 2%, 10% and 25% whey protein (pH 4 and pH 7), using a co-linear treatment chamber and parallel plate configurations. Independent of solid contents, a similar microbial inactivation level was observed using a co-linear continuous treatment chamber. Nevertheless, a co-linear configuration is characterized by a relatively high resistance that results in inhomogeneous treatment within the product (Jaeger et al., 2009). The design and setup of a suitable PEF treatment chamber will be a key factor during the development of homogeneous treatment models (Alkhafaji and Farid, 2007). The amount of commercial PEF equipment at the industrial scale is constantly increasing by aiming for better conservation of nutritive values and organoleptic properties. In recent years the commercial applications of PEF technology transitioned from lab- and pilot-scale to industrial equipment reported by a number of companies worldwide (Coolwave Processing B.v., Diversified Technolgies, Inc., Energy Pulse Systems Lda., Heat and Control, Inc., Elea GmbH, Inotec, KEA-Tec GmbH, Steribeam, Scandinova, Wek-tec, Energy Pulse Systems, Montena Technology, Pulsemaster, Thomson-CSF, Diversified Technologies, IXL, PurePulse, etc). PEF has still not been widely employed by food manufacturers because the cost of PEF processing per unit operations is higher than that of thermal processing and the treated products may need to be degassed (Krahl et al., 2009). Depending on the PEF generator characteristics, the total costs for liquid processing are 10–80€ per ton of food (Roohinejad et al., 2018). The major challenge for the beverage industry is the development of new processing units, which ultimately provide an economic advantage compared to conventional thermal processing. As the consumption of indigenous cereal beverages is on the rise, so is the concern about the safety and hygiene of these beverages and their potential health risk of foodborne diseases all around the world. The presence of pathogens in traditional fermented foods is well documented, and the extent of the safety of these products is determined in part by processors' knowledge, attitudes and production practices. In some countries, unbroken cold storage is not available for the general public, and as a result, locally produced beverages suffer from short shelf life. PEF units that are affordable, moderate in size, portable, economic to maintain and relatively less sophisticated to operate offer convenience for small and medium-sized food producers in many countries. Major critical considerations that need to be addressed are the competitiveness of the PEF application with regard to cost and efficiency, and its appropriateness for diverse food contents. It is expected that new modifications of the technical properties of PEF processing equipment will increase production efficiency while lowering the cost per food unit. Researchers and developers also need to pay attention to the establishment of relevant national food safety laws and regulations.

Hurdle applications in which PEF is combined with various treatments lead to an increase in cell membrane permeability and eventually cause higher microbial death (Amiali and Ngadi, 2012). Depending on the required inactivation of the target microflora and the beverage composition, it may be advantageous to combine PEF treatment with physical (ultraviolet light, microwave, high-intensity light pulses, ultrasound, etc.), chemical (antimicrobial constituents, pH, high-pressure CO_2, etc.) or even mild thermal treatments. For instance, although PEF at near-ambient temperatures has no effect on spore destruction, the combination of elevated temperatures with PEF has been reported to be effective in the inactivation of *B. subtilis* spores (Siemer et al., 2014). Even further, PEF pretreatment of cereals combined with other processes offers the potential to improve fermentation properties such as increasing water-soluble carbohydrates and proteins during the brewing/fermentation process. In future studies, hurdle strategies where PEF is combined with other non-thermal processes and the efficiency of proper chambers for hurdle treatments on a commercial scale will need to be explored in detail.

2.8 CONCLUSION

There is a complex relationship between human intestinal microflora and issues such as malnutrition, obesity, diabetes, insulin resistance and chronic inflammation, which can lead to chronic diseases and even cancer. It is estimated that 400 kinds of probiotic foods are available worldwide, and 80% of these products are probiotic fermented dairy products, while fermented cereal products on the market are very rare. Fermented cereal beverages are a rich source of probiotics, prebiotics and dietary fiber. There are many kinds of cereals, each with its own uniqueness and distinctive characteristics. Consumers and the food market are eager to explore new products that are safe, delicious, and nutritious. Conventionally, pasteurization and UHT treatment are commonly applied to preserve suspension and microbial stabilities while extending the shelf life of grain-based beverages. PEF is the most promising non-thermal alternative option with combined effects in the extraction and product preservation of grain-based beverages and will likely be increasingly commercialized in the coming decade.

REFERENCES

Aboagye, G., Gbolnyo-Cass, S., Kortei, N. K., and Annan, T. 2020. Microbial evaluation and some proposed good manufacturing practices of locally prepared malted corn drink ("asaana") and *Hibiscus sabdarifa* calyxes oextract ("sobolo") beverages sold at a university cafeteria in Ghana. *Scientific African* 8:330–339.

Aboulfazli, F., Shori, A. B., and Baba, A. S. 2016. Effects of the replacement of cow milk with vegetable milk on probiotics and nutritional profile of fermented ice cream. *LWT – Food and Science and Technology* 70:261–270.

Adebayo-Tayo, B., and Fashogbon, R. 2020. In vitro antioxidant, antibacterial, in vivo immunomodulatory, antitumor, and hematological potential of exopolysaccharide produced by wild type and mutant *Lactobacillus delbureckii* subsp. *bulgaricus*. *Heliyon* 62:3268.

Adesulu-Dahunsi, A. T., Dahunsi, S. O., and Olayanju, A. 2020. Synergistic microbial interactions between lactic acid bacteria and yeasts during production of Nigerian indigenous fermented foods and beverages. *Food Control* 110:106963.

Adinsi, L., Akissoe, H. N., Vieira-Dalodé, H. N., and Anihouvi, G. 2014. Sensory evaluation and consumer acceptability of a beverage made from malted and fermented cereal: Case of Gowe from Benin. *Food Science and Nutrition* 3:1–17.

Ajiboye, T. O., Iliasu, G. A., Adeleye, A. O. et al. 2014. Nutritional and antioxidant dispositions of sorghum/millet-based beverages indigenous to Nigeria. *Food Science and Nutrition* 2(5):597–604.

Akins-Lewenthal, D. 2012. Supply chain management to maintain microbiological integrity of processed grain ingredients. *Cereal Foods World* 57(3):115–117.

Al Daccache, M., Koubaa, M., Salameh, D., Vorobiev, E., Maroun, R. G., and Louka, N. 2020. Control of the sugar/ethanol conversion rate during moderate pulsed electric field-assisted fermentation of a Hanseniaspora sp. strain to produce low-alcohol cider. *Innovative Food Science and Emerging Technologies* 59:102258.

Alalwan, T. A., Mandeel, Q. A., and Al-Sarhani, L. 2017. Traditional plant-based foods and beverages in Bahrain. *Journal of Ethnic Foods* 4(4):274–283.

Alexandre, E. M. C., Pinto, C. A., Moreira, S. A., Pintado, M., and Saraiva, J. A. 2019. Nonthermal food processing/preservation technologies. In: *Saving Food*, ed. Charis M. Galanakis, 141–169. Academic Press.

Alkhafaji, S. R., and Farid, M. 2007. An investigation on pulsed electric fields technology using new treatment chamber design. *Innovative Food Science and Emerging Technologies* 8(2):205–212.

Aloys, N., and Angeline, N. 2009. Traditional fermented foods and beverages in Burundi. *Food Research International* 42(5–6):588–594.

Amiali, M., and Ngadi, M. O. 2012. Microbial decontamination of food by pulsed electric fields (PEFs). In: *Microbial Decontamination in the Food Industry*, eds. A. Demirci, and M. O. Ngadi, 407–449. Woodhead Publishing.

Andersson, A. A., Andersson, R., Piironen, V., et al. 2013. Contents of dietary fibre components and their relation to associated bioactive components in whole grain wheat samples from the HEALTHGRAIN diversity screen. *Food Chemistry* 136(3–4):1243–1248.

Anunciação, P. C., de Morais Cardoso, L., Gomes, J. V. P., et al. 2017. Comparing sorghum and wheat whole grain breakfast cereals: Sensorial acceptance and bioactive compound content. *Food Chemistry* 221:984–989.

Apprich, S., Tirpanalan, O., Hell, J., et al. 2014. Wheat bran-based biorefinery 2: Valorization of products. *LWT – Food Science and Technology* 56(2):222–231.

Asrani, P., Patial, V., and Asrani, R. K. 2019. *Production of Fermented Beverages: Shedding Light on Indian Culture and Traditions, Production and Management of Beverages*, 409–437. Woodhead Publishing.

Azzurra, and Paola. 2009. Consumers behaviors and attitudes toward healthy food products: The Case of organic and functional foods. *113th EAAE Seminar: A Resilient European Food Industry and Food Chain in a Challenging World*: pp. 1–15.

Baier, A. K., Bußler, S., and Knorr, D. 2015. Potential of high isostatic pressure and pulsed electric fields to improve mass transport in pea tissue. *Food Research International* 76:66–73.

Bansal, S., Mangal, M., Sharma, S. K., Yadav, D. N., and Gupta, R. K. 2016. Optimization of process conditions for developing yoghurt like probiotic product from peanut. *LWT – Food and Science and Technology* 73:6–12.

Barba, F. J., Koubaa, M., do Prado-Silva, L., Orlien, V., and de Souza Sant'Ana, A. 2017. Mild processing applied to the inactivation of the main foodborne bacterial pathogens: A review. *Trends in Food Science and Technology* 66:20–35.

Barba, F. J., Parniakov, O., Pereira, S. A., et al. 2015. Current applications and new opportunities for the use of pulsed electric fields in food science and industry. *Food Research International* 77:773–798.

Barbosa-Cánovas, G. V., Góngora-Nieto, M. M., Pothakamury, U. R., and Swanson, B. G. 1999. PEF inactivation of vegetative cells, spores, and enzymes in foods. In: *Preservation of Foods with Pulsed Electric Fields*, eds. G. V. Barbosa-Cánovas, U. R. Pothakamury, and M. M. Góngora-Nieto, 108–152. Academic Press.

Barbosa-Pereira, L., Guglielmetti, A., and Zeppa, G. 2018. Pulsed electric field assisted extraction of bioactive compounds from cocoa bean shell and coffee silverskin. *Food and Bioprocess Technology* 11(4):818–835.

Behera, S. S., and Panda, S. K. 2020. Ethnic and industrial probiotic foods and beverages: Efficacy and acceptance. *Current Opinion in Food Science* 32:29–36.

Beveridge, J. R., Wall, K., MacGregor, S. J., Anderson, J. G., and Rowan, N. J. 2004. Pulsed electric field inactivation of spoilage microorganisms in alcoholic beverages. *Proceedings of the IEEE* 92(7):1138–1143.

Bevilacqua, A., Casanova, F. P., Petruzzi, L., Sinigaglia, M., and Corbo, M. R. 2016. Using physical approaches for the attenuation of lactic acid bacteria in an organic rice beverage. *Food Microbiology* 53(B):1–8.

Boussetta, N., Soichi, E., Lanoisellé, J. L., and Vorobiev, E. 2014. Valorization of oilseed residues: Extraction of polyphenols from flaxseed hulls by pulsed electric fields. *Industrial Crops and Products* 52:347–353.

Brianceau, S., Turk, M., Vitrac, X., and Vorobiev, E. 2015. Combined densification and pulsed electric field treatment for selective polyphenols recovery from fermented grape pomace. *Innovative Food Science and Emerging Technologies* 29:2–8.

Buszewska-Forajta, M. 2020. Mycotoxins, invisible danger of feedstuff with toxic effect on animals. *Toxicon* 182:34–53.

Costa, K. K. F. D., Júnior, M. S. S., Rosa, S. I. R., Caliari, M., and Pimentel, T. C. 2017. Changes of probiotic fermented drink obtained from soy and rice byproducts during cold storage. *LWT – Food and Science and Technology* 78:23–30.

Cregenzán-Alberti, O., Halpin, R. M., Whyte, P., Lyng, J. G., and Noci, F. 2015. Study of the suitability of the central composite design to predict the inactivation kinetics by pulsed electric fields (PEF) in *Escherichia coli, Staphylococcus aureus* and *Pseudomonas fluorescens* in milk. *Food and Bioproducts Processing* 95:313–322.

Dankovtsev, A. V., Vostrikov, S. V., and Markina, N. S. 2002. Fermentation studies on the traditional Russian drink "Sourish Shchi". *Journal of the Institute of Brewing* 108(4):474–477.

Davoodi, H., Esmaeili, S., and Mortazavian, A. M. 2013. Effects of milk and milk products consumption on cancer: A review. *Comprehensive Review in Food Science and Food Safety* 12(3):249–264.

Delsart, C., Grimi, N., Boussetta, N., et al. 2016. Impact of pulsed-electric field and high-voltage electrical discharges on red wine microbial stabilization and quality characteristics. *Journal of Applied Microbiology* 120(1):152–164.

Dermesonlouoglou, E., Chalkia, A., Dimopoulos, G., and Taoukis, P. 2018. Combined effect of pulsed electric field and osmotic dehydration pre-treatments on mass transfer and quality of air dried goji berry. *Innovative Food Science and Emerging Technologies* 49:106–115.

Dicko, M. H., Gruppen, H., Traoré, A. S., Voragen, A. G. J., and van Berkel, W. J. H. 2006. Sorghum grain as human food in Africa: Relevance of content of starch and amylase activities. *African Journal of Biotechnology* 5:384–395.

Dimopoulos, G., Stefanou, N., Andreou, V., and Taoukis, P. 2018. Effect of pulsed electric fields on the production of yeast extract by autolysis. *Innovative Food Science and Emerging Technologies* 48:287–295.

Dongmo, S. N., Procopio, S., Sacher, B., and Becker, T. 2016. Flavor of lactic acid fermented malt based beverages: Current status and perspectives. *Trends in Food Science and Technology* 54:37–51.

Duque, S. M. M., Leong, S. Y., Agyei, D., Singh, J., Larsen, N., and Oey, I. 2020a. Modifications in the physicochemical properties of flour "fractions" after Pulsed Electric Fields treatment of thermally processed oat. *Innovative Food Science and Emerging Technologies* 64:102406.

Duque, S. M. M., Leong, S. Y., Agyei, D., Singh, J., Larsen, N., and Oey, I. 2020b. Understanding the impact of Pulsed Electric Fields treatment on the thermal and pasting properties of raw and thermally processed oat flours. *Food Research International* 129:108839.

Dymek, K., Dejmek, P., Panarese, V., et al. 2012. Effect of pulsed electric field on the germination of barley seeds. *LWT-Food Science and Technology* 47(1):161–166.

Evans, J., Fiore, R., Pedersen, J. A., and Frøst, M. B. 2015. Place-based taste: Geography as a starting point for deliciousness. *Flavour* 4:7.

Ewe, J. A., Wan-Abdullah, W. N., Alias, A. K., and Liong, M. T. 2012. Enhanced growth of lactobacilli and bio-conversion of isoflavones in biotin-supplemented soymilk by electroporation. *International Journal of Food Sciences and Nutrition* 63(5):580–596.

Fauster, T., Ostermeier, R., Scheibelberger, R., and Jäger, H. 2021. Pulsed electric field (PEF) application in the potato industry. In: *Innovative Food Processing Technologies*, eds. K. Knoerzer, and K. Muthukumarappan, 253–270. Elsevier.

Ferdousi, R., Rouhi, M., Mohammadi, R., Mortazavian, A. M., Khosravi-Darani, K., and Rad, A. H. 2013. Evaluation of probiotic survivability in yogurt exposed to cold chain interruption. *Iranian Journal of Pharmaceutical Research: IJPR* 12(Suppl):139–144.

Flander, L., Salmenkallio-Marttila, M., Suortti, T., and Autio, K. 2007. Optimization of ingredients and baking process for improved wholemeal oat bread quality. *LWT-Food Science and Technology* 40(5):860–870.

Galanakis, C. M., Barba, F. J., and Prasad, K. N. 2015. Cost and safety issues of emerging technologies against conventional techniques. In: *Food Waste Recovery: Processing Technologies and Industrial Techniques*, ed. C. M. Galakanis, 321–336. Academic Press.

Ganeva, V., Angelova, B., Galutzov, B., Goltsev, V., and Zhiponova, M. 2020. Extraction of proteins and other intracellular bioactive compounds from baker's yeasts by pulsed electric field treatment. *Frontiers in Bioengineering and Biotechnology* 8:1433–1439.

Garcia, D., Gómez, N., Mañas, P., Raso, J., and Pagán, R. 2007. Pulsed electric fields cause bacterial envelopes permeabilization depending on the treatment intensity, the treatment medium pH and the microorganism investigated. *International Journal of Food Microbiology* 113(2):219–227.

Gasparin, F. S. R., Teles, J. M., and Araújo, S. C. 2010. Alergia à proteína do leite de vaca versus intolerância à lactose: As diferenças e semelhanças. *Revista Saúde & Pesquisa* 3:107–114.

Gassem, M. A. A. 2002. A microbiological study of sobia: A fermented beverage in the western province of Saudi Arabia. *World Journal of Microbiology and Biotechnology* 18(3):173–177.

Genc, M., Zorba, M., and Ova, G. 2002. Determination of rheological properties of boza by using physical and sensory analysis. *Journal of Food Engineering* 52(1):95–98.

Ghosh, K., Maity, C., Adak, A., et al. 2014. Ethnic preparation of haria, a rice-based fermented beverage, in the province of lateritic West Bengal, India. *Ethnobotanical Res. Appl.* 12:039–049.

Giteru, S. G., Oey, I., and Ali, M. A. 2018. Feasibility of using pulsed electric fields to modify biomacromolecules: A review. *Trends in Food Science and Technology* 72:91–113.

Gobbetti, M., De Angelis, M., Corsetti, A., and Di Cagno, R. 2005. Biochemistry and physiology of sourdough lactic acid bacteria 16(1–3):57–69.

Gonze-Amaro, R. M., Figueroa-Cárdenas, J. D., Perales, H., and Santiago-Ramos, D. 2015. Maize races on functional and nutritional quality of tejate: A maize-cacao beverage. *LWT – Food Science and Technology* 63(2):1008–1015.

Granato, D., Masson, M. L., and Ribeiro, J. C. B. 2012. Sensory acceptability and physical stability evaluation of a prebiotic soy-based dessert developed with passion fruit juice. *Food Science and Technology* 32(1):119–126.

Greenwood, D., Threapleton, D., Evans, C., Cleghorn, C., Nykjaer, C., Woodhead, C., and Burley, V. 2014. Association between sugar-sweetened and artificially sweetened soft drinks and type 2 diabetes. Systemic review and dose response meta-analysis of prospective studies. *British Journal of Nutrition* 112(5):725–734.

Guderjan, M., Töpfl, S., Angersbach, A., and Knorr, D. 2005. Impact of pulsed electric field treatment on the recovery and quality of plant oils. *Journal of Food Engineering* 67(3):281–287.

Gupta, M., and Bajaj, B. K. 2017. Development of fermented oat flour beverage as a potential probiotic vehicle. *Food Bioscience* 20:104–109.

Gupta, R. K., Gangoliya, S. S., and Singh, N. K. 2015. Reduction of phytic acid and enhancement of bioavailable micronutrients in food grains. *Journal of Food Science and Technology* 52(2):676–684.

Han, Z., Zeng, X. A., Yu, S. J., Zhang, B. S., and Chen, X. D. 2009. Effects of pulsed electric fields (PEF) treatment on physicochemical properties of potato starch. *Innovative Food Science and Emerging Technologies* 10(4):481–485.

Heini, R. L., Kaukovirta-Norja, A., and Poutanen, K. 2011. Flavor in processing new oat foods. *Cereal Foods World (CFW)* 56(1):21–26.

Hernández-Hernández, H. M., Moreno-Vilet, L., and Villanueva-Rodríguez, S. J. 2019. Current status of emerging food processing technologies in Latin America: Novel non-thermal processing. *Innovative Food Science and Emerging Technologies* 58:102233.

Hong, J., Chen, R., Zeng, X.A., and Han, Z. 2015. Effect of pulsed electric fields assisted acetylation on morphological, structural and functional characteristics of potato starch. *Food Chemistry* 192:15–24.

Huang, K., Jiang, T., Wang, W., Gai, L., and Wang, J. 2014. A comparison of pulsed electric field resistance for three microorganisms with different biological factors in grape juice via numerical simulation. *Food and Bioprocess Technology* 7(7):1981–1995.

Jaeger, H., Meneses, N., and Knorr, D. 2009. Impact of PEF treatment inhomogeneity such as electric field distribution, flow characteristics and temperature effects on the inactivation of E. coli and milk alkaline phosphatase. *Innovative Food Science and Emerging Technologies* 10(4):470–480.

Janositz, A., Noack, A. K., and Knorr, D. 2011. Pulsed electric fields and their impact on the diffusion characteristics of potato slices. *LWT-Food Science and Technology* 44(9):1939–1945.

Jargin, S. V. 2009. Kvass: A possible contributor to chronic alcoholism in the former Soviet Union—Alcohol content should be indicated on labels and in advertising. *Alcohol and Alcoholism* 44(5):529.

Jeske, S., Zannini, E., and Arendt, E. K. 2018. Past, present and future: The strength of plant-based dairy substitutes based on gluten-free raw materials. *Food Research International* 110:42–51.

Jin, W., Wang, Z., Peng, D., et al. 2020. Effect of pulsed electric field on assembly structure of α-amylase and pectin electrostatic complexes. *Food Hydrocolloids* 101:105547.

Kanwar, S. S., Gupta, M., Katoch, C., and Kanwar, P. 2011. Cereal based traditional alcoholic beverages of Lahaul and Spiti area of Himachal Pradesh. *Indian Journal of Traditional Knowledge* 10:251–257.

Katina, K., and Poutanen, K. 2013. Nutritional aspects of cereal fermentation with lactic acid bacteria and yeast. *Handbook on Sourdough Biotechnology*: pp. 229–244.

Khan, A. A., Randhawa, M. A., Carne, A., et al. 2018. Effect of low and high pulsed electric field processing on macro and micro minerals in beef and chicken. *Innovative Food Science and Emerging Technologies* 45:273–279.

Kim, S. M., Chung, H. J., and Lim, S. T. 2014. Effect of various heat treatments on rancidity and some bioactive compounds of rice bran. *Journal of Cereal Science* 60(1):243–248.

Kitabatake, N., Gimbi, D. M., and Oi, Y. 2003. Traditional non-alcoholic beverage, Togwa, in East Africa, produced from maize flour and germinated finger millet. *International Journal of Food Science and Nutrition* 54(6):447–455.

Krahl, M., Zarnkow, M., Stürmer, F., and Becker, T. 2009. Flavor stability of alternative malt-based beverages. *Master Brewers Association of the America Technical Quarterly* 46:1–7.

Kumari, B., Tiwari, B. K., Walsh, D., et al. 2019. Impact of pulsed electric field pre-treatment on nutritional and polyphenolic contents and bioactivities of light and dark brewer's spent grains. *Innovative Food Science and Emerging Technologies* 54:200–210.

Lacerda Ramos, C., and Schwan Freitas, R. 2017. Technological and nutritional aspects of indigenous Latin America fermented foods. *Current Opinion in Food Science* 13:97–102.

Lanciotti, R., Patrignani, F., Iucci, L., Saracino, P., and Guerzoni, M. E. 2007. Potential of high pressure homogenization in the control and enhancement of proteolytic and fermentative activities of some Lactobacillus species. *Food Chemistry* 102(2):542–550.

Lee, S., and Kim, M. 2019. Leuconostoc mesenteroides MKSR isolated from kimchi possesses α-glucosidase inhibitory activity, antioxidant activity, and cholesterol-lowering effects. *LWT* 116:108570.

Li, Q., Wu, Q. Y., Jiang, W., et al. 2019. Effect of pulsed electric field on structural properties and digestibility of starches with different crystalline type in solid state. *Carbohydrate Polymers* 207:362–370.

Li, S. Q., and Zhang, Q. H. 2004. Inactivation of E. coli 8739 in enriched soymilk using pulsed electric fields. *Journal of Food Science* 69(7):169–174.

Li, X., and Farid, M. 2016. A review on recent development in non-conventional food sterilization technologies. *Journal of Food Engineering* 182:33–45.

Li, Y., Chen, Z., and Mo, H. 2007. Effects of pulsed electric fields on physicochemical properties of soybean protein isolates. *LWT-Food Science and Technology* 40(7):1167–1175.

Liu, T., Burritt, D. J., Eyres, G. T., and Oey, I. 2018. Pulsed electric field processing reduces the oxalate content of oca (Oxalis tuberosa) tubers while retaining starch grains and the general structural integrity of tubers. *Food Chemistry* 245:890–898.

Lohani, U. C., and Muthukumarappan, K. 2017. Process optimization for antioxidant enriched sorghum flour and apple pomace based extrudates using liquid CO2 assisted extrusion. *LWT* 86:544–554.

Lopez-Gamez, G., Elez-Martínez, P., Martín-Belloso, O., and Soliva-Fortuny, R. 2020. Pulsed electric fields affect endogenous enzyme activities, respiration and biosynthesis of phenolic compounds in carrots. *Postharvest Biology and Technology* 168:111284.

Luana, N., Rossana, C., Curiel, J. A., Kaisa, P., Marco, G., and Rizzello, C. G. 2014. Manufacture and characterization of a yogurt-like beverage made with oat flakes fermented by selected lactic acid bacteria. *International Journal of Food Microbiology* 185:17–26.

Lye, H., Khoo, Y., Karim, A., Rusul, G., and Liong, T. 2012. Growth properties and cholesterol removal ability of electroporated Lactobacillus acidophilus BT 1088. *Journal of Microbiology and Biotechnology* 22(7):981–989.

Maniglia, B. C., Castanha, N., Rojas, M. L., and Augusto, P. E. 2020. Emerging technologies to enhance starch performance. *Current Opinion in Food Science* 37:26.

Martinez, J. M., Cebrián, G., Álvarez, I., and Raso, J. 2016. Release of mannoproteins during Saccharomyces cerevisiae autolysis induced by pulsed electric field. *Frontiers in Microbiology* 7:1435.

Martínez, J. M., Delso, C., Aguilar, D., Cebrián, G., Álvarez, I., and Raso, J. 2018. Factors influencing autolysis of Saccharomyces cerevisiae cells induced by pulsed electric fields. *Food Microbiology* 73:67–72.

Martín-García, B., Tylewicz, U., Verardo, V., Pasini, F., Gómez-Caravaca, D., Fiorenza, M., and Dalla Rosa, M. 2020. Pulsed electric field (PEF) as pre-treatment to improve the phenolic compounds recovery from brewers' spent grains Caboni, Rosa, M.D. *Innovative Food Science and Emerging Technologies* 64:102402.

Martínez-Villaluenga, C., Peñas, E., and Hernández-Ledesma, B. 2020. Pseudocereal grains: Nutritional value, health benefits and current applications for the development of Gluten-free foods. *Food and Chemical Toxicology* 137:111178.

Matsugo, S., Sakamoto, T., and Wada, N. 2018. 20Liposoluble anti oxidative components in Japanese traditional fermented food "Amazake" made from brown rice. *Journal of Bioprocessing and Biotechniques* 8:1.

Mattar, J. R., Turk, M. F., Nonus, M., Lebovka, N. I., El Zakhem, H., and Vorobiev, E. 2014. Stimulation of Saccharomyces cerevisiae cultures by pulsed electric fields. *Food and Bioprocess Technology* 7(11):3328–3335.

Mattar, J. R., Turk, M. F., Nonus, M., Lebovka, N. I., El Zakhem, H., and Vorobiev, E. 2015. S. cerevisiae fermentation activity after moderate pulsed electric field pre-treatments. *Bioelectrochemistry* 103:92–97.

Mattison, K. W., Dubin, P. L., and Brittain, I. J. 1998. Complex formation between bovine serum albumin and strong polyelectrolytes: Effect of Polymer Charge Density. *The Journal of Physical Chemistry B* 102:3830–3836.

McKevith, B. 2004. Nutritional aspects of cereals. *Nutrition Bulletin* 29(2):111–142.

Melini, V., Panfili, G., Fratianni, A., and Acquistucci, R. 2019. Bioactive compounds in rice on Italian market: Pigmented varieties as a source of carotenoids, total phenolic compounds and anthocyanins, before and after cooking. *Food Chemistry* 277:119–127.

Menezes, A. G. T., Ramos, C. L., Dias, D. R., and Schwan, R. F. 2018. Combination of probiotic yeast and lactic acid bacteria as starter culture to produce maize-based beverages. *Food Research International* 111:187–197.

Miguel, M. G. D. C. P., Santos, M. R. R. M., Duarte, W. F., de Almeida, E. G., and Schwan, R. F. 2012. Physicochemical and microbiological characterization of corn and rice "calugi" produced by Brazilian Amerindian people. *Food Research International* 49(1):524–532.

Milani, E. A., Alkhafaji, S., and Silva, F. V. 2015. Pulsed electric field continuous pasteurization of different types of beers. *Food Control* 50:223–229.

Misihairabgwi, J., and Cheikhyoussef, A. 2017. Traditional fermented foods and beverages of Namibia. *Journal of Ethnic Foods* 4(3):145–153.

Morales, J. C., and Zepeda, R. A. G. 2017. Effect of different corn processing techniques in the nutritional composition of nixtamalized corn tortillas. *Journal of Nutrition and Food Sciences* 7(2):2.

Morales-De La Peña, M., Salvia-Trujillo, L., Rojas-Graü, M. A., and Martín-Belloso, O. 2010. Isoflavone profile of a high intensity pulsed electric field or thermally treated fruit juice-soymilk beverage stored under refrigeration. *Innovative Food Science and Emerging Technologies* 11(4):604–610.

Mora-Rochin, S., Gutiérrez-Uribe, J. A., Serna-Saldivar, S. O., Sánchez-Peña, P., Reyes-Moreno, C., and Milán-Carrillo, J. 2010. Phenolic content and antioxidant activity of tortillas produced from pigmented maize processed by conventional nixtamalization or extrusion cooking. *Journal of Cereal Science* 52(3):502–508.

Mugochi, T., Mutukumira, T. and Zvauya, R. 2001. Comparison of sensory characteristics of traditional Zimbabwean non-alcoholic cereal beverages, masvusvu and mangisi with mahewu, a commercial cereal product. *Ecology of Food and Nutrition* 40:299–309.

Mukisa, I. M., Porcellato, D., Byaruhanga, Y. B., Muyanja, C. M. B. K., and Rudi, K. 2012. The dominant microbial community associated with fermentation of Obushera (sorghum and millet beverages) determined by culture-dependent and culture-independent methods. *International Journal of Food Microbiology* 160:1–10.

Muyanja, C., Birungi, S., Ahimbisibwe, M., Semanda, J.,. and Namugumya, B. S. 2010. Traditional processing, microbiological and physicochemical changes during fermentation of malwa. *African Journal of Food, Agriculture, and Nutrition and Development* 10:1664.

Mwesigye, P. K., and Okurut, T. O. 1995. A survey of the production and consumption of traditional alcoholic beverages in Uganda. *Process Biochemistry* 30(6):497–501.

Najim, N., and Aryana, K. J. 2013. A mild pulsed electric field condition that improves acid tolerance, growth, and protease activity of Lactobacillus acidophilus LA-K and Lactobacillus delbrueckii subspecies bulgaricus LB-12. *Journal of Dairy Science* 96(6):3424–3434.

Nazir, M., Arif, S., Khan, R. S., Nazir, W., Khalid, N., and Maqsood, S. 2019. Opportunities and challenges for functional and medicinal beverages: Current and future trends. *Trends in Food Science and Technology* 88:513–526.

Nemo, R., and Bacha, K. 2020. Microbial, physicochemical and proximate analysis of selected Ethiopian traditional fermented beverages. *LWT* 131:109713.

Novik, G., and Savich, V. 2020. Beneficial microbiota. Probiotics and pharmaceutical products in functional nutrition and medicine. *Microbes and Infection* 22(1):8–18.

Nout, M. J. R. 2009. Rich nutrition from the poorest – Cereal fermentations in Africa and Asia. *Food Microbiology* 26(7):685–692.

Nowacka, M., Tappi, S., Wiktor, A., et al. 2019. The impact of pulsed electric field on the extraction of bioactive compounds from beetroot. *Foods* 8(7):244.

Oh, Y. J., and Jung, D. S. 2015. Evaluation of probiotic properties of Lactobacillus and Pediococcus strains isolated from Omegisool, a traditionally fermented millet alcoholic beverage in Korea. *LWT-Food Science and Technology* 63(1):437–444.

Oliveira, P. M., Zannini, E., and Arendt, E. K. 2014. Cereal fungal infection, mycotoxins, and lactic acid bacteria mediated bioprotection: From crop farming to cereal products. *Food Microbiology* 37:78–95.

Onyango, C., Noetzold, H., Bley, T., and Henle, T. 2004. Proximate composition and digestibility of fermented and extruded uji from maize–finger millet blend. *LWT-Food Science and Technology* 37(8):827–832.

Panghal, A., Janghu, S., Virkar, K., Gat, Y., Kumar, V., and Chhikara, N. 2018. Potential non-dairy probiotic products – A healthy approach. *Food Bioscience* 21:80–89.

Park, C. S., and Lee, T. S. 2002. Quality characteristics of takju prepared by wheat flour nuruks. *Korean Journal of Food Science and Technology* 34(2):296–302.

Parniakov, O., Roselló-Soto, E., Barba, F., Grimi, N., Lebovka, N., and Vorobiev, E. 2015. New approaches for the effective valorization of papaya seeds: Extraction of proteins, phenolic compounds, carbohydrates, and isothiocyanates assisted by pulsed electric energy. *Food Research International* 77:711–717.

Pashazadeh, B., Elhamirad, A. H., Hajnajari, H., Sharayei, P., and Armin, M. 2020. Optimization of the pulsed electric field-assisted extraction of functional compounds from cinnamon. *Biocatalysis and Agricultural Biotechnology* 23:101461.

Perez-Armendariz, B., and Cardoso-Ugarte, G. A. 2020. Traditional Fermented Beverages in Mexico: Biotechnological, Nutritional, and Functional Approaches. *Food Research International* 136:109307.

Peterson, D. M., Emmons, Ch. L., and Hibbs, A. H. 2001. Phenolic antioxidants and antioxidant activity in pearling fractions of oat groats. *Journal of Cereal Science* 33(1):97–103.

Peyer, L. C., Axel, C., Lynch, K. M., Zannini, E., Jacob, F., and Arendt, E. K. 2016a. Inhibition of Fusarium culmorum by carboxylic acids released from lactic acid bacteria in a barley malt substrate. *Food Control* 69:227–236.

Peyer, L. C., Zannini, E., and Arendt, E. K. 2016b. Lactic acid bacteria as sensory biomodulators for fermented cereal-based beverages. *Trends in Food Science and Technology* 54:17–25.

Phiri, S., Schoustra, S. E., van den Heuvel, J., Smid, E. J., Shindano, J., and Linnemann, A. R. 2020. How processing methods affect the microbial community composition in a cereal-based fermented beverage. *LWT* 128:109451.

Poutanen, K., Shepherd, R., Shewry, P. R., Delcour, J. A., Bjorck, I., and Van Der Kamp, J. W. 2008. Beyond whole grain: The European HEALTHGRAIN project aims at healthier cereal foods. *Cereal Foods World* 53(1):32–35.

Prado, F. C., Parada, J. C., Pandey, A., and Soccol, C. R. 2007. Trends in non-dairy probiotic beverages. *Food Research International* 41:111–123.

Puerari, C., Magalhães-Guedes, K. T., and Schwan, R. F. 2015. Physicochemical and microbiological characterization of chicha, a rice-based fermented beverage produced by Umutina Brazilian Amerindians. *Food Microbiology* 46:210–217.

Puertolas, E., Koubaa, M., and Barba, F. J. 2016. An overview of the impact of electrotechnologies for the recovery of oil and high-value compounds from vegetable oil industry: Energy and economic cost implications. *Food Research International* 80:19–26.

Qian, J. Y., Gu, Y. P., Jiang, W., and Chen, W. 2014. Inactivating effect of pulsed electric field on lipase in brown rice. *Innovative Food Science and Emerging Technologies* 22:89–94.

Rafiq, S. I., Singh, S., and Saxena, D. C. 2016. Effect of heat-moisture and acid treatment on physicochemical, pasting, thermal and morphological properties of Horse Chestnut (Aesculus indica) starch. *Food Hydrocolloids* 57:103–113.

Raso, J., Condón, S., and Álvarez, I. 2014. Non-thermal processing| Pulsed electric field. In: *Encyclopedia of Food Microbiology*, eds. C. A. Batt, and M. Tortorello, 966–973. Academic Press.

Ray, S., Bagyaraj, D. J., Thilagar, G., and Tamang, J. P. 2016. Preparation of Chyang, an ethnic fermented beverage of the Himalayas, using different raw cereals. *Journal of Ethnic Foods* 3:297–299.

Ribas-Agustí, A., Martín-Belloso, O., Soliva-Fortuny, R., and Elez-Martínez, P. 2019. Influence of pulsed electric fields processing on the bioaccessible and non-bioaccessible fractions of apple phenolic compounds. *Journal of Functional Foods* 59:206–214.

Rifna, E. J., Singh, S. K., Chakraborty, S., and Dwivedi, M. 2019. Effect of thermal and non-thermal techniques for microbial safety in food powder: Recent advances. *Food Research International* 126:108654.

Rodríguez-Roque, M. J., de Ancos, B., Sánchez-Moreno, C., Cano, M. P., Elez-Martínez, P., and Martín-Belloso, O. 2015. Impact of food matrix and processing on the in vitro bioaccessibility of vitamin C, phenolic compounds, and hydrophilic antioxidant activity from fruit juice-based beverages. *Journal of Functional Foods* 14:33–43.

Roohinejad, S., Parniakov, O., Nikmaram, N., Greiner, R., and Koubaa, M. 2018. Energy saving food processing. In: *Sustainable Food Systems from Agriculture to Industry*, ed. C. M. Galanakis, 191–243. Academic Press.

Rooney, L. W., and Waniska, R. D. 2000. Sorghum food and industrial utilization. In: Smith C. W., Frederiksen R. A., editors. *Sorghum: Origin, History, Technology, and Production*. New York: John Wiley & Sons, Inc. pp. 689–729.

Roselló-Soto, E., Barba, F. J., Parniakov, O., Galanakis, C. M., Lebovka, N., Grimi, N., and Vorobiev, E. 2015. High voltage electrical discharges, pulsed electric field, and ultrasound assisted extraction of protein and phenolic compounds from olive kernel. *Food and Bioprocess Technology* 8(4):885–894.

Roselló-Soto, E., Poojary, M. M., Barba, F. J., Lorenzo, J. M., Mañes, J., and Moltó, J. C. 2018. Tiger nut and its by-products valorization: From extraction of oil and valuable compounds to development of new healthy products. *Innovative Food Science and Emerging Technologies* 45:306–312.

Sack, M., Sigler, J., Frenzel, S., et al. 2010. Research on industrial-scale electroporation devices fostering the extraction of substances from biological tissue. *Food Engineering Reviews* 2(2):147–156.

Saldaña, G., Puértolas, E., López, N., García, D., Álvarez, I., and Raso, J. 2009. Comparing the PEF resistance and occurrence of sublethal injury on different strains of Escherichia coli, Salmonella Typhimurium, Listeria monocytogenes and Staphylococcus aureus in media of pH 4 and 7. *Innovative Food Science and Emerging Technologies* 10(2):160–165.

Salmeron, I. 2017. Fermented cereal beverages: From probiotic, prebiotic and synbiotic towards nanoscience designed healthy drinks. *Letters in Applied Microbiology* 65(2):114–124.

Salvia-Trujillo, L., Morales-de la Peña, M., Rojas-Graü, A., Welti-Chanes, J., and Martín-Belloso, O. 2017. Mineral and fatty acid profile of high intensity pulsed electric fields or thermally treated fruit juice-milk beverages stored under refrigeration. *Food Control* 80:236–243.

Sanchez-Vega, R., Mujica-Paz, H., Marquez-Melenez, R., Ngadi, M. O., and Ortega-Rivas, E. 2009. Enzyme inactivation on apple juice treated by ultrapasteurization and pulsed electric fields technology. *Journal of Food Processing and Preservation* 33(4):486–499.

Sanni, A. I. 1989. Chemical studies on sekete beer. *Food Chemistry* 33(3):187–191.

Santos, C. C. A., da Silva Libeck, B., and Schwan, R. F. 2014. Co-culture fermentation of peanut-soy milk for the development of a novel functional beverage. *International Journal of Food Microbiology* 186:32–41.

Sanz-Puig, M., Santos-Carvalho, L., Cunha, L. M., Pina-Pérez, M. C., Martínez, A., and Rodrigo, D. 2016. Effect of pulsed electric fields (PEF) combined with natural antimicrobial by-products against S. *Typhimurium. Innovative Food Science and Emerging Technologies* 37:322–328.

Schottroff, F., Johnson, K., Johnson, N. B., Bédard, M. F., and Jaeger, H. 2020. Challenges and limitations for the decontamination of high solids protein solutions at neutral pH using pulsed electric fields. *Journal of Food Engineering* 268:109737.

Selma, M. V., Fernández, P. S., Valero, M., and Salmeron, M. C. 2003. Control of Enterobacter aerogenes by high-intensity, pulsed electric fields in horchata, a Spanish low-acid vegetable beverage. *Food Microbiology* 20(1):105–110.

Selma, M. V., Salmerón, M. C., Valero, M., and Fernández, P. S. 2006. Efficacy of pulsed electric fields for Listeria monocytogenes inactivation and control in horchata. *Journal of Food Safety* 26(2):137–149.

Sethi, S., Tyagi, S. K., and Anurag, R. K. 2016. Plant-based milk alternatives an emerging segment of functional beverages: A review. *Journal of Food Science and Technology* 53(9):3408–3423.

Sha, S. P., Anupama, A., Pradhan, P., Prasad, G. S., and Tamang, J. P. 2016. Identification of yeasts by polymerase-chain-reaction-mediated denaturing gradient gel electrophoresis in marcha, an ethnic amylolytic starter of India. *Journal of Ethnic Foods* 3(4):292–296.

Sharma, P., Trivedi, N., and Gat, Y. 2017. Development of functional fermented whey-oat based product using probiotic bacteria. *Biotechnology* 7(4):272.

Shokribousjein, Z., Deckers, S. M., Gebruers, K., et al. 2011. Hydrophobins, beer foaming and gushing. *Cerevisia* 35(4):85–101.

Siemer, C., Toepfl, S., and Heinz, V. 2014. Inactivation of Bacillus subtilis spores by pulsed electric fields (PEF) in combination with thermal energy II. Modeling thermal inactivation of B. subtilis spores during PEF processing in combination with thermal energy. *Food Control* 39:244–250.

Silva, A. R., Silva, M. M., and Ribeiro, B. D. 2020. Health issues and technological aspects of plant-based alternative milk. *Food Research International* 131:108972.

Soni, S., and Dey, G. 2014. Perspectives on global fermented foods. *British Food Journal* 116(11):1767–1787.

Soukand, R., Pieroni, A., Biró, M., et al. 2015. An ethnobotanical perspective on traditional fermented plant foods and beverages in Eastern Europe. *Journal of Ethnopharmacology* 170:284–296.

Stanton, C., Gardiner, G., Meehan, H., Collins, K., Fitzgerald, G., Lynch, P. B., and Ross, R. P. 2001. Market potential for probiotics. *The American Journal of Clinical Nutrition* 73(2) Supplement:476–483.

Sullivan, P., Arendt, E., and Gallagher, E. 2013. The increasing use of barley and barley by-products in the production of healthier baked goods. *Trends in Food Science and Technology* 29(2):124–134.

Taylor, J. R., and Duodu, K. G. 2019. Traditional sorghum and millet food and beverage products and their technologies. In: *Sorghum and Millets*, eds. J. R. N. Taylor, and K. G. Duodu, 259–292. AACC International Press.

Taylor, J. R., and Kruger, J. 2019. Sorghum and millets: Food and beverage nutritional attributes. In: *Sorghum and Millets*, eds. J. R. N. Taylor, and K. G. Duodu, 171–224. AACC International Press.

Taylor, J. R. N. 2016. Fermentation: Foods and nonalcoholic beverages. In: *Encyclopedia of Food Grains*, eds. C. Wrigley, H. Corke, K. Seetharaman, and J. Faubion, 183–192. Academic Press.

Terefe, N., Yang, Y., Knoerzer, K., Buckow, R., and Versteeg, C. 2010. High pressure and thermal inactivation kinetics of polyphenol oxidase and peroxidase in strawberry puree. *Innovative Food Science and Emerging Technologies* 11(1):52–60.

Terefe, S. N., Buckow, R., and Versteeg, C. 2014. Quality-related enzymes in fruit and vegetable products: Effects of novel food processing technologies. Quality-related enzymes in fruit and vegetable products: Effects of novel food processing technologies, part 1: High-pressure processing – Part 1. *Critical Reviews in Food Science and Nutrition* 54(1):24–63.

Timmermans, R. A. H., Mastwijk, H. C., Berendsen, L. B. J. M., Nederhoff, A. L., Matser, A. M., Van Boekel, M. A. J. S., and Nierop Groot, M. N. 2019. Moderate intensity Pulsed Electric Fields (PEF) as alternative mild preservation technology for fruit juice. *International Journal of Food Microbiology* 298:63–73.

Toepfl, S., Siemer, C., and Heinz, V. 2014. Effect of high-intensity electric field pulses on solid foods. In: *Emerging Technologies for Food Processing*, ed. D. Sun, 147–154. Academic Press.

Ulmer, H. M., Heinz, V., Gänzle, M. G., Knorr, D., and Vogel, R. F. 2002. Effects of pulsed electric fields on inactivation and metabolic activity of Lactobacillus plantarum in model beer. *Journal of Applied Microbiology* 93(2):326–335.

Vorobiev, E., and Lebovka, N. 2021. Effects of pulsed electric energy in food and agricultural products: A review of recent research advances. In: *Innovative Food Processing Technologies*, eds. K. Knoerzer, and K. Muthukumarappan, 173–198. Elsevier.

Vivas, N., Waniska, R., and Rooney, L. 1987. Thin porrides (Atole) from maize and sorghum. *Cereal Chemistry* 64:390–394.

Walkling-Ribeiro, M., Rodríguez-González, O., Jayaram, S. H., and Griffiths, M. W. 2011. Processing temperature, alcohol and carbonation levels and their impact on pulsed electric fields (PEF) mitigation of selected characteristic microorganisms in beer. *Food Research International* 44(8):2524–2533.

Wang, L. H., Pyatkovskyy, T., Yousef, A., Zeng, X. A., and Sastry, S. K. 2020. Mechanism of Bacillus subtilis spore inactivation induced by moderate electric fields. *Innovative Food Science and Emerging Technologies* 62:102349.

Wang, M. S., Wang, L. H., Bekhit, A. E. D. A., et al. 2018. A review of sublethal effects of pulsed electric field on cells in food processing. *Journal of Food Engineering* 223:32–41.

Wang, S., Li, C., Copeland, L., Niu, Q., and Wang, S. 2015. Starch retrogradation: A comprehensive review. *Comprehensive Reviews in Food Science and Food Safety* 14(5):568–585.

Wang, T., He, F., and Chen, G. 2014. Improving bioaccessibility and bioavailability of phenolic compounds in cereal grains through processing technologies: A concise review. *Journal of Functional Foods* 7:101–111.

Westling, M., Danielsson Tham, M. L., Jass, J., Nilsen, A., Öström, Å., and Tham, W. 2016. Contribution of Enterobacteriaceae to sensory characteristics in soft cheeses made from raw milk. *Procedia Food Science* 7:17–20.

Wu, X., Li, F., Wu, W. Wang, Z., Zheng, F., Lu X., Li Z., and Ren, Q. 2020. Effects of rice bran rancidity on the oxidation and structural characteristics of rice bran protein. *LWT* 120:108943.

Xu, J., Wu, H., Wang, Z., Zheng, F., Lu, X., Li, Z., and Ren, Q. 2018. Microbial dynamics and metabolite changes in Chinese Rice Wine fermentation from sorghum with different tannin content. *Scientific Reports* 8(1):4639.

Yu, H., and Bogue, J. 2013. Concept optimisation of fermented functional cereal beverages. *British Food Journal* 115(4):541–563.

Zeng, F., Gao, Q. Y., Han, Z., Zeng, X. A., and Yu, S. J. 2016. Structural properties and digestibility of pulsed electric field treated waxy rice starch. *Food Chemistry* 194:1313–1319.

Zhang, L., Li, C. Q., Jiang, W., Wu, M., Rao, S. Q., and Qian, J. Y. 2019. Pulsed electric field as a means to elevate activity and expression of α-amylase in barley (*Hordeum vulgare* L.) malting. *Food and Bioprocess Technology* 12(6):1010–1020.

Zhang, L., Wang, L. J., Jiang, W., and Qian, J. Y. 2017a. Effect of pulsed electric field on functional and structural properties of canola protein by pretreating seeds to elevate oil yield. *LWT* 84:73–81.

Zhang, Z. H., Wang, L. H., Zeng, X. A., Han, Z., and Wang, M. S. 2017b. Effect of pulsed electric fields (PEFs) on the pigments extracted from spinach (Spinacia oleracea L.). *Innovative Food Science and Emerging Technologies* 43:26–34.

Zhao, W., Yang, R., Wang, M., and Lu, R. 2009a. Effects of pulsed electric fields on bioactive components, colour and flavor of green tea infusions. *International Journal of Food Science and Technology* 44:321–327.

Zhao, W., Yang, R., Tang, Y., Zhang, W., and Hua, X. 2009b. Investigation of the protein–protein aggregation of egg white proteins under pulsed electric fields. *Journal of Agriculture and Food Chemistry* 57:3571–3577.

Zhou, M. X., Holmes, M. G., Robards, K., and Helliwell, S. 1998. Fatty acid composition of lipids of Australian oats. *Journal of Cereal Science* 28(3):311–319.

Chapter 3

Innovative Approaches in High-Pressure Processing for the Extraction of Antioxidants from Grains

Zehra Gulsunoglu-Konuskan, Zeynep Tacer-Caba and Meral Kilic-Akyilmaz

CONTENTS

3.1 INTRODUCTION

Oxidation is important in order for living organisms to produce the necessary energy for biological processes. Reactive oxygen species (ROS) are generated *in vivo* during oxidation and are used in cell signaling, apoptosis, gene expression and the transport of ions. However, when these ROS molecules are generated in excess or cellular defenses are inadequate, biomolecules like protein, lipids and nucleic acids can be injured by the oxidative stress process. Therefore, oxidative damage plays an important role in aging, cancer, coronary heart diseases, Alzheimer's disease, neurodegenerative disorders, cataracts and inflammation (Babbar et al., 2011; Dai & Mumper, 2010). Most living organisms have enzymatic and non-enzymatic antioxidant protection and repair mechanisms that have been built to protect them from oxidative stress. However, these native antioxidative systems are usually not enough to avoid oxidative damage in the human body. The consumption of antioxidants mitigates the oxidative damage on the human body.

Grains can be classified as cereals and pseudocereals, which are staple foods and supply significant amounts of energy, macro- and micronutrients to the human diet. The major grains consumed in the human diet are wheat, rice, maize, oats, rye, barley, sorghum, millet, amaranth and teff. In Asia, rice is the most consumed grain, while wheat is the major grain consumed in Europe and the USA. Grains and their processed products contain diverse antioxidants including carotenoids, tocopherols, phenolic acids, alkylresorcinols, anthocyanins, flavonoids, proanthocyanidins and lignans (Luthria, Lu, & John, 2015). Lignans are a group of dietary phytoestrogens found in a variety of grains like corn, oats, rye and wheat. When one consumes these lignans, they are converted to enterodiol and enterolactone, which have strong antioxidant activity. The bran and germ of grains contain the majority of bioactive phytochemicals (phenolic acids, alkylresorcinols and lignans) (Luthria et al., 2015).

Cereals and pseudocereals, especially colored ones, contain many phenolic compounds (Figure 3.1). The concentration and diversity of phenolic compounds and also nutrient composition have been affected by several factors such as genotype, soil, environmental and climatic conditions (Ofosu et al., 2020). Phenolic compounds have the ability of free radical scavenging, thus providing a wide range of beneficial health effects as a result of their antioxidant properties. They are synthesized by plants as secondary metabolites. Phenolic compounds can be divided into four general groups: phenolic acids (gallic, protocatechuic, caffeic and rosmarinic acids),

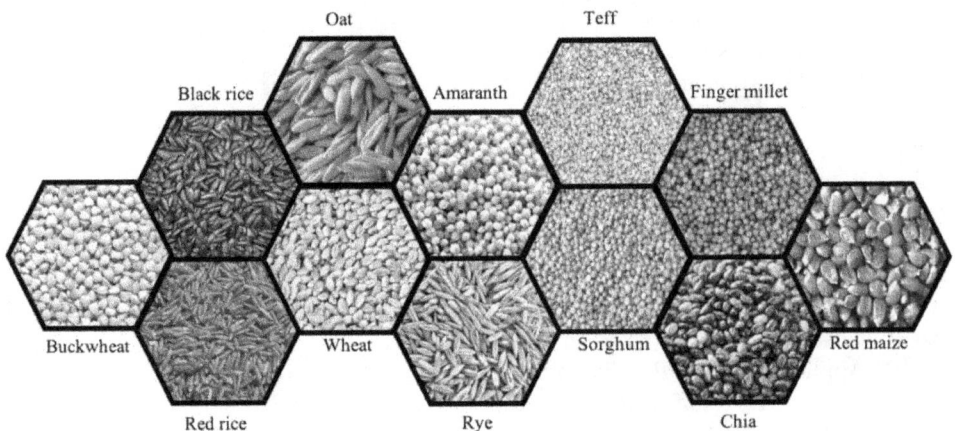

Figure 3.1   Some cereal and pseudocereal grains that are rich sources of antioxidants.

flavonoids (quercetin and catechin), tannins and less common stilbenes and lignans. The most common phenolic compounds are phenolic acids and flavonoids. The dominant phenolic acid in grains is ferulic acid followed by hydroxycinnamic, gallic, vanillic and *p*-coumaric acids (Jideani et al., 2014).

Phenolic compounds can be present in soluble free, soluble conjugated and insoluble or bound forms. Soluble forms of phenolic compounds can be easily extracted by solvent extraction techniques using different organic solvents like methanol, ethanol and acetone, while the extraction of bound phenolic compounds requires a pretreatment for their release. Insoluble bound phenolics can be conjugated to structural components of the cell wall such as cellulose, hemicellulose, lignin, pectin and structural proteins (Acosta-Estrada et al., 2014). The release of bound phenolics is possible by acid or alkali treatments, while hot acidic conditions can destroy some phenolic compounds. As the grains contain mostly bound phenolics in bran, extraction of them requires a pretreatment for their dissolution.

Conventional techniques of extraction have some disadvantages such as long extraction time, high cost and toxicity, use of high-quantity solvent and difficult recovery of the solvent. Because of these limitations, novel techniques including ultrasonic extraction, microwave-assisted extraction, supercritical fluid extraction and high-pressure processing are getting more attention for the extraction of phenolic compounds. The advantages of these methods are reduced extraction time and solvents and high reproducibility (Tanase et al., 2019).

High-pressure processing (HPP) is a novel technology that has been developed as an alternative to the thermal processing of foods. HPP is applied at pressures of 10–1,000 MPa which can disrupt the cellular structure and increase the bioaccessibility of bioactive compounds from foods (Kim et al., 2015). HPP can be used for the extraction of antioxidants due to its effect on cellular structure. In this chapter, the antioxidants present in cereals, the HPP technique and the extraction of antioxidants by HPP are reviewed.

3.2 ANTIOXIDANTS FOUND IN GRAINS

3.2.1 Cereal Grains

Cereals have been known to play a crucial role in meeting the food demands of the human population since ancient times. Cereals such as maize, wheat and rice account for around 80% of food intake and are fortified to enhance vitamins and other essential micro-nutrients. The most-consumed cereals are maize (*Zea mays*), rice (*Oryza sativa*), wheat (*Triticum* spp.), barley (*Hordeum vulgare*), sorghum (*Sorghum bicolor*), millet, oats (*Avena sativa*) and rye (*Secale cereale*) (Zhu, 2019). Cereals are important for the human diet due to their contribution of energy, protein, minerals and antioxidants such as carotenoids, tocopherols and phenolic compounds (Hussain et al., 2012). The antioxidant activity of cereal and pseudocereal grains determined by different methods is presented in Table 3.1.

3.2.1.1 Wheat

Wheat is a common cereal crop worldwide with a production of about 3.5 billion tons over the period from 2015 to 2018 and is used to manufacture various food products (FAO, 2020). Although wheat is mostly viewed as a source of calories, it has also been proven that wheat has many bioactive phytochemicals including amino acids, minerals, vitamins, dietary fiber, phenolics, carotenoids, tocopherols, alkylresorcinols, benzoxazinoids, phytosterols

TABLE 3.1   ANTIOXIDANT ACTIVITY OF CEREAL AND PSEUDOCEREAL GRAINS

Grain Type	Form	Antioxidant Method*	Antioxidant Activity	References
Amaranth	Soluble	DPPH	0.13–0.97 µmol TE/g	Alvarez-Jubete et al. (2010); Inglett et al. (2015); Peiretti et al. (2017)
	Soluble	FRAP	0.55 mg TE/ g	
	Soluble	TEAC	0.069 mmol TE/g	
	Insoluble-bound	DPPH	9.88–9.92 µmol TE/g	
Barley	Soluble	FRAP	0.19–0.23 Fe⁺² mmol/g	Lahouar et al. (2014)
	Soluble	ABTS	3.13–3.72 µmol TE/g	
Buckwheat	Soluble	DPPH	6.2 mg TE/ g	Alvarez-Jubete et al. (2010); Inglett et al. (2011); Inglett et al. (2015)
	Soluble	FRAP	4.36 mg TE/g	
	Insoluble-bound	DPPH	10.25–10.40 µmol TE/g	
Chia seed	Hydrophilic	ORAC	517.3 µmol TE/g	Da Silva Marineli (2014)
	Lipophilic	ORAC	6.48 µmol TE/g	
	Soluble	DPPH	436.61 µmol TE/g	
	Soluble	FRAP	405.7 µmol TE/g	
Maize (blue)	Soluble	DPPH	49.2 µmol TE/g	Herrera-Sotero et al. (2017)
	Soluble	TBARS	792 µg/mL	
Maize (red)	Soluble	FRAP	3.28 µmol TE/g	Capocchi et al. (2017)
	Soluble	CUPRAC	9.23 µmol TE/g	
Maize (yellow)	Soluble	FRAP	2.46 µmol TE/g	Capocchi et al. (2017)
	Soluble	CUPRAC	7.85 µmol TE/g	
Maize (sweet)	Soluble	DPPH	85.01–281.97 mg FW/mL	Zhang et al. (2017)
	Insoluble-bound	DPPH	117.48–314.61 mg FW/mL	
Millet	Soluble	DPPH	0.64–11.2 µmol FAE/g	Xiang et al. (2019a); Xiang et al. (2019b); Chandrasekara and Shahidi (2011)
	Soluble	SOSC	3.74–17.7 µmol FAE/g	
	Soluble	ORAC	109.36–218.20 µmol TE/g	
	Soluble	ABTS	15.59–35.39 µmol TE/g	
	Esterified	DPPH	1.19–2.81 µmol FAE/g	
	Esterified	SOSC	1.43–7.58 µmol FAE/g	
	Esterified	ORAC	20.64–26.72 µmol TE/g	
	Etherified	DPPH	0.41–2.24 µmol FAE/g	
	Etherified	SOSC	1.38–2.87 µmol FAE/g	
	Etherified	ORAC	32.19–45.78 µmol TE/g	
	Insoluble-bound	DPPH	4.06–22.7 µmol FAE/g	
	Insoluble-bound	SOSC	3.51–83.4 µmol FAE/g	
	Insoluble-bound	ORAC	95.27–792.7 µmol TE/g	
	Insoluble-bound	ABTS	6.56–14.70 µmol TE/g	

(Continued)

TABLE 3.1 (CONTINUED)   ANTIOXIDANT ACTIVITY OF CEREAL AND PSEUDOCEREAL GRAINS

Grain Type	Form	Antioxidant Method*	Antioxidant Activity	References
Oat	Soluble	FRAP	194.3–212.9 µmol Fe²⁺/g	Rakic et al. (2014); Chen et al. (2018)
	Soluble	DPPH	0.35–0.39 mg/mL	
	Soluble	ORAC	17.07–25.62 µmol TE/g	
	Soluble	CAA	25.31–33.38 µmol QE/100 g	
Quinoa	Soluble	DPPH	57.7 mg TE/100 g	Alvarez-Jubete et al. (2010); Ballester-Sanchez et al. (2020)
	Soluble	FRAP	92.1 mg TE/100 g	
	Insoluble-bound	DPPH	63 µmol TE/g	
	Insoluble-bound	FRAP	43 µmol TE/g	
Rye	Soluble	ABTS	302.3–1570 mg TE/kg	Kulichova et al. (2019)
	Soluble	DPPH	120.7–3041.3 mg TE/kg	
	Soluble	FRAP	2277.5–5365 mg TE/kg	
Rice (white)	Soluble	DPPH	0.05–0.19 µmol TE/g	Pang et al. (2018); Štastná et al. (2019)
	Soluble	ABTS	0.13–0.25 µmol TE/g	
	Soluble conjugated	DPPH	0.25 mg TE/g	
	Soluble conjugated	ABTS	0.11 mg TE/g	
	Insoluble-bound	DPPH	1.00–1.65 µmol TE/g	
	Insoluble-bound	ABTS	2.56–3.28 µmol TE/g	
Rice (red)	Soluble	DPPH	0.19–1.66 µmol TE/g	Pang et al. (2018); Štastná et al. (2019)
	Soluble	ABTS	0.47–2.66 µmol TE/g	
	Soluble conjugated	DPPH	0.42 mg TE/g	
	Soluble conjugated	ABTS	0.12 mg TE/g	
	Insoluble-bound	DPPH	1.42–1.65 µmol TE/g	
	Insoluble-bound	ABTS	2.95–4.15 µmol TE/g	
Rice (black)	Soluble	DPPH	1.04–3.72 µmol TE/g	Pang et al. (2018); Štastná et al. (2019)
	Soluble	ABTS	1.89–6.19 µmol TE/g	
	Soluble conjugated	DPPH	0.39 mg TE/g	
	Soluble conjugated	ABTS	0.14 mg TE/g	
	Insoluble-bound	DPPH	1.6–3.4 µmol TE/g	
	Insoluble-bound	ABTS	3.22–7.20 µmol TE/g	
Sorghum (white)	Soluble	DPPH	0.33 mg TE/g	Rao et al. (2018)
	Soluble	FRAP	2.31 mg TE/g	
Sorghum (red)	Soluble	DPPH	0.41–1.17 mg TE/g	Rao et al. (2018); Irondi et al. (2019)
	Soluble	ABTS	5.47 mmol TE/g	
	Soluble	TEAC	19.83 µg/mL	
	Soluble	Reducing power	72.41 mg GAE/g	
	Soluble	FRAP	2.72–4.83 mg TE/g	

(Continued)

TABLE 3.1 (CONTINUED)   ANTIOXIDANT ACTIVITY OF CEREAL AND PSEUDOCEREAL GRAINS

Grain Type	Form	Antioxidant Method*	Antioxidant Activity	References
Sorghum (black)	Soluble	DPPH	18.04 mg TE/g	Rao et al. (2018)
	Soluble	FRAP	20.92 mg TE/g	
Sorghum (brown)	Soluble	DPPH	21.02 mg TE/g	Rao et al. (2018)
	Soluble	FRAP	4.62 mg TE/g	
Teff	Soluble	DPPH	2.37–4.88 µmol TE/g	Inglett et al. (2015)
	Insoluble-bound	DPPH	10.28–10.42 µmol TE/g	
Teff (white)	Soluble	DPPH	2.9–5.7 µmol TE/g	Shumoy and Raes (2016); Šťastná et al. (2019)
	Soluble	FRAP	6.2–11.5 µmol Fe^{+2}/g	
	Soluble conjugated	DPPH	0.22 mg TE/g	
	Soluble conjugated	ABTS	0.27 mg TE/g	
	Insoluble-bound	DPPH	21.1–136 µmol TE/g	
	Insoluble-bound	ABTS	0.48 mg TE/g	
	Insoluble-bound	FRAP	36.0–50.2 µmol Fe^{+2}/g	
Teff (brown)	Soluble	DPPH	5.4–6.5 µmol TE/g	Shumoy and Raes (2016); Šťastná et al. (2019)
	Soluble	ABTS	0.46 mg TE/g	
	Soluble	FRAP	12.3–15.7 µmol Fe^{+2}/g	
	Soluble conjugated	DPPH	0.44 mg TE/g	
	Soluble conjugated	ABTS	0.26 mg TE/g	
	Insoluble-bound	DPPH	87.8–93.7 µmol TE/g	
	Insoluble-bound	ABTS	0.52 mg TE/g	
	Insoluble-bound	FRAP	56.7–63.4 µmol Fe^{+2}/g	
Wheat	Soluble	DPPH	0.44 mg TE/g	Alvarez-Jubete et al. (2010); Leoncini et al. (2012)
	Soluble	FRAP	1.10 mg TE/g	
	Insoluble-bound	DPPH	5.61–8.05 µmol TE/g	
	Insoluble-bound	FRAP	0.95–1.11 mmol Fe^{+2}/100g	

* SOSC: singlet oxygen scavenging capacity; DPPH: 2,2-diphenyl-1-picrylhydrazyl assay; ABTS: 2,2'-azino-bis-3-ethylbenzthiazoline-6-sulphonic acid; FRAP: ferric reducing antioxidant power; CUPRAC: cupric reducing antioxidant capacity; TEAC: trolox equivalent antioxidant capacity; TBARS: thiobarbituric acid reactive substance; ORAC: oxygen radical absorbance capacity; CAA: cellular antioxidant activity.

and lignans (Luthria et al., 2015; Shewry, 2018). Bioactive compounds in wheat show high free-radical scavenging activity and anticancer, anticholesterolemic and antidiabetic activities (Luthria et al., 2015).

Phenolic compounds were found in both soluble and bound forms in wheat (Khosravi & Razavi, 2020). Ferulic acid accounted for most of the phenolic compounds and it constitutes 70–90% of the total phenolic content in wheat. In addition to ferulic acid, dihydroferulic, sinapic, caffeic, vanillic, syringic, salicylic, o-coumaric, p-coumaric and hydroxybenzoic acids can also be found in wheat in both free and bound forms (Luthria et al., 2015; Khosravi & Razavi, 2020). Flavonoids such as apigenin, chrysoeriol, kaempferol, quercetin and luteolin, especially in dark-colored wheat, were mostly in conjugated form with different sugars (Khosravi & Razavi, 2020). Anthocyanins like cyanidin, malvidin, petunidin, delphinidin and their sugar derivatives were also found in mostly dark-colored wheat, but they are less concentrated than phenolic acids (Khosravi & Razavi, 2020).

Germ and bran fractions (pericarp, testa and alurone layer) of wheat generally contain high amounts of antioxidants (Luthria et al., 2015). The phenolic compounds of the aleurone layer were hydroxycinnamic acids including ferulic acid; the majority were bound to arabinoxylans, diferulic acid, sinapic acid and p-coumaric acid. Syringic and vanillic acids, lignans and lignins were also identified (Prückler et al., 2014). Moreover, whole wheat grains also contain moderate sources of vitamin E and α-, β-, δ- and γ-tocopherols (Luthria et al., 2015). According to Hussain et al. (2012), 20% of the daily vitamin E requirement can be provided by the consumption of wheat and wheat products.

3.2.1.2 Rice

Rice is one of the most commonly consumed staple foods providing nutrients for more than half of the world's population. Rice is a source of proteins, minerals, vitamins and antioxidants, which are phenolic acids, anthocyanins, proanthocyanidins, tocopherols and oryzanol. Rice bran also contains dietary fiber, vitamin B, tocopherols, tocotrienols, oryzanols and other phenolic compounds (Bordiga et al., 2014). Most of the antioxidants are primarily found in the bran, which is generally discarded as waste (Pang et al., 2018). The antioxidants give some biological functions to rice in the reduction of the risk of some diseases (Bordiga et al., 2014; Jideani et al., 2014; Alves et al., 2016). Recently, consumers' interest has tended to rise in the consumption of pigmented rice varieties due to their high content of bioactive compounds.

The antioxidant activity of rice originates from phenolic acids, anthocyanins and proanthocyanidins. Quercetin and isorhamnetin are flavonol aglycons, and feruloylquinic acids were the dominant monoacyl quinic acids detected in rice samples (Bordiga et al., 2014; Zaupa et al., 2015). The most common flavan-3-ol monomers including (+)-catechin, (–)-epicatechin, (–)-gallocatechin and (–)-epigallocatechin were identified in rice samples (Bordiga et al., 2014; Zaupa et al., 2015).

White rice was found to contain mainly phenolic acids such as protocatechuic, ferulic acid O-dihexoside, sinapic acid O-dihexoside, ferulic, diferulic, triferulic, sinapic, p-coumaric and m-coumaric acids (Zaupa et al., 2015). The main phenolics in black and red rice varieties were identified as ferulic, p-coumaric and vanillic acids (Pang et al., 2018). Red rice is characterized by the presence of procyanidins, whereas black rice is characterized by the presence of anthocyanins, mainly cyanidin-3-glucoside and peonidin-3-glucoside with a small amount of cyanidin-3-rutinoside, malvidin-3-glucoside and peonidin-3-rutinoside (Bordiga et al., 2014; Pang et al., 2018). Black rice had a high content of phenolics, flavonoids and anthocyanins and showed higher antioxidant activity when compared to white and red rice (Bordiga et al., 2014; Pang et al., 2018).

3.2.1.3 Rye

Rye is one of the most cultivated and consumed grains and is mainly used in bread manufacture. Rye is rich in bioactive compounds such as dietary fiber, alkylresorcinols, folate, tocols, phenolic acids and sterols with antioxidant properties. However, these compounds can be different between individual rye cultivars (Kulichová et al., 2019).

The most abundant phenolic acid found in rye is ferulic acid, comprising 74% of the bound phenolic acid fraction, followed by sinapic and *p*-coumaric acids (Bondia-Pons et al., 2009). Additionally, syringic, caffeic, vanillic and *p*-hydroxybenzoic acids were also present at low concentrations. The concentration of phenolic compounds in rye milling fractions can be listed in descending order as follows: rye bran, rye grain, rye flour and rye bread (Bondia-Pons et al., 2009). Rye and rye products were shown to inhibit low-density lipoprotein (LDL) oxidation (Andreasen et al., 2001). Moreover, rye bran was found to have high antioxidant activity as a consequence of high amounts of monomeric and dimeric hydroxycinnamates (Andreasen et al., 2001).

3.2.1.4 Sorghum

Sorghum is a cereal crop originating in sub-Saharan Africa (Luthria & Liu, 2013). Sorghum is a good source of carbohydrates, protein, vitamins, minerals and phenolic compounds including phenolic acids, flavonoids and tannins (Irondi et al., 2019). Phenolic compounds in sorghum were generally found in bound form and their concentrations were influenced by cultivar and growing conditions (Luthria & Liu, 2013; Rao et al., 2018). Compared with other grains such as barley, wheat, rye and millet, sorghum is popular for its rich polyphenol content (Ofosu et al., 2020).

The main phenolic compounds found in sorghum grains were phenolic acids including gallic, chlorogenic, caffeic, ellagic and *p*-coumaric acids (Irondi et al., 2019), sinapic and ferulic acids (Luthria & Liu, 2013) and flavonoids including quercetin, luteolin and apigenin (Irondi et al., 2019). Catechin, taxifolin, glycitein, naringenin, ononin and hispidulin were also reported by Ofosu et al. (2020). Sorghum bran was found to contain mainly ferulic acid followed by diferulic, *p*-coumaric, caffeic and sinapic acids (Chiremba et al., 2012).

Sorghum grains have colored pericarps including white, black, red and brown (Rao et al., 2018). The colored sorghum contains unique anthocyanins that can be used as food colorants (Luthria & Liu, 2013). A unique bioactive compound present in sorghum is 3-deoxyanthocyanidin. Condensed tannins (Rao et al., 2018) and cyanidins (Ofosu et al., 2020) were also identified.

3.2.1.5 Maize

Maize (*Zea mays* L.) is a rich source of antiodixants including phenolics and carotenoids that possesses higher antioxidant activity compared to wheat, oats and rice (Capocchi et al., 2017). Maize species have a large genetic diversity and are identified by different shapes and colors from white to yellow, red, purple and blue (Herrera-Sotero et al., 2017). The concentrations of bioactive compounds depend on the variety of maize (Zhang et al., 2017).

Identified phenolic acids from maize are *p*-hydroxybenzoic, *p*-coumaric, ferulic, trans-ferulic, diferulic, vanillic, caffeic, protocatechuic and syringic acids (Das & Singh, 2015; Zhang et al., 2017). Moreover, cyanidin-3-glucoside, quercetin and kaempferol were also found in maize (Das & Singh, 2015). The maize bran was also found to contain phenolic compounds, mainly ferulic acid followed by diferulic, *p*-coumaric and sinapic acid in bound form (Chiremba et al., 2012). When compared with the yellow ones, dark-colored maize varieties contain higher health-promoting antioxidants such as carotenoids, tocopherols and anthocyanins (Capocchi et al., 2017). Red maize was reported to have a higher amount of anthocyanins followed by dark blue (purple), light blue and multi-colored maize (Zilic et al., 2012). Herrera-Sotero et al. (2017) reported 20 compounds in red maize as anthocyanins, which were mainly cyanidin derivatives. Phenolic compounds identified in purple maize are chlorogenic, caffeic, ferulic acids, rutin, morin, quercetin, naringenin and kaempferol (Ramos-Escudero et al., 2012) and several anthocyanins including cyanidin, pelargonidin, peonidin, cyanidin-3,5-diglucoside, cyanidin-3-glucoside, pelargonidin-3-glucoside, cyanidin-3-malo-nylglucoside, peonidin-3-glucoside, cyanidin-3-(6"-malonylglucoside) and cyanidin-3-succi-nylglucoside (Lao & Giusti, 2016).

3.2.1.6 Millet
Millet is a small, seeded grain that is produced in the world's tropical and semi-arid regions (Chandrasekara & Shahidi, 2011). Millet has unique health benefits due to its rich nutrient content and it is used as an ingredient in multigrain and gluten-free cereal products (Pradeep & Sreerama, 2018).

Whole millet grains were found to contain hydroxybenzoic acid and their derivatives (gallic, protocatechuic, p-hydroxybenzoic, gentisic, vanillic and syringic acids) and hydroxycinnamic acids (chlorogenic, caffeic, trans-cinnamic, p-coumaric, sinapic, trans-ferulic and cis-ferulic acids) as well as flavonoids (catechin, gallocatechin, epigallocatechin, epicatechin, taxifolin, vitexin, luteolin, tricin, myricetin, kaempherol, quercetin and apigenin) (Chandrasekara & Shahidi, 2011). The major phenolic compounds present in millet varied depending on the variety. Mainly protocatechuic, ferulic and caffeic acids, catechin and epicatechin were found in finger millet (Chandrasekara & Shahidi, 2011; Xiang et al., 2019a). The free phenolic fraction of foxtail millet was comprised of vanillic, ferulic, sinapic and caffeic acids, apigenin, luteolin, naringenin, kaempferol (Chandrasekara & Shahidi, 2011; Pradeep & Sreerama, 2018), N', N"-di-p-coumaroylspermidine and N'-p-coumaroyl-N"-feruloylspermidine (Xiang et al., 2019b). Pradeep and Sreerama (2018) reported vanillic, ferulic, p-coumaric acids, catechin, luteolin and apigenin as bound phenolic compounds in foxtail millet. N', N"-dicaffeoylspermidine, p-coumaric acid, kaempferol-C-hexoside, kaempferolC- and O-dihexoside were also reported in bound form in foxtail millet (Xiang et al., 2019b). Little millet was also shown to have a similar phenolic profile to that of foxtail millet (Pradeep & Sreerama, 2018).

3.2.1.7 Barley
Barley (*Hordeum vulgare* L.) is one of the ancient cereal crops. Barley is attracting worldwide attention because of its high content of biologically active compounds such as protein, dietary fiber, vitamins and phenolic compounds (Mareček et al., 2017). Studies have shown that the consumption of foods rich in whole barley can provide protection from the development of some diseases like hyperlipidemia, diabetes and atherosclerosis (Ge et al., 2020).

Tocopherols, lignans, flavonoids and phenolic acids are major antioxidants found in barley grains. These phenolic antioxidants were found to be mainly located in the husk and bran fractions of barley in free and bound forms (Lahouar et al., 2014; Mareček et al., 2017). The husk of barley was found to be rich in phenolic compounds including syringic and p-coumaric acids, and a small amount of protocatechuic acid, (+)-catechin, naringin, naringenin, gallic acid, hyperoside, rutin, 4,5-di-O-caffeoylquinic acid, cirsiliol and apegenin were also present in barley husks (Hajji et al., 2018). The phenolic composition and antioxidant activity can vary depending on the grain color and genotype. Ferulic acid is the predominant phenolic acid found in white barley, while chlorogenic acid is the main phenolic acid in yellow, blue and black barley. Some phenolic acids also found in colored barley are gallic, protocatechuic, caffeic, hydroxybenzoic and p-coumaric acids. The main flavonoids present in all types of barley were reported as vitexin followed by quercitrin, isovitexin, afzelin, isoquercitrin, epicatechin, hesperidin, (+)-catechin and rutin, and a small amount of phloretin, daidzein, psoralen, naringenin, phlorizin and glycitein were also identified (Ge et al., 2020).

3.2.1.8 Oat

Oat is mainly consumed as whole grain and it has two main species including *Avena sativa* L. and *Avena nuda* L. *Avena sativa* L. is widely distributed in North America, Europe and Asia, while *Avena nuda* L. is more common in China. The phenolic composition and antioxidant activity of oat can be influenced by cultivar and location (Chen et al., 2018). The oat grain is comprised of the hull and kernel (groat). The whole oat usually consists of 25–35% hull, depending on environmental and genetic factors. The color of the hull changes with the variety, the most common colors being white, yellow, reddish-brown and black (Varga et al., 2018). Oat displays antioxidant activity and therapeutic effect due to its high phytochemical content such as β-glucan, tocols, flavonoids, phytic acid and phenolic compounds (Rakic et al., 2014; Chen et al., 2018).

The main phenolic compounds found in oat were identified as gallic, protocatechuic, *p*-coumaric, hydroxybenzoic, vanillic, caffeic and ferulic acids (Rakic et al., 2014; Chen et al., 2018). Besides these common antioxidants, *N*-cinnamoylanthranilic acids called avenanthramides, which are unique and low-molecular-weight antioxidant groups, were only found in oats and more than 35 avenanthramides were identified in oats (Chen et al., 2018). Avenanthramides are described as soluble amides of cinnamic acids or avenalumic acids (*p*-coumaric, caffeic, ferulic and sinapic acids) with anthranilic acid (or a hydroxylated and/or methoxylated derivative of anthranilic acid) (Varga et al., 2018). Avenanthramides were shown to have 10–30 times higher radical scavenging activity compared with caffeic acid, ferulic acid and vanillin in *in vitro* experiments (Chen et al., 2018).

3.2.2 Pseudocereal Grains

Pseudocereals are structurally similar to the cereals and are used in similar ways as foods. Common pseudocereals or gluten-free cereals are buckwheat (*Fagopyrum* spp.), quinoa (*Chenopodium quinoa* Willd), amaranth (*Amaranthus* spp.), teff (*Eragrostis* spp.) and chia (*Salvia hispanica* L.). The pseudocereals are not suitable for use in bread manufacturing, however, they can be used in diet therapy for celiac disease and gluten sensitivity (Zhu, 2019). Moreover, increased consumption of pseudocereals in a daily diet can improve antioxidant intake due to the diverse bioactive compounds found in pseudocereals.

3.2.2.1 Teff

Teff (*Eragrostis tef*) is an ancient agricultural crop belonging to the *Poaceae* family that originated and diversified in Ethiopia (Satheesh & Fanta, 2018). Although the consumption and awareness of teff are limited, the usage of teff in food systems is getting more popular because of its nutritional and bioactive components and lack of gluten. Because of its small size, teff is always milled with bran and germ, which makes teff an important source of bioactive compounds. Moreover, teff is a good source of essential amino acids, carbohydrates, minerals and dietary fiber (Inglett et al., 2015; Shumoy & Raes, 2016; Satheesh & Fanta, 2018).

Teff has varieties with different colors from white to brown, which indicates differences in phenolic profile and antioxidant activity. The major phenolic compounds found in soluble form in white teff are epigallocatechin, rutin, protocatechin ethyl acid, chlorogenic, gallic, *p*-coumaric and sinapic acids, while brown teff contains mainly epigallocatechin, rutin, *p*-coumaric, gallic and sinapic acids. The bound fraction contains epigallocatechin, catechin, quercetin, ferulic and sinapic acids in white and brown teff. Unlike white teff, brown teff also contains cinnamic acid. The concentrations of phenolic compounds and antioxidant activity in brown teff were found to be higher than those in white teff (Shumoy & Raes, 2016; Stastna et al., 2019).

3.2.2.2 Quinoa

Quinoa (*Chenopodium quinoa*) belongs to the *Chenopodiacea* family and it is endemic in the Andean region from Colombia to the north side of Argentina and the south of Chile (Satheesh & Fanta, 2018). Quinoa can adapt to hard environmental conditions like highlands, salinity, drought and frost due to its broad genetic diversity (Park et al., 2017). Quinoa is also nutritious with higher amounts of vitamins, protein (lysine and methionine), dietary fiber, minerals and omega-6 essential fatty acid than staple cereals (Inglett et al., 2015). In addition, it can also be safely used for the treatment of celiac disease. The health benefits of quinoa were reported to be the regulation of levels of serum cholesterol and glucose, low-density lipoproteins and triglycerides (Vilcacundo & Hernández-Ledesma, 2017). Quinoa has antioxidative compounds that can protect against several conditions like cancer, allergies, inflammatory diseases, obesity, diabetes and cardiovascular diseases (Inglett et al., 2015; Satheesh & Fanta, 2018).

Quercetin and kaempferol glycosides are the major polyphenols found in quinoa seeds (Alvarez-Jubete et al., 2010). Caffeic, ferulic, *p*-coumaric, gallic and protocatechuic acids are also common phenolic acids. Regarding free phenolics, epigallocatechin, rutin and sinapic acid are found in red quinoa, while ferulic acid is abundant in black quinoa. Protocatechuic, vanillic and ferulic acids were detected in quinoa in a soluble conjugated phenolic acids fraction. Epigallocatechin is the main phenolic compound present in bound form in red and black quinoa, whereas cinnamic acid is detected only in black quinoa. Lipophilic antioxidants such as tocopherols and carotenoids are also present in quinoa and contribute to antioxidant activity (Tang & Tsao, 2017; Stastna et al., 2019).

3.2.2.3 Buckwheat

Buckwheat (*Fagopyrum esculentum* Moench), is a common pseudocereal that contains high amounts of manganese, magnesium, dietary fiber and phenolic compounds (Inglett et al., 2015). Due to its high phenolic content and antioxidant activity, buckwheat was shown to exert several health benefits such as reducing the risk of chronic diseases and regulating the levels of total triglycerides and total cholesterol in serum and the liver, as well as blood sugar and blood pressure (Guo et al., 2012). Inglett et al. (2015) reported that when compared with cereals like oats and barley, whole buckwheat had 2–5 times higher amount of phenolic compounds, and buckwheat bran and hulls had 2–7 times higher antioxidant activity than those of barley, triticale and oats.

The phenolic composition of buckwheat consists of mainly rutin (quercetin-3-rutinoside), quercetin, catechin, hiperin, quercitrin, epicatechin, orientin, isoorientin, vitexin and isovitexin (Verardo et al., 2010). Luteolin, apigenin glycosides, caffeic acid and syringic acid are flavone glycosides that are present in buckwheat (Alvarez-Jubete et al., 2010). Other phenolics tentatively identified in buckwheat by Verardo et al. (2010) were 2-Hydroxy-3-*O*-β-D-glucopyranosyl-benzoic acid, 1-*O*-caffeoyl-6-*O*-α-rhamnopyranosyl-β-glycopyranoside and epicatechin-3-(3″-*O*-methyl) gallate.

Phenolic acids are mostly found in free form in the bran of buckwheat. Major phenolics in the bran are *p*-hydroxybenzoic acid, caffeic acid, chlorogenic acid and protocatechuic acid (Guo et al., 2012). The bound phenolics of buckwheat were reported as 3-methylcatechol, *p*-coumaric, gallic, caffeic acid, catechin, epicatechin, benzoic acid, epicatechin-gallate, isoquercetin and rutin (Guo et al., 2012). The amount and distribution of phenolics can be different based on the variety of buckwheat (Inglett et al., 2011).

3.2.2.4 Amaranth

Amaranth (*Amaranthus caudatus*) is an ancient pseudocereal grain that is attracting increasing attention due to its unique nutritional components and potential health benefits. It can be adapted to all regions of the world from tropical to temperate climates as a consequence of its high genetic variability (Tang & Tsao, 2017). Amaranth is a good source of the essential amino acid, lysine, which is low in other grains, and also vitamins like thiamine, niacin, riboflavin, folate and minerals including calcium, iron, magnesium, phosphorus, zinc and manganese (Inglett et al., 2015).

A number of studies reported that an amaranth-rich diet may reduce blood pressure and cholesterol. Moreover, anti-atherosclerotic, anthelmintic, antinociceptive, antipyretic, anticancer, antiallergenic and hepatoprotective activities, as well as stimulation of the immune system, were reported (Martinez-Lopez et al., 2020).

Amaranth was found to have high amounts of antioxidative phenolic compounds like phenolic acids, including gallic, *p*-hydroxybenzoic, vanillic and its derivatives, caffeic, protocatechuic and flavonoids like rutin (Alvarez-Jubete et al., 2010; Pasko et al., 2011; Peiretti et al., 2017). In the free fraction of amaranth, the main phenolics were identified as gallic, protocatechuic and *p*-hydroxybenzoic acids, while in the bound form ferulic acids, mainly trans-ferulic and cis-ferulic, were found (Tang & Tsao, 2017).

3.2.2.5 Chia

Chia (*Salvia hispanica* L.) belongs to the *Lamiaceae* family and grows in southern Mexico and northern Guatemala. However, chia is also cultivated in Australia, Bolivia, Columbia, Peru, Argentina, North America, and Europe (Grancieri, Martino, & Gonzalez de Mejia, 2019). The chia seed is a good source of protein, dietary fiber, minerals, oil and phenolic compounds (da Silva Marineli et al., 2014). Chia seed possesses some health benefits like cardiac and hepatic protective effects, anti-aging and anti-carcinogenic activity due to its antioxidant compounds such as tocopherols, phytosterols, carotenoids and phenolic compounds including chlorogenic acid, caffeic acid, myricetin, quercetin and kaempferol (Ullah et al., 2016).

The major phenolic compounds identified in chia seed were quercetin, kaempferol, rosmarinic, caffeic, chlorogenic, protocatechuic and gallic acids and daidzein (Reyes-Caudillo, Tecante, & Valdivia-López, 2008; Martinez-Cruz & Paredes-Lopez, 2014). Different phenolic compounds were also identified in chia seed as danshensu (3-[3,4-dihydroxyphenyl] lactic acid) and its derivatives, such as salvianolic acids (Oliveira-Alves et al., 2017). The antioxidant activity of chia was reported to depend on the variety (Martinez-Cruz & Paredes-Lopez, 2014).

3.3 HIGH-PRESSURE PROCESSING (HPP)

3.3.1 The Working Principle of HPP

HPP is a novel technology that involves the application of hydrostatic or hydrodynamic high pressure on food products with specially designed equipment. HPP is a commonly used term that refers to the application of static high pressure via a pressure-transmitting liquid, commonly water. Dynamic high-pressure application is carried out in high pressure homogenizers (HPH). HPP can be utilized for processing liquid or solid foods, while HPH is only applied to liquid foods with or without particulate solids. Higher pressures up to 800 MPa can be applied in HPP compared with up to 400 MPa in HPH. Heating of the food product can occur in both processes but the level of temperature increase is much higher in HPH. While pressure, temperature and time are operating parameters in HPP, valve geometry and number of passages or cycles are additional parameters considered in HPH.

The effect of HPP is explained by Le Chatelier's principle which states that when a disturbance is applied to a system at equilibrium, the system responds in a way that minimizes the disturbance. Therefore, the reactions that lead to an increase in volume will be inhibited while the reactions causing a reduction in volume (e.g. phase transitions, change in molecular configuration and chemical reactions) will be stimulated by the application of high pressure (Tao et al., 2014). In addition, pressure is instantaneously and homogeneously transmitted throughout a sample if the sample is in direct contact with the pressure-transmitting medium according to the isostatic rule (Martínez-Monteagudo & Balasubramaniam, 2016). This means that high pressure will be applied to the whole product uniformly in a short time independent of sample size and geometry, which is an advantage over thermal processing. Therefore, HPP can reduce processing time and energy and the risk of over-processing food products (Farkas, 2016).

Reduction in volume by hydrostatic pressure changes the distance between molecules by influencing distance-dependent interactions, van der Waals and hydrophobic, electrostatic and hydrogen bonds (Martínez-Monteagudo & Balasubramaniam, 2016). As high pressure does not affect covalent bonds, HPP can have an impact on macromolecules with non-covalent bonds. Therefore, it can modify the structure of microorganisms, enzymes, protein and starch, while it has no or negligible effect on small molecules like flavor compounds, pigments and nutrients in food products (Shoquin et al., 2004). The impact of HP on non-covalent bonds causes loss of enzyme activity and damage to cell membranes that could result in the death of bacteria, yeast, molds and parasites (Farkas & Hoover, 2000).

3.3.2 Equipment and Operational Parameters in HPP

The principal components of an HPP unit are a pressure chamber, closures or plugs to seal the chamber, a frame (yoke or wire-wound) to cover the chamber, a pressure intensifier and a pump (Balasubramaniam, Barbosa-Cánovas, & Lelieveld, 2016). Pressures ranging from 100 MPa to 800 MPa are applied by electrohydraulic pressure intensifiers mounted on the chamber at or below room temperature for a few minutes depending on the purpose of the application. Commercial pressure vessels have typical volumes between 35 and 525 L while pilot systems have between 1 and 30 L volume (Balasubramaniam et al., 2016). The typical cycle time for the process is about 10 min (Farkas & Hoover, 2000; Martínez-Monteagudo & Balasubramaniam, 2016). The temperature of the food is increased by the application of high pressure due to compression (adiabatic) heating, however, the temperature is reduced back to its initial value after decompression. The temperature increases by 3°C per 100 MPa for water and high-moisture foods and up to 8–9°C per 100 MPa for lipid-containing foods (Martínez-Monteagudo & Balasubramaniam, 2016).

Batch and semi-continuous systems can be used for the application of HPP. Batch systems can be used for liquid or solid foods. The product must be packed with flexible polymer packaging that allows a change in volume by 10–20% (Tao et al., 2014). Flexible polymeric materials that can resist compression and decompression such as polyethylene terephthalate, polyethylene, polypropylene, ethylene vinyl alcohol copolymer and their combinations are used as packaging materials (Juliano et al., 2010). Coating materials such as aluminum oxide and silicon oxide or metalization by a thin layer of aluminum on the surface can be applied to improve the barrier properties of the polymeric films (Mensitieri, Scherillo, & Iannace, 2013). Packaging material should also protect the food from light, oxygen and moisture. Vacuum packaging is recommended for HPP for uniform treatment and to avoid air-related problems in processing and storage.

In batch systems, pre-packaged products in a basket are placed in the pressure chamber, the chamber is closed with plugs and then the pressure-transmitting fluid, commonly water in large-scale systems and water/glycol or oil in small-scale ones, is admitted by a low-pressure pump into the chamber. The contents of the chamber are compressed to the target pressure by pressure intensifiers and held for the target time. After application, the pressure is released, the chamber is opened and the contents are unloaded. Cycle time for a high-pressure application includes time for holding and depressurization. The capacity of an HP unit is determined by the cycle time and the loading factor (i.e. the percentage of the vessel volume that is used for holding the packaged product) (Tao et al., 2014). Multiple units can be used in a sequence based on plant capacity.

Pumpable fluid food products can be processed by semi-continuous HPP systems. The product and the pressure-transmitting fluid are separated by a free-floating piston (Tao et al., 2014). After the process, the product is discharged into a sterile holding tank and then packaged in an aseptic filling unit. A recently developed semi-continuous system allows the processing of bulk liquids at a capacity of 10,000 L/h with two chambers with a volume of 525 L (1,050 bulk, www.hyperbaric.com, 2018). The liquid in a feed tank is filled into a bladder or a processing bag in the chamber that has a volume of 90% of the chamber volume. Water is pumped in as pressure-transmitting fluid surrounding the bladder in the chamber and the liquid is pressurized at 600 MPa for 2–3 min. After HPP, the liquid is transferred into a sterile processing tank and fed to a packaging unit. The company claims a reduction in investment cost, energy, labor and wear of the equipment compared to systems with in-pack equipment. Continuous HPP systems have been developed where the fluid is pressurized while flowing in a tube with certain dimensions by a high-pressure pump (Lelieveld & Hoogland, 2016). However, this system is not used as commonly as batch and semi-continuous systems.

HPH is an extension of conventional homogenization technology applied at elevated pressures up to 400 MPa. The system involves a high-pressure pump and valves or nozzles made from abrasion-resistant materials. The valves or nozzles block the flow of fluid and create additional pressure, shear, turbulence, cavitation and impingement on the fluid beyond those caused by conventional homogenization (Harte, 2016). In addition, dissipation of the kinetic energy of the fluid in the homogenizer chamber results in the breakage of particles and an irreversible increase in the temperature of the fluid up to 18°C per 100 MPa (Martínez-Monteagudo & Balasubramaniam, 2016). These effects cause a reduction in the size of droplets or particles in a liquid and the stabilization of dispersed systems.

3.3.3 Utilization of HPP in the Food Industry

The aim of food processing is to preserve food materials by converting them into safe, nutritious and stable products that can be reached by consumers at any season and location. The process parameters are optimized so that pathogenic and spoilage microorganisms and enzymes are inactivated and, at the same time, the nutritional value and sensory properties of the foods are protected. However, food safety is achieved at the expense of some unavoidable degradation in nutritional and sensory properties by conventional thermal processing. HPP is one of the novel technologies that has been developed as an alternative to thermal processing with the aim of preserving and protecting nutritional and sensory quality and extending the shelf life of food products. The ability of HPP to modify physicochemical properties of food components also allowed its use for modifying food texture and producing new functional food ingredients. In addition, HPP can reduce or eliminate the use of chemical additives in some cases by its effect on the structure and safety of foods. Furthermore, the combination of HPP with other food processing methods such as freezing, thawing, blanching, heating and extraction can intensify their effects.

The first application of high pressure in food processing was reported in 1894. Bert Hite from the Agriculture Research Station in Morgantown, West Virginia applied pressures between 200–680 MPa to milk and found out that spoilage microorganisms can be inactivated by pressure (Farkas & Hoover, 2000; Tao et al., 2014). Hayashi and coworkers restarted studies on the application of high pressure for foods in 1986 (Hayashi, 1996). However, due to lack of industrial equipment and proper packaging materials, the use of HPP in the industry started in 1992 in Japan and then the USA, and the numbers of companies using HPP equipment have increased greatly in the 2000s (Tao et al., 2014). In Japan, jam was the first HPP-treated commercial product introduced to the market followed by fruit jellies and sauces. Since then, many products treated with HPP including juices and beverages, vegetable products, meat products, seafood and fish have been commercialized in Japan, the USA and Europe. Currently, batch HPP at 600 MPa is successfully applied for pasteurization of a wide variety of food products.

HPP can inactivate vegetative pathogens and spoilage microorganisms with no or negligible effect on physicochemical properties like nutrients, color and flavor (Balasubramaniam et al., 2016; Bevilacqua et al., 2019). On the other hand, inactivation of microbial spores requires the combination of HPP with heat treatment (HPTS) at temperatures above 90°C. In 2009, the FDA certified an industrial pressure-assisted thermal sterilization (PATS) process for treating mashed potatoes at 600 MPa and 121°C (Stewart, Dunne, & Keener, 2016). Sterilization of low-acid foods with HPP can be achieved with less heat load compared to conventional retort processing. There are some limitations that have to be resolved before industrial use of HPTS, including limited knowledge on the inactivation mechanisms of highly resistant bacterial spores and technical limitations, such as process homogeneity, discontinuous/batch processing and process complexity (Sevenich & Mathys, 2018). Uniform temperature distribution is critical in HPTS in equipment design and implementation. A large number of small vessels was tried as an option. Another more feasible option was called pressure-enhanced sterilization (PES) using an insulated carrier filled with hot water as a cover of a packaged product with current HPP systems where carrier fluid can transmit pressure and heat the contents homogeneously (Balasubramaniam et al., 2016; Sevenich & Mathys, 2018). Preheating to a temperature lower than 121°C is an advantage of the PES system.

The influence of high pressure on non-covalent bonds has been investigated as a way to modify and functionality of macromolecules in foods. Secondary, tertiary and quaternary structures of proteins involving hydrophobic, hydrogen and ionic bonds can be changed upon application of HPP, however, the effect of HPH is lesser due to lower pressures and shorter time applied (Huang, Hsu, Yang, & Wang, 2014; Harte, 2016). The solubility of a protein can be lowered by high pressure, causing unfolding and exposure of sulfhydryl groups leading to protein aggregation (Perez-Andres et al., 2018; Queirós, Saraiva, & da Silva, 2018). For example, proteins of amaranth were unfolded by HPP depending on pressure level (200–400 MPa/5 min) and concentration of protein, amount of sulfhydryl groups were increased and solubility of proteins was decreased with pressure (Condés et al., 2012; Condés, Añón, & Maurí, 2015). Unfolding and aggregation of proteins can enhance their structural-functional properties in gels, emulsions and foams. The effect of high pressure on protein has been found to depend on the type and structure of the protein, pH, ionic strength and the composition of the medium and conditions of HPP. Structural modification can be enhanced when HPP is used in combination with pH and heat and it can also improve the action of hydrolytic enzymes (Queirós et al., 2018). HPP can be used in the improvement of structural-functional and bioactive properties of food proteins.

Modification of protein conformation by HPP has also been investigated as a potential method for reducing the allergenicity of food proteins (Ekezie, Cheng, & Sun, 2018; Dong, Wang, & Raghavan, 2020). The allergenicity of proteins can be reduced by protein denaturation, the release of membrane-bound proteins or increased susceptibility to enzymatic hydrolysis. However, the type of protein and the matrix were found to be important for this effect. Allergic rice protein fractions were dissolved and released by high-pressure treatment at pressures between 100–400 MPa (Kato et al., 2000). The effect of high pressure was explained by partial destruction of the endosperm, permeation of water and an increase in the solubility of proteins where pressures above 500 MPa did not change the results. In addition, pressure treatment at 100 MPa improved the quality of cooked rice, improving brightness of color, flavor and texture (Watanabe et al., 1991).

HPP can change structural-functional properties of starch, which is a major structure-developing component in grain-based products. Gelatinization and retrogradation of starch play an important role in the textural quality and stability of a food product. Pressures between 300–600 MPa can induce gelatinization of starch depending on the source and type of starch, pressure level and holding time (Hayashi, 1987; Hayashi & Hayashida, 1989; Hayashi, 1996; Stolt et al., 2001; Grgić et al., 2019). HPP caused swelling and gelatinization of starch granules with no change in granule shape and limited release of amylose compared to the thermal process (Stute et al., 1996; Stolt et al., 2001; Oh et al., 2008; Kieffer et al., 2007; Yamamoto & Buckow, 2016). Retrogradation was observed after HPP but it was found slower than that of heat-treated starch (Katopo, Song, & Jane, 2002; Hu et al., 2011). In addition, HPP can assist in the modification of starch when applied with chemical modification (Hu et al., 2011; Yamamoto & Buckow, 2016). Effects of HP on starch and protein were utilized for improvement of the texture of gluten-free batters. HP treatment improved the elasticity of batters made from wheat, white rice and teff by starch gelatinization and protein cross-linking (Vallons et al., 2010, 2011; Deora, Deswal, & Mishra, 2014).

HPP can increase the rate of osmotic dehydration and drying of certain food products as HP disintegrates cell walls and tissue structure and increases the permeability of cell membranes (Rastogi et al., 2007). Salt reduction in processed food products such as cured meats was also possible by application of HPP (Clariana et al., 2011; Hayes & Allen, 2011). The use of HPP for reduction of food contaminants that are derived from thermal processing has been investigated. The Maillard reaction and formation of acrylamide were reduced by application of HPP compared to thermal treatment at the same temperature and time in model systems (De Vleeschouwer et al., 2010). On the other hand, incomplete inactivation of enzymes can cause degradation of nutrients and sensory properties during storage time (Jambrak et al., 2018).

The application of HPH is mainly limited to size reduction operations compared to the wider use of HPP, although it is a much older technology. HPH was first introduced to the food industry by Augusto Gaulin in 1899 for the emulsification of milk (Sevenich & Mathys, 2018). This technology has been used since then for stabilization of food dispersions, especially emulsions. HPH stabilizes dispersions by reducing the size of droplets or particles and modifying the rheological properties of the mixtures. HPH can achieve much higher pressures in the range of 100–450 MPa by using specially designed valves and nozzles than those applied in conventional homogenization (5–50 MPa). Disruption of vegetative microbial cells can be affected by cavitation, shear stress, turbulence, impingement, and high pressure. Pasteurization and sterilization can be accomplished according to the process parameters including initial temperature, pressure and valve geometry (Martínez-Monteagudo & Balasubramaniam, 2016). Combination of HPH with preheating can result in temperatures between 120 and 150°C in the cavity, which could disrupt microbial spores in a very short time (Sevenich & Mathys, 2018). Microbial inactivation by HPH was attributed mostly to the heating component rather than pressure and shear effects in studies made with fluid milk (Cavender & Kerr, 2011; Harte, 2016; Lelieveld & Hoogland, 2016). In addition, cellular disruption allows the release of intracellular components, and changes in the structure and conformation of polymers such as starch, protein, carbohydrates, enzymes and microorganisms occur depending on temperature and pressure (Mesa et al., 2020). Modified components can be used as new food ingredients with desired functionalities. Temperature increase during HPH operations adversely affects heat-sensitive nutrients and bioactive components.

3.4 EXTRACTION OF ANTIOXIDANTS FROM GRAINS

Extraction is the separation of a targeted component by using a proper solvent from a complex matrix. Solvent extraction is the basic conventional technology for extraction of antioxidants from food matrices. Limitations of conventional solvent extraction led to investigations on applications of novel technologies for the improvement of extraction. Choosing the most effective method of extraction is vital for achieving maximum yield and stability for the compound of interest.

3.4.1 Conventional Techniques

Dating back to 1879, Soxhlet is the oldest extraction method by diffusion, without any shear stress applied to the solid matrix. Extraction of oil or juice by pressing the raw material, solvent reflux and solvent extraction are examples of other highly used traditional extraction applications in history (Rodrigues & Fernandes, 2017). Considering its positive attributes such as ease of use, efficiency and wide applicability, solvent extraction is the most common method used for the extraction of antioxidants (Dai & Mumper, 2010).

In conventional extraction systems, three main mechanisms drive the process. Mass transfer is the main mechanism that runs the extraction process. The extraction solvent is the second element that is used to solubilize the target compound. Finally, heat is the facilitator of the diffusion and the solvent permeator into the solid matrix (Rodrigues & Fernandes, 2017). An increase in extraction temperature can enhance both solubility and the rate of mass transfer. Moreover, high temperature also decreases the viscosity and the surface tension, therefore aiding in extraction, although long extraction times and high temperatures increase the chance of oxidation of antioxidants and hence decrease the yield of antioxidants in the extract (Dai & Mumper, 2010).

The solubility of antioxidants in different solvents depends on their chemical structure (e.g. aglycones-free phenolic acids, esters, glycosides), their interaction with other components to form bound complexes and the polarity of the solvent used for extraction (Garcia-Salas et al., 2010; Naczk & Shahidi, 2006). Generally, water and a changing amount of aqueous ethanol (e.g. 80%), methanol (80%) or acetone (50 to 80%) are utilized, although propanol, ethyl acetate, dimethylformamide and their combinations may also be used to extract antioxidants from food matrices (Guido & Moreira, 2017; Naczk & Shahidi, 2006). Additional steps such as removing non-phenolic substances such as waxes, terpenes, fats and chlorophylls may also be required. Moreover, acidic or alkaline solvent systems are also used (Zhou & Yu, 2004). The yield of extraction is affected by the extraction method and by parameters such as the solute to solvent ratio, type of solvent, pH, temperature, time and chemical and physical characteristics of the matrix (Chandrasekara et al., 2016; Naczk & Shahidi, 2006). The next stage after extraction is determination of the activity and stability of antioxidants in the extract. In addition, the bioaccessibility and bioavailability of antioxidants are evaluated by *in vitro* and *in vivo* methods.

3.4.2 Novel Techniques

Conventional extraction is considered a time-consuming processing operation with low efficiency. Moreover, the long extraction times and high temperature applied in conventional extraction techniques increase the chance of oxidation of antioxidants. Almost all of the traditional extraction methods have these common properties, therefore alternative techniques have been sought to overcome these drawbacks. For this purpose, novel technologies such as HPP, subcritical fluid extraction, supercritical fluid extraction (SFE), microwave, pressurized liquid extraction, pulsed electric field and high voltage electrical discharge and ultrasound have been applied for extraction (Xu et al., 2017; Saini et al., 2020). Each novel technology uses different mechanisms of action in extraction. Novel techniques of extraction are also called green technologies as they cause less toxicity to the environment by reducing the use of chemicals and energy (Moreno-Najera et al., 2020). In addition, novel extraction technologies can improve the stability of antioxidants during extraction (Setyaningsih et al., 2016).

Ultrasound technology uses microflows through acoustic cavitation formed by alternating compression and expansion cycles that damage the plant cell membrane and release targeted compounds into the solvent. Ultrasound-assisted extraction has been utilized in the extraction of antioxidative compounds from cereals. In an effort to increase the efficiency of water as a solvent, ultrasound has been applied in the extraction of phenolics from brewer's spent grain (Alonso-Riaño et al., 2020). Extraction at 47°C and 21.7 mL water/g dry brewer's spent grain was found to maximize the yield and ultrasound assistance was shown to improve the efficiency of the extraction (Alonso-Riaño et al., 2020). For ultrasound-assisted extraction of phenolics from rice grains, optimum conditions were reported as the use of 80% methanol in water as a solvent at pH 4.25 and 45°C and ultrasound application at an amplitude of 47% (max 200 W) for 25 min (Setyaningsih et al., 2019).

SFE is based on the use of a supercritical fluid that is above its critical temperature and pressure. The most specific property of this method is that, in this supercritical state of the fluid, some physical properties of the fluid (e.g. viscosity, density and diffusivity) change between those of gaseous and liquid states (Bezerra et al., 2019). This renders the fluid an effective solvent in extraction without leaving any residue in the end product. Therefore, SFE is widely used instead of solvent extraction for extracting antioxidants (Roselló-Soto et al., 2019). SFE has several advantages such as lower temperatures and energy consumption, and high product quality and purity. On the other hand, SFE is limited to compounds of low or medium polarity (Garcia-Salas et al., 2010). Inertness, nontoxicity, high solubility and suitability to heat-sensitive compounds make carbon dioxide the most used solvent in SFE for the extraction of apolar and moderately polar compounds (Bezerra et al., 2019).

In microwave-assisted extraction (MAE), microwaves, electromagnetic radiation with a wavelength from 0.001 m to 1 m, pass through the medium. During this passage, the energy of the microwaves is absorbed by the matrix (Zhang et al., 2011). The evaporation of food moisture caused by the heat produced creates high pressure against the cell walls, changes the physical properties of the biological tissues and enables better penetration of the extraction solvent through the matrix, improving the yield. A lower amount of solvent consumption is another advantage of MAE (Joana Gil-Chávez et al., 2013). However, degradation of antioxidants due to the thermal effects of microwaves and the use of only solvents that are able to absorb microwaves are disadvantages of MAE (Xu et al., 2017).

Enzyme-assisted extraction (EAE) is a potential green extraction method because it requires mild extraction conditions and has minimal environmental effects. Enzymes have high specificity and high efficiency; therefore, they help in the release of bioactive compounds in the plant matrix by degrading the cellular components and structural integrity of the plant cell wall. Cellulase, pectinase, hemicellulase and glucosidase are extensively used in EAE. EAE techniques have been shown to improve the extraction efficiencies for antioxidants including phenolics, flavonoids, anthocyanins and carotenoids. Different carbohydrase enzymes were used in the extraction of antioxidants from rice bran (Kim & Lim 2016). Carbohydrase increased the number of phenolics and among them, pentopan resulted in the highest level of cell wall hydrolysis (Kim & Lim 2016).

3.4.3 Application of HPP in Extraction

3.4.3.1 Effect of HPP on Extraction

HPP can be applied for extraction of nutrients, flavor, essential oil, colorant and functional components from foods and for enhancement of their bioaccessibility and bioavailability. As high pressure has no effect on covalent bonds and the process can be applied at low temperatures for a short time, the activity and functionality of components are expected to be preserved more compared to conventional extraction methods usually performed at elevated temperatures with alkali or acid, in some cases for a long time. Even relatively low pressures around 100–200 MPa result in the crystallization of phospholipids in cell membranes and an increase in membrane permeability at room temperature (Matser & Timmerman, 2016). High pressure can break down weak chemical bonds causing cellular disintegration and rupture of the cell wall. As a result, the permeability of the cell wall is increased and intracellular components are released from the matrix (Jung, 2016). A solvent is enforced to the cell by the pressure gradient created by a high-pressure application in a short time and the solubility of the component is increased (Khan, Aslam, & Makroo, 2019). HPP can be applied as a pretreatment for cellular destruction before extraction or during extraction with the solvent (HPE) (Barba et al., 2015). Pressure, temperature, duration and number of cycles are operational factors that affect the efficiency of extraction by HPP in addition to the type of solvent, solvent to solid ratio, compound and matrix (Jun, 2013).

The use of hydrolytic enzymes together with HPP can improve the efficiency of extraction by releasing bioactive compounds from the cellular structure (Nadar, Rao, & Rathod, 2018). The pressures applied in HPP and HPE are higher compared to the pressures between 10–100 MPa applied in supercritical extraction, thus they can increase the yield and rate of extraction (Shouqin, Junjie, & Changzheng, 2004; Nadar et al., 2018).

The efficiency of HPP in extraction is especially related to its effect on the cellular matrix and cell wall. HPH can also increase extraction efficiency by its multiple mechanisms of action on cells and cell walls. The disintegration of the matrix by HPP can enhance the extraction yield for bound bioactive components. On the other hand, de-compartmentation of cellular components and incomplete inactivation of oxidative enzymes can cause degradation of bioactive compounds during and after extraction (Chakraborty et al., 2017). In some cases, HPP can cause degradation of the compound of interest by reactions favored under high pressure such as trans- to cis-isomerization in carotenoids or anthocyanin condensation reactions forming pyran rings (Jung, 2016). The presence of vitamin C also influences the activity of a specific bioactive compound (Jung, 2016).

Food wastes have been used for the recovery of nutrients and functional components in order to better utilize natural resources. Grain wastes in particular are a rich source of prebiotics, fiber, phenolic antioxidants and cellulose that could be recovered and presented for industrial use. HPP, by its effect on cellular structures in food materials, can increase the rate of mass transfer and efficiency and reduce time, amount of solvent, impurities in the extract and energy in an extraction compared to conventional solvent extraction. Enhanced cell permeability can also improve the bioaccessibility of micronutrients (Meng et al., 2019; Scepankova et al., 2018). The more efficient use of environmentally friendly solvents such as water is also possible in extraction by HPP.

3.4.3.2 Application of HPP for Extraction of Antioxidants from Grains

HPP has been utilized for the treatment of grains as an alternative to thermal treatment and its effect on the recovered amount and activity of antioxidants has been determined. Different studies revealed that the level of pressure applied in HPP is important for the yield of phenolic compounds and antioxidant activity from grains. HPP was applied to vacuum-packaged black rice (30 g/42 mL distilled water) after holding in water for 1 h and the amount of phenolics and DPPH radical scavenging activity was measured after cooking at 118°C for 15 min by Meng et al. (2019). Application of pressure at 400 and 500 MPa at 25°C for 15 min resulted in higher amounts of extractable phenolics, flavonoids, anthocyanins and antioxidant activity in cooked black rice compared to those of untreated and HPP-treated samples (Meng et al., 2019). There were decreases in phenolic content and antioxidant activity after storage at 4°C for 40 d in all samples but the decrease was lesser in HPP-treated samples at a pressure of 400 and 500 MPa. In another study, HPP at 500 MPa and 18°C for 10 min increased the antioxidant activity of germinated brown rice by 8.87%, 12.72% and 57.71% in different antioxidant capacity assays of FRAP, DPPH and ABTS, respectively, in comparison to those of an untreated sample (Xia et al., 2017). The authors also reported enhanced *in vitro* bioaccessibility of antioxidants in brown rice by HPP.

The DPPH scavenging activity of buckwheat flour (10 g sample:100 mL water) was not affected to a great extent by HPP treatment at 600 MPa and 20°C or at 45°C for 15 min in two cycles compared to that of an untreated sample (Zhou et al. 2015). However, the iron chelating capacity was increased by 11% by HPP treatment of buckwheat flour, which was explained by a possible release of bound phenolic compounds from the cell wall structure (Zhou et al. 2015).

Pressure treatment at 30 MPa at 37°C for 24 or 48 h increased the amounts of γ-oryzanol, tocopherol and tocotrienol in germinated rough rice depending on the duration of germination and pressure application (Kim et al., 2015). The authors concluded that pressure application in combination with germination accelerates the biosynthesis of physiological metabolic material as a result of an enhanced enzymatic reaction rate and significantly increases the extractability of phenolics (Kim et al., 2015). A similar trend was observed by Kim et al. (2017) in HPP-treated two-day-germinated rough rice where pressure from 10 MPa to 100 MPa was applied at 37°C for 24 h. More phenolic acids were recovered from the germinated rough rice by increasing levels of pressure in HPP and about a threefold increase was obtained by application of HPP at 100 MPa compared to that from an untreated sample. Tocopherol and tocotrienols were also extracted with a higher yield by HPP treatment up to 30 MPa, however, their extracted concentrations declined under HPP treatment at 50 and 100 MPa compared to those in an untreated sample (Kim et al., 2017).

The effects of pressure level (200–600 MPa), temperature (20–60°C) and time (30–120 min) of HPP applied during soaking on the phenolic content and antioxidant activity of germinated foxtail millet flour were investigated by Sharma et al. (2018). About a 17% increase in total phenolic content was obtained by HPP at 20°C with the extended application of high-pressure levels. An increase in temperature to 40 and 60°C caused a reduction in total phenolic content by 20% and 26%, respectively, under HPP at 600 MPa for 120 min. This reduction was attributed to the oxidation of free phenolics present in the pericarp and aleurone layer of the grain. On the other hand, the antioxidant activity of fox millet by FRAP assay was increased by pressure level, time and temperature, reaching 105.33% of an untreated sample under HPP at 600 MPa for 120 min. This was attributed to the enhanced extractability of antioxidants and the possible formation of antioxidants by a Maillard reaction (Sharma et al., 2018).

Saikaew et al. (2018) reported that the total phenolic content and antioxidant activity of purple corn were preserved better by HPP at room temperature and pressure in the range of 250–700 MPa for 30–45 min than by steam treatment, which is the common cooking method. However, the preservation effect of HPP was reduced by increasing pressure from 250 to 550 MPa, and a further increase in pressure to 700 MPa allowed partial recovery of the phenolics and anti-oxidant activity of untreated kernels. Up to 66% reduction in anthocyanin content was deter-mined in the sample treated with HPP at 550 MPa for 45 min while the other pressure–time combinations and steam treatment provided higher anthocyanin content. Degradation of phe-nolic compounds at pressures lower than 700 MPa was attributed to incomplete inactivation of oxidative enzymes including lipoxygenase and peroxidase and to a lesser extent polyphenol oxidase in corn (Saikaew et al., 2018). In addition, the authors suggested that high pressure can induce some degradation reactions such as polymerization with proteins, condensation reac-tions through covalent association with other phenolics or ionization.

Djulis was treated by Sun et al. (2019) with HPP at 500 MPa and 10°C or 30°C for 3 min as an alternative method to traditional thermal methods including cooking in boiling water for 20 min and pasteurization at 65°C for 30 min. Pressure applications resulted in higher reten-tion of phenolics and flavonoid compounds and antioxidant activity compared to thermal treat-ments. In addition, the hull of the grain was found to contain a much higher amount of phenolics and antioxidant activity than the seed.

High pressure can be applied during solvent extraction in high-pressure extraction (HPE). This approach is utilized with extraction solvents (including water) that are normally less effi-cient at low temperatures to reach higher yields than that of conventional extraction (Dai & Mumper, 2010; Joana Gil-Chávez et al., 2013; Mendiola et al., 2007). Although there is no study about the use of HPE in the extraction of grains, pressurized liquid extraction at pressures lower than those in HPP was applied for the analysis of phenolic compounds in various rice grains in a study by Setyaningsih et al. (2016). In this study, phenolic antioxidants were recovered at a maximum level from rice powder (2.5 g) into 60% ethyl acetate in methanol, an extractor with a volume of 11 mL operating at 20 MPa at 190°C in three cycles of 10 min. The extraction tempera-ture, solvent and extraction time were found to have a significant effect on the recovery of phe-nolic compounds. Pressurized liquid extraction is carried out under nitrogen at a high pressure that allows rapid penetration of solvent into the matrix and greater contact between solvent and matrix, and at a high temperature that disrupts phenolics–matrix bonds and increases solubil-ity in the extraction solvent (Setyaningsih et al., 2016). As higher pressures are applied in HPP compared to pressurized liquid extraction, extraction can be achieved at lower temperatures in a shorter time by HPP and the heat-induced degradation of antioxidants can be prevented.

Park et al. (2016) studied HPP alone at 500 MPa and 20°C for 6 min and in combination with enzymatic hydrolysis for the extraction of rice husks. The extract obtained by the combination of enzymatic hydrolysis and HPP treatment with 0.5% Celluclast yielded higher antioxidant activity and total amount of phenolics than those of the untreated, only HPP-treated and lower-enzyme-treated rice husk samples (Park et al., 2016).

Studies have shown that antioxidants and antioxidant activity in grains can be preserved better by HPP compared to traditional thermal treatments. The effects of HPP on antioxidants and matrices have been found to depend on applied pressure, temperature and treatment time. Although HPP can provide significant savings in time, solvents and energy, some degradations of antioxidants can also occur by pressure-induced chemical reactions and residual enzymatic activity. Therefore, further research studies on the determination of optimum operating condi-tions are required for the extraction of a specific antioxidant from a specific grain matrix.

3.5 CONCLUSIONS

HPP provides advantages over thermal processing in the preservation of nutritional value and sensory quality of food products, providing fresh-tasting and safe foods with a long shelf life. A major disadvantage of HPP technology for the food industry is the high initial investment cost that needs to be overcome in the future. However, increasing consumer demand for minimally processed foods and ingredients could drive the use of this technology by the food industry. Further developments in HPP equipment and processes will accelerate the utilization of this novel technology for the benefit of consumers, manufacturers and the environment.

HPP and HPH can improve yield and efficiency in extractions by increasing the permeation of solvents and diffusion of intracellular components, as they are effective in cellular disruption. These novel technologies can provide significant savings in time, energy and solvents in the extraction processes. However, the target compound and food matrix determine the effectiveness of HPP on a particular extraction. Therefore, the effect of HP on the antioxidant, the matrix and the incomplete inactivation of enzymes should be considered in applications to ensure the stability of the antioxidant. In studies to date, grains and especially bran fractions were shown to have a significant potential for health-promoting antioxidants. However, there is a lack of studies on the extraction of the antioxidants from grain matrices by HPP that would provide data for industrial applications. Mechanisms and factors affecting the yield, activity and stability of antioxidants require special attention in the HPP-assisted extractions of grains. Therefore, more research studies are required before the use of HPP technology for the extraction of diverse antioxidants from grain matrices.

REFERENCES

Acosta-Estrada, B.A., Gutierrez-Uribe, J.A., & Serna-Saldivar, S.O. (2014). Bound phenolics in foods, a review. *Food Chemistry, 152*, 46–55.

Alonso-Riaño, P., Sanz-Diez, M.T., Blanco, B., Beltran, S., Trigueros, E., & Benito-Roman, O. (2020). Water ultrasound-assisted extraction of polyphenol compounds from brewer's spent grain: Kinetic study, extract characterization, and concentration. *Antioxidants, 9*(3), 265.

Alvarez-Jubete, L., Wijngaard, H., Arendt, E.K., & Gallagher, E. (2010). Polyphenol composition and in vitro antioxidant activity of amaranth, quinoa buckwheat and wheat as affected by sprouting and baking. *Food Chemistry, 119*(2), 770–778.

Alves, G.H., Ferreira, C.D., Vivian, P.G., Monks, J.L.F., Elias, M.C., Vanier, N.L., & de Oliveira, M. (2016). The revisited levels of free and bound phenolics in rice: Effects of the extraction procedure. *Food Chemistry, 208*, 116–123.

Andreasen, M.F., Landbo, A.K., Christensen, L.P., Hansen, Å., & Meyer, A.S. (2001). Antioxidant effects of phenolic rye (*Secale cereale* L.) extracts, monomeric hydroxycinnamates, and ferulic acid dehydrodimers on human low-density lipoproteins. *Journal of Agricultural and Food Chemistry, 49*(8), 4090–4096.

Babbar, N., Oberoi, H.S., Uppal, D.S., & Patil, R.T. (2011). Total phenolic content and antioxidant capacity of extracts obtained from six important fruit residues. *Food Research International, 44*(1), 391–396. doi:10.1016/j.foodres.2010.10.001.

Balasubramaniam, V.B., Barbosa-Canovas, G.V., & Lelieveld, H.L.M. (2016). High-pressure processing equipment for the food industry. In: V. Balasubramaniam, G. Barbosa-Cánovas, & H. Lelieveld (Eds.), *High Pressure Processing of Food. Food Engineering Series* (pp. 39–65). Springer.

Ballester-Sánchez, J., Fernández-Espinar, M.T., & Haros, C.M. (2020). Isolation of red quinoa fibre by wet and dry milling and application as a potential functional bakery ingredient. *Food Hydrocolloids, 101*(105513), 1–9.

Barba, F.J., Terefe, N.S., Buckow, R., Knorr, D., & Orlien, V. (2015). New opportunities and perspectives of high pressure treatment to improve health and safety attributes of foods. A review. *Food Research International*, *77*, 725–742.

Bevilacqua, A., Campaniello, D., Speranza, B., Altieri, C., Sinigaglia, M., & Corbo, M.R. (2019). Two nonthermal technologies for food safety and quality-ultrasound and high pressure homogenization: Effects on microorganisms, advances, and possibilities: A review. *Journal of Food Protection*, *82*(12), 2049–2064. doi:10.4315/0362-028x.jfp-19-059.

Bezerra, F.W.F., De Oliveira, M.S., Bezerra, P.N., Cunha, V.M.B., Silva, M.P., Da Costa, W.A., Pinto, R.H.H., Cordeiro, R.M., Da Cruz, J.N., Chaves Neto, A.M.J., & Carvalho, R.N. (2019). Extraction of bioactive compounds. In: A.M. Inamuddin, Asiri, & A.M. Isloor (Eds.), *Green Sustainable Process for Chemical and Environmental Engineering and Science: Supercritical Carbon Dioxide as Green Solvent*. Elsevier Inc., 149–167. doi:10.1016/B978-0-12-817388-6.00008-8.

Bondia-Pons, I., Aura, A.M., Vuorela, S., Kolehmainen, M., Mykkänen, H., & Poutanen, K. (2009). Rye phenolics in nutrition and health. *Journal of Cereal Science*, *49*(3), 323–336.

Bordiga, M., Gomez-Alonso, S., Locatelli, M., Travaglia, F., Coïsson, J.D., Hermosin-Gutierrez, I., & Arlorio, M. (2014). Phenolics characterization and antioxidant activity of six different pigmented *Oryza sativa* L. cultivars grown in Piedmont (Italy). *Food Research International*, *65*, 282–290.

Capocchi, A., Bottega, S., Spanò, C., & Fontanini, D. (2017). Phytochemicals and antioxidant capacity in four Italian traditional maize (*Zea mays* L.) varieties. *International Journal of Food Sciences and Nutrition*, *68*(5), 515–524.

Cavender, G.A., & Kerr, W.L. (2011). Inactivation of vegetative cells by continuous high-pressure processing: New insights on the contribution of thermal effects and release device. *Journal of Food Science*, *76*(7), E525–E529.

Chakraborty, S., Hulle, N.R.S., Jabeen, K., & Rao, P.S. (2017). Effect of combined high pressure-temperature treatments on bioactive compounds in fruit purees. In: J.J. Moreno (Ed.), *Innovative Processing Technologies for Foods with Bioactive Compounds* (pp. 105–130). CRC Press.

Chandrasekara, A., Rasek, O.A., John, J.A., Chandrasekara, N., & Shahidi, F. (2016). Solvent and extraction conditions control the assayable phenolic content and antioxidant activities of seeds of black beans, canola and millet. *Journal of the American Oil Chemists' Society*, *93*(2), 275–283.

Chandrasekara, A., & Shahidi, F. (2011). Determination of antioxidant activity in free and hydrolyzed fractions of millet grains and characterization of their phenolic profiles by HPLC-DAD-ESI-MSn. *Journal of Functional Foods*, *3*(3), 144–158.

Chen, C., Wang, L., Wang, R., Luo, X., Li, Y., Li, J., Li, Y., & Chen, Z. (2018). Phenolic contents, cellular antioxidant activity and antiproliferative capacity of different varieties of oats. *Food Chemistry*, *239*, 260–267.

Chiremba, C., Taylor, J.R., Rooney, L.W., & Beta, T. (2012). Phenolic acid content of sorghum and maize cultivars varying in hardness. *Food Chemistry*, *134*(1), 81–88.

Clariana, M., Guerrero, L., Sarraga, C., Diaz, I., Valero, A., & Garcia-Regueiro, J.A. (2011). Influence of high pressure application on the nutritional, sensory and microbiological characteristics of sliced skin vacuum packed dry-cured ham. Effects along the storage period. *Innovative Food Science and Emerging Technologies*, *12*(4), 456–465.

Condés, M.C., Speroni, F., Maurí, A., & Añón, M.C. (2012). Physicochemical and structural properties of amaranth protein isolates treated with high pressure. *Innovative Food Science and Emerging Technologies*, *14*, 11–17.

Condés, M.C., Añón, M.C., & Maurí, A.N. (2015). Amaranth protein films prepared with high-pressure treated proteins. *Journal of Food Engineering*, *166*, 38–44.

da Silva Marineli, R., Moraes, É.A., Lenquiste, S.A., Godoy, A.T., Eberlin, M.N., & Maróstica Jr, M.R. (2014). Chemical characterization and antioxidant potential of Chilean chia seeds and oil (*Salvia hispanica* L.). *LWT-Food Science and Technology*, *59*(2), 1304–1310.

Dai, J., & Mumper, R.J. (2010). Plant phenolics: Extraction, analysis and their antioxidant and anticancer properties. *Molecules*, *15*(10), 7313–7352. doi:10.3390/molecules15107313.

Das, A.K., & Singh, V. (2015). Antioxidative free and bound phenolic constituents in pericarp, germ and endosperm of Indian dent (*Zea mays* var. *indentata*) and flint (*Zea mays* var. *indurata*) maize. *Journal of Functional Foods*, *13*, 363–374.

de Vleeschouwer, K., van der Plancken, I., van Loey, A., & Hendrickx, M.E. (2010). The effect of high pressure-high temperature processing conditions on acrylamide formation and other Maillard reaction compounds. *Journal of Agricultural and Food and Chemistry*, *58*(22), 11740–11748. doi:10.1021/jf102697b.

Deora, N.S., Deswal, A., & Mishra, H.N. (2014). Alternative approaches towards gluten-free dough development: Recent trends. *Food Engineering Reviews*, *6*(3), 89–104. doi:10.1007/s12393-014-9079-6.

Dong, X., Wang, J., & Raghavan, V. (2020). Critical reviews and recent advances of novel non-thermal processing techniques on the modification of food allergens. *Critical Reviews in Food Science and Nutrition*, *2020*, 1–15. doi:10.1080/10408398.2020.1722942.

Ekezie, F.G.C., Cheng, J.H., & Sun, D.W. (2018). Effects of nonthermal food processing technologies on food allergens: A review of recent research advances. *Trends in Food Science and Technology*, *74*, 12–25.

FAO & FAOSTAT. (2020). Food and agriculture organization of the United Nations (FAO). http://www.fao.org/faostat/en/#data/QC.

Farkas, D. (2016). A short history of research and development efforts leading to the commercialization of high-pressure processing of food. In: V. Balasubramaniam, G. Barbosa-Cánovas, & H. Lelieveld (Eds.), *High Pressure Processing of Food. Food Engineering Series* (pp. 19–36). Springer.

Farkas, D.V., & Hoover, D.G. (2000). High pressure processing. *Journal of Food Science*, *65*(4), 47–64.

Garcia-Salas, P., Morales-Soto, A., Segura-Carretero, A., & Fernández-Gutiérrez, A. (2010). Phenolic-compound-extraction systems for fruit and vegetable samples. *Molecules*, *15*(12), 8813–8826.

Ge, X., Jing, L., Zhao, K., Su, C., Zhang, B., Zhang, Q., Han, L., Yu, X., & Li, W. (2020). The phenolic compounds profile, quantitative analysis and antioxidant activity of four naked barley grains with different color. *Food Chemistry*, *335*, 127655.

Grancieri, M., Martino, H.S.D., & Gonzalez de Mejia, E. (2019). Chia seed (*Salvia hispanica* L.) as a source of proteins and bioactive peptides with health benefits: A review. *Comprehensive Reviews in Food Science and Food Safety*, *18*(2), 480–499.

Grgić, I., Ačkar, D., Barišić, V., Vlainić, M., Knežević, N., & Knežević, Z.M. (2019). Nonthermal methods for starch modification-A review. *Journal of Food Processing and Preservation*, *43*(12), e14242. doi:10.1111/jfpp.14242.

Guido, L.F., & Moreira, M.M. (2017). Techniques for extraction of brewer's spent grain polyphenols: A review. *Food and Bioprocess Technology*, *10*(7), 1192–1209.

Guo, X.D., Wu, C.S., Ma, Y.J., Parry, J., Xu, Y.Y., Liu, H., & Wang, M. (2012). Comparison of milling fractions of Tartary buckwheat for their phenolics and antioxidant properties. *Food Research International*, *49*(1), 53–59.

Hajji, T., Mansouri, S., Vecino-Bello, X., Cruz-Freire, J.M., Rezgui, S., & Ferchichi, A. (2018). Identification and characterization of phenolic compounds extracted from barley husks by LC-MS and antioxidant activity in vitro. *Journal of Cereal Science*, *81*, 83–90.

Harte, F. (2016). Food processing by high-pressure homogenization. In: V. Balasubramaniam, G. Barbosa-Cánovas, & H. Lelieveld (Eds.), *High Pressure Processing of Food. Food Engineering Series* (pp. 123–142). Springer.

Hayashi, R.(1987). Possibility of high pressure technology for cooking, sterilization, processing and storage of foods. *Syokuhin to Kaihatsu*, *22*(7), 55–62 (in Japanese).

Hayashi, R. (1996). An overview of the use of high pressure in bioscience and biotechnology. In: R. Hayashi & C. Balny (Eds.). *High Pressure Bioscience and Biotechnology, Progress in Biotechnology* (pp. 1–6). Elsevier.

Hayashi, R., & Hayashida, A. (1989). Increased amylase digestibility of pressure-treated starch. *Agricultural and Biological Chemistry*, *53*, 2543–2544.

Hayes, J., & Allen, P. (2011, August 7–12). Monitoring the effects of high pressure processing, salt levels and refrigeration on the sensory and technological properties of pork sausages. *157th International Congress of Meat Science and Technology*, Ghent, Belgium. http://icomst-proceedings.helsinki.fi/papers/2011_36_05.pdf.

Herrera-Sotero, M.Y., Cruz-Hernández, C.D., Trujillo-Carretero, C., Rodríguez-Dorantes, M., García-Galindo, H.S., Chávez-Servia, J.L., Oliart-Ros, R.M., & Guzmán-Gerónimo, R.I. (2017). Antioxidant and antiproliferative activity of blue corn and tortilla from native maize. *Chemistry Central Journal, 11*(1), 1–8.

Hu, X., Xu, X., Jin, Z., Tian, Y., Bai, Y., & Xie, Z. (2011). Retrogradation properties of rice starch gelatinized by heat and high hydrostatic pressure (HHP). *Journal of Food Engineering, 106*(3), 262–266.

Huang, H.W., Hsu, C.P., Yang, B.B., & Wang, C.Y. (2014). Potential utility of high-pressure processing to address the risk of food allergen concerns. *Comprehensive Reviews in Food Science and Food Safety, 13*(1), 78–90.

Hussain, A., Larsson, H., Olsson, M.E., Kuktaite, R., Grausgruber, H., & Johansson, E. (2012). Is organically produced wheat a source of tocopherols and tocotrienols for health food? *Food Chemistry, 132*(4), 1789–1795.

Inglett, G.E., Chen, D., Berhow, M., & Lee, S. (2011). Antioxidant activity of commercial buckwheat flours and their free and bound phenolic compositions. *Food Chemistry, 125*(3), 923–929.

Inglett, G.E., Chen, D., & Liu, S.X. (2015). Antioxidant activities of selective gluten free ancient grains. *Food and Nutrition Sciences, 6*(07), 612.

Irondi, E.A., Adegoke, B.M., Effion, E.S., Oyewo, S.O., Alamu, E.O., & Boligon, A.A. (2019). Enzymes inhibitory property, antioxidant activity and phenolics profile of raw and roasted red sorghum grains in vitro. *Food Science and Human Wellness, 8*(2), 142–148.

Jambrak, A.R., Vukušić, T., Donsi, F., Paniwnyk, L., & Djekic, I. (2018). Three pillars of novel nonthermal food technologies: Food safety, quality, and environment. *Journal of Food Quality.* doi:10.1155/2018/8619707, http://www.ncbi.nlm.nih.gov/pubmed/8619707.

Jideani, A.I., Silungwe, H., Takalani, T., Anyasi, T.A., Udeh, H., & Omolola, A. (2014). Antioxidant-rich natural grain products and human health. In: O. Oguntibeju (Ed.), *Antioxidant-Antidiabetic Agents and Human Health* (pp. 167–187). Tech Publisher.

Joana Gil-Chávez, G., Villa, J.A., Fernando Ayala-Zavala, J., Basilio Heredia, J., Sepulveda, D., Yahia, E.M., & González-Aguilar, G.A. (2013). Technologies for extraction and production of bioactive compounds to be used as nutraceuticals and food ingredients: An overview. *Comprehensive Reviews in Food Science and Food Safety, 12*(1), 5–23. doi:10.1111/1541-4337.12005.

Juliano, P., Koutchma, T., Sui, Q.A., Barbosa-Canovas, G.V., & Sadler, G. (2010). Polymeric-based food packaging for high-pressure processing. *Food Engineering Reviews, 2*(4), 274–297.

Jun, X. (2013). High-pressure processing as emergent technology for the extraction of bioactive ingredients from plant materials. *Critical Reviews in Food Science and Nutrition, 53*(8), 837–852. doi:10.1080/104 08398.2011.561380.

Jung, S. (2016). Applications and opportunities for pressure-assisted extraction. In: V. Balasubramaniam, G. Barbosa-Cánovas, & H. Lelieveld (Eds.), *High Pressure Processing of Food. Food Engineering Series* (pp. 173–192). Springer.

Kato, T., Katayama, E., Matsubara, S., Omi, Y., & Matsuda, T. (2000). Release of allergenic proteins from rice grains induced by high hydrostatic pressure. *Journal of Agricultural and Food Chemistry, 48*(8), 3124–3129.

Katopo, H., Song, Y., & Jane, J. (2002). Effect and mechanism of ultrahigh hydrostatic pressure on the structure and properties of starches. *Carbohydrate Polymers, 47*(3), 233–244.

Khan, S.A., Aslam, R., & Makroo, H.A. (2019). High pressure extraction and its application in the extraction of bio-active compounds: A review. *Journal of Food Process Engineering, 42*(1), e12896. doi:10.1111/jfpe.12896.

Khosravi, A., & Razavi, S.H. (2020). The role of bioconversion processes to enhance polyphenol bioaccessibility in rice bioaccessibility of polyphenols in rice. *Food Bioscience, 33.* doi:10.1016/j.fbio.2020.100605, http://www.ncbi.nlm.nih.gov/pubmed/100605.

Kieffer, R., Schurer, F., Kohler, P., & Wieser, H. (2007). Effect of hydrostatic pressure and temperature on the chemical and functional properties of wheat gluten: Studies on gluten, gliadin and glutenin. *Journal of Cereal Science, 45*(3), 285–292.

Kim, M.Y., Lee, S.H., Jang, G.Y., Park, H.J., Li, M., Kim, S., Lee, Y.R., Noh, Y.H., Lee, J., & Jeong, H.S. (2015). Effects of high hydrostatic pressure treatment on the enhancement of functional components of germinated rough rice (*Oryza sativa* L.). *Food Chemistry*, *166*, 86–92.

Kim, S.M., & Lim, S.T. (2016). Enhanced antioxidant activity of rice bran extract by carbohydrase treatment. *Journal of Cereal Science*, *68*, 116–121.

Kim, M.Y., Lee, S.H., Jang, G.Y., Li, M., Lee, Y.R., Lee, J., & Jeong, H.S. (2017). Changes of phenolic-acids and vitamin E profiles on germinated rough rice (*Oryza sativa* L.) treated by high hydrostatic pressure. *Food Chemistry*, *217*, 106–111.

Kulichová, K., Sokol, J., Nemeček, P., Maliarová, M., Maliar, T., Havrlentová, M., & Kraic, J. (2019). Phenolic compounds and biological activities of rye (*Secale cereale* L.) grains. *Open Chemistry*, *17*(1), 988–999.

Lahouar, L., El Arem, A., Ghrairi, F., Chahdoura, H., Salem, H.B., El Felah, M., & Achour, L. (2014). Phytochemical content and antioxidant properties of diverse varieties of whole barley (*Hordeum vulgare* L.) grown in Tunisia. *Food Chemistry*, *145*, 578–583.

Lao, F., & Giusti, M.M. (2016). Quantification of purple corn (*Zea mays* L.) anthocyanins using spectrophotometric and HPLC approaches: Method comparison and correlation. *Food Analytical Methods*, *9*(5), 1367–1380.

Lelieveld, H.L.M., & Hoogland, H. (2016). Continuous high-pressure processing to extend product shelf life. In: V. Balasubramaniam, G. Barbosa-Cánovas, & H. Lelieveld (Eds.), *High Pressure Processing of Food. Food Engineering Series* (pp. 67–72). Springer.

Leoncini, E., Prata, C., Malaguti, M., Marotti, I., Segura-Carretero, A., Catizone, P., Dinelli, G., & Hrelia, S. (2012). Phytochemical profile and nutraceutical value of old and modern common wheat cultivars. *PLoS ONE*, *7*(9), e45997.

Luthria, D.L., & Liu, K. (2013). Localization of phenolic acids and antioxidant activity in sorghum kernels. *Journal of Functional Foods*, *5*(4), 1751–1760.

Luthria, D.L., Lu, Y., & John, K.M. (2015). Bioactive phytochemicals in wheat: Extraction, analysis, processing, and functional properties. *Journal of Functional Foods*, *18*, 910–925.

Mareček, V., Mikyška, A., Hampel, D., Čejka, P., Neuwirthová, J., Malachová, A., & Cerkal, R. (2017). ABTS and DPPH methods as a tool for studying antioxidant capacity of spring barley and malt. *Journal of Cereal Science*, *73*, 40–45.

Martinez-Cruz, O., & Paredes-López, O. (2014). Phytochemical profile and nutraceutical potential of chia seeds (*Salvia hispanica* L.) by ultra high performance liquid chromatography. *Journal of Chromatography. Part A*, *1346*, 43–48.

Martinez-Lopez, A., Millan-Linares, M.C., Rodriguez-Martin, N.M., Millan, F., & Montserrat-de la Paz, S. (2020). Nutraceutical value of kiwicha (*Amaranthus caudatus* L.). *Journal of Functional Foods*, *65*. http://www.ncbi.nlm.nih.gov/pubmed/103735.

Martínez-Monteagudo, S.I., & Balasubramaniam, V.B. (2016). Fundamentals and applications of high-pressure processing technology. In: V. Balasubramaniam, G. Barbosa-Cánovas, & H. Lelieveld (Eds.), *High Pressure Processing of Food. Food Engineering Series* (pp. 3–17). Springer.

Matser, A., & Timmermans, R. (2016). High-pressure effects on fruits and vegetables. In: V. Balasubramaniam, G. Barbosa-Cánovas, & H. Lelieveld (Eds.), *High Pressure Processing of Food. Food Engineering Series* (pp. 541–552). Springer.

Mendiola, J.A., Herrero, M., Cifuentes, A., & Ibañez, E. (2007). Use of compressed fluids for sample preparation: Food applications. *Journal of Chromatography. Part A*, *1152*(1), 234–246. doi:10.1016/j.chroma.2007.02.046.

Meng, L., Zhang, W., Zhou, X., Wu, Z., Hui, A., He, Y., Gao, H., & Chen, P. (2019). Effect of high hydrostatic pressure on the bioactive compounds, antioxidant activity and in vitro digestibility of cooked black rice during refrigerated storage. *Journal of Cereal Science*, *86*, 54–59. doi:10.1016/j.jcs.2019.01.005.

Mensitieri, G., Scherillo, G., & Iannace, S. (2013). Flexible packaging structures for high pressure treatments. *Innovative Food Science and Emerging Technologies*, *17*, 12–21.

Mesa, J., Hinestroza-Córdoba, L.I., Barrera, C., Seguí, L., Betoret, E., & Betoret, N. (2020). High homogenization pressures to improve food quality, functionality and sustainability. *Molecules*, *25*(14), 3305. doi:10.3390/molecules25143305.

Moreno-Najera, L.C., Ragazzo-Sanchez, J.A., Gaston Pena, C.R., & Calderon-Santoyo, M. (2020). Green technologies for the extraction of proteins from jackfruit leaves (*Artocarpus heterophyllus* Lam). *Food Science and Biotechnology, 29*(12), 1675–1684. doi:10.1007/s10068-020-00825-4.

Naczk, M., & Shahidi, F. (2006). Phenolics in cereals, fruits and vegetables: Occurrence, extraction and analysis. *Journal of Pharmaceutical and Biomedical Analysis, 41*(5), 1523–1542. doi:10.1016/j.jpba.2006.04.002.

Nadar, S.N., Rao, P., & Rathod, V.K. (2018). Enzyme assisted extraction of biomolecules as an approach to novel extraction technology: A review. *Food Research International, 108*, 309–330.

Ofosu, F.K., Elahi, F., Daliri, E.B.M., Tyagi, A., Chen, X.Q., Chelliah, R., Kim, J.H., Han, S.I., & Oh, D.H. (2020). UHPLC-ESI-QTOF-MS/MS characterization, antioxidant and antidiabetic properties of sorghum grains. *Food Chemistry, 337*, 127788.

Oh, H.E., Pinder, D.N., Hemar, Y., Anema, S.G., & Wong, M. (2008). Effect of high-pressure treatment on various starch-in-water suspensions. *Food Hydrocolloids, 22*(1), 150–155.

Oliveira-Alves, S.C., Vendramini-Costa, D.B., Cazarin, C.B.B., Júnior, M.R.M., Ferreira, J.P.B., Silva, A.B., Prado, M.A., & Bronze, M.R. (2017). Characterization of phenolic compounds in chia (*Salvia hispanica* L.) seeds, fiber flour and oil. *Food Chemistry, 232*, 295–305.

Pang, Y., Ahmed, S., Xu, Y., Beta, T., Zhu, Z., Shao, Y., & Bao, J. (2018). Bound phenolic compounds and antioxidant properties of whole grain and bran of white, red and black rice. *Food Chemistry, 240*, 212–221.

Park, C.Y., Kim, S., Lee, D., Park, D.J., & Imm, J.Y. (2016). Enzyme and high pressure assisted extraction of tricin from rice hull and biological activities of rice hull extract. *Food Science and Biotechnology, 25*(1), 159–164.

Park, J.H., Lee, Y.J., Kim, Y.H., & Yoon, K.S. (2017). Antioxidant and antimicrobial activities of quinoa (*Chenopodium quinoa* Willd.) seeds cultivated in Korea. *Preventive Nutrition and Food Science, 22*(3), 195.

Pasko, P., Bartoń, H., Zagrodzki, P., Chłopicka, J., Iżewska, A., Gawlik, M., Gawlik, M., & Gorinstein, S. (2011). Effect of amaranth seeds in diet on oxidative status in plasma and selected tissues of high fructose-fed rats. *Food Chemistry, 126*(1), 85–90.

Peiretti, P.G., Meineri, G., Gai, F., Longato, E., & Amarowicz, R. (2017). Antioxidative activities and phenolic compounds of pumpkin (*Cucurbita pepo*) seeds and amaranth (*Amaranthus caudatus*) grain extracts. *Natural Product Research, 31*(18), 2178–2182.

Perez-Andrés, J.M., Charoux, C.M.G., Cullen, P.J., & Tiwari, B.K. (2018). Chemical modifications of lipids and proteins by nonthermal food processing technologies. *Journal of Agricultural and Food and Chemistry, 66*(20), 5041–5054.

Pradeep, P.M., & Sreerama, Y.N. (2018). Phenolic antioxidants of foxtail and little millet cultivars and their inhibitory effects on α-amylase and α-glucosidase activities. *Food Chemistry, 247*, 46–55.

Prückler, M., Siebenhandl-Ehn, S., Apprich, S., Hoeltinger, S., Haas, C., Schmid, E., & Kneifel, W. (2014). Wheat bran-based biorefinery I: Composition of wheat bran and strategies of functionalization. *LWT-Food Science and Technology, 56*(2), 211–221.

Queirós, R.P., Saraiva, J.A., & da Silva, J.A.L. (2018). Tailoring structure and technological properties of plant proteins using high hydrostatic pressure. *Critical Reviews in Food Science and Nutrition, 58*(9), 1538–1556.

Rakic, S., Janković, S., Marčetić, M., Živković, D., & Kuzevski, J. (2014). The impact of storage on the primary and secondary metabolites, antioxidant activity and digestibility of oat grains (*Avena sativa*). *Journal of Functional Foods, 7*, 373–380.

Ramos-Escudero, F., Muñoz, A.M., Alvarado-Ortíz, C., Alvarado, A., & Yánez, J.A. (2012). Purple corn (*Zea mays* L.) phenolic compounds profile and its assessment as an agent against oxidative stress in isolated mouse organs. *Journal of Medicinal Food, 15*(2), 206–215.

Rao, S., Santhakumar, A.B., Chinkwo, K.A., Wu, G., Johnson, S.K., & Blanchard, C.L. (2018). Characterization of phenolic compounds and antioxidant activity in sorghum grains. *Journal of Cereal Science, 84*, 103–111.

Rastogi, N.K., Raghavarao, K.S.M.S., Balasubramaniam, V.M., Niranjan, K., & Knorr, D. (2007). Opportunities and challenges in high pressure processing of foods. *Critical Reviews in Food Science and Nutrition, 47*(1), 69–112.

Reyes-Caudillo, E., Tecante, A., & Valdivia-López, M.A. (2008). Dietary fibre content and antioxidant activity of phenolic compounds present in Mexican chia (*Salvia hispanica* L.) seeds. *Food Chemistry*, *107*(2), 656–663.

Rodrigues, S., & Fernandes, F.A.N. (2017). Extraction processes assisted by ultrasound. In: D. Bermudez-Aguirre (Ed.), *Ultrasound: Advances for Food Processing and Preservation* (pp. 351–368). Elsevier.

Roselló-Soto, E., Thirumdas, R., Lorenzo, J.M., Munekata, P.E.S., Putnik, P., Roohinejad, S., Mallikarjunan, K., & Barba, F.J. (2019). An integrated strategy between gastronomic science, food science and technology, and nutrition in the development of healthy food products. In: F.J. Barba, J.M.A. Saraiva, G. Cravotto, & J.M. Lorenzo (Eds.), *Innovative Thermal and Non-Thermal Processing, Bioaccessibility and Bioavailability of Nutrients and Bioactive Compounds* (pp. 3–21). Elsevier.

Saikaew, K., Lertrat, K., Meenune, M., & Tangwongchai, R. (2018). Effect of high-pressure processing on colour, phytochemical contents and antioxidant activities of purple waxy corn (*Zea mays* L. var. *ceratina*) kernels. *Food Chemistry*, *243*, 328–337.

Saini, P., Kumar, N., Kumar, S., Mwaurah, P.W., Panghal, A., Attkan, A.K., Singh, V.K., Garg, M.K., & Singh, V. (2020). Bioactive compounds, nutritional benefits and food applications of colored wheat: A comprehensive review. *Critical Reviews in Food Science and Nutrition*, 1–14. doi:10.1080/10408398.2020.1793727.

Satheesh, N., Fanta, S.W., & Yildiz, F. (2018). Review on structural, nutritional and anti-nutritional composition of teff (*Eragrostis tef*) in comparison with quinoa (*Chenopodium quinoa* Willd.). *Cogent Food and Agriculture*, *4*(1). http://www.ncbi.nlm.nih.gov/pubmed/1546942.

Scepankova, H., Martins, M., Estevinho, L., Delgadillo, I., & Saraiva, J.A. (2018). Enhancement of bioactivity of natural extracts by non-thermal high hydrostatic pressure extraction. *Plant Foods for Human Nutrition*, *73*(4), 253–267.

Setyaningsih, W., Saputro, I.E., Carrera, C.A., & Palma, M. (2019). Optimisation of an ultrasound-assisted extraction method for the simultaneous determination of phenolics in rice grains. *Food Chemistry*, *288*, 221–227. doi:10.1016/j.foodchem.2019.02.107.

Setyaningsih, W., Saputro, I.E., Palma, M., & Barroso, C.G. (2016). Pressurized liquid extraction of phenolic compounds from rice (*Oryza sativa*) grains. *Food Chemistry*, *192*, 452–459.

Sevenich, R., & Mathys, A. (2018). Continuous versus discontinuous ultra-high-pressure systems for food sterilization with focus on ultra-high-pressure homogenization and high-pressure thermal sterilization: A review. *Comprehensive Reviews in Food Science and Food Safety*, *17*(3), 646–662.

Sharma, N., Goyal, S.K., Alam, T., Fatma, S., Chaoruangrit, A., & Niranjan, K. (2018). Effect of high pressure soaking on water absorption, gelatinization, and biochemical properties of germinated and non-germinated foxtail millet grains. *Journal of Cereal Science*, *83*, 162–170. doi:10.1016/j.jcs.2018.08.013.

Shewry, P.R. (2018). Do ancient types of wheat have health benefits compared with modern bread wheat? *Journal of Cereal Science*, *79*, 469–476.

Shouqin, Z., Junjie, Z., & Changzhen, W. (2004). Novel high pressure extraction technology. *International Journal of Pharmaceutics*, *278*(2), 471–474.

Shumoy, H., & Raes, K. (2016). Antioxidant potentials and phenolic composition of Tef varieties: An indigenous Ethiopian cereal. *Cereal Chemistry Journal*, *93*(5), 465–470.

Stastna, K., Mrázková, M., Sumczynski, D., Cındık, B., & Yalçın, E. (2019). The nutritional value of non-traditional gluten-free flakes and their antioxidant activity. *Antioxidants*, *8*(11), 565.

Stewart, C.M., Dunne, C.P., & Keener, L. (2016). Pressure-assisted thermal sterilization validation. In: V. Balasubramaniam, G. Barbosa-Cánovas, & H. Lelieveld (Eds.), *High Pressure Processing of Food. Food Engineering Series* (pp. 687–716). Springer.

Stolt, M., Oinonen, S., & Autio, K. (2001). Effect of high pressure on the physical properties of barley starch. *Innovative Food Science and Emerging Technologies*, *1*(3), 167–175.

Stute, R., Kingler, R.W., Boguslawski, S., Eshtiaghi, M.N., Knorr, D., & Knorr, D. (1996). Effects of high pressure treatment on starches. *Starch – Starke*, *48*(11–12), 399–408.

Sun, L.C., Sridhar, K., Tsai, P.J., & Chou, C.S. (2019). Effect of traditional thermal and high-pressure processing (HPP) methods on the color stability and antioxidant capacities of Djulis (*Chenopodium formosanum* Koidz.). *LWT- Food Science and Technology*, *109*, 342–349. doi:10.1016/j.lwt.2019.04.049.

Tanase, C., Coşarcă, S., & Muntean, D.L. (2019). A critical review of phenolic compounds extracted from the bark of woody vascular plants and their potential biological activity. *Molecules*, *24*(6), 1182.

Tang, Y., & Tsao, R. (2017). Phytochemicals in quinoa and amaranth grains and their antioxidant, anti-inflammatory, and potential health beneficial effects: A review. *Molecular Nutrition and Food Research*, *61*(7). http://www.ncbi.nlm.nih.gov/pubmed/1600767.

Tao, Y., Sun, D.W., Hogan, E., & Kelly, A.L. (2014). High-pressure processing of foods: An overview. In: D.W. Sun (Ed.), *Emerging Technologies for Food Processing* (pp. 3–24). Elsevier.

Ullah, R., Nadeem, M., Khalique, A., Imran, M., Mehmood, S., Javid, A., & Hussain, J. (2016). Nutritional and therapeutic perspectives of Chia (*Salvia hispanica* L.): A review. *Journal of Food Science and Technology*, *53*(4), 1750–1758.

Vallons, K.J.R., Ryan, L.A.M., & Arendt, E.K. (2011). Promoting structure formation by high pressure in gluten-free flours. *LWT – Food Science and Technology*, *44*(7), 1672–1680. doi:10.1016/j.lwt.2010.11.024.

Vallons, K.R., Ryan, L.M., Koehler, P., & Arendt, E. (2010). High pressure-treated sorghum flour as a functional ingredient in the production of sorghum bread. *European Food Research and Technology*, *231*(5), 711–717. doi:10.1007/s00217-010-1316-5.

Varga, M., Jójárt, R., Fónad, P., Mihály, R., & Palágyi, A. (2018). Phenolic composition and antioxidant activity of colored oats. *Food Chemistry*, *268*, 153–161.

Verardo, V., Arráez-Román, D., Segura-Carretero, A., Marconi, E., Fernández-Gutiérrez, A., & Caboni, M.F. (2010). Identification of buckwheat phenolic compounds by reverse phase high performance liquid chromatography–electrospray ionization-time of flight-mass spectrometry (RP-HPLC–ESI-TOF-MS). *Journal of Cereal Science*, *52*(2), 170–176.

Vilcacundo, R., & Hernández-Ledesma, B. (2017). Nutritional and biological value of quinoa (*Chenopodium quinoa* Willd.). *Current Opinion in Food Science*, *14*, 1–6.

Watanabe, M., Arai, E., Honma, K., & Fuke, S. (1991). Improving the cooking properties of aged rice grains by pressurising and enzymatic treatment. *Agricultural and Biological Chemistry*, *55*, 2725–2731.

Xia, Q., Wang, L., Xu, C., Mei, J., & Li, Y. (2017). Effects of germination and high hydrostatic pressure processing on mineral elements, amino acids and antioxidants in vitro bioaccessibility, as well as starch digestibility in brown rice (*Oryza sativa* L.). *Food Chemistry*, *214*, 533–542. doi:10.1016/j.foodchem.2016.07.114.

Xiang, J., Li, W., Ndolo, V.U., & Beta, T. (2019a). A comparative study of the phenolic compounds and in vitro antioxidant capacity of finger millets from different growing regions in Malawi. *Journal of Cereal Science*, *87*, 143–149.

Xiang, J., Zhang, M., Apea-Bah, F.B., & Beta, T. (2019b). Hydroxycinnamic acid amide (HCAA) derivatives, flavonoid C-glycosides, phenolic acids and antioxidant properties of foxtail millet. *Food Chemistry*, *295*, 214–223.

Xu, D.P., Li, Y., Meng, X., Zhou, T., Zhou, Y., Zheng, J., Zhang, J.J., & Li, H.B. (2017). Natural antioxidants in foods and medicinal plants: Extraction, assessment and resources. *International Journal of Molecular Sciences*, *18*(1), 96. doi:10.3390/ijms18010096.

Yamamoto, K., & Buckow, R. (2016). Pressure gelatinization of starch. In: V. Balasubramaniam, G. Barbosa-Cánovas, & H. Lelieveld (Eds.), *High Pressure Processing of Food. Food Engineering Series* (pp. 433–460). Springer.

Zaupa, M., Calani, L., Del Rio, D., Brighenti, F., & Pellegrini, N. (2015). Characterization of total antioxidant capacity and (poly) phenolic compounds of differently pigmented rice varieties and their changes during domestic cooking. *Food Chemistry*, *187*, 338–347.

Zhang, H.F., Yang, X.H., & Wang, Y. (2011). Microwave assisted extraction of secondary metabolites from plants: Current status and future directions. *Trends in Food Science and Technology*, *22*(12), 672–688.

Zhang, R., Huang, L., Deng, Y., Chi, J., Zhang, Y., Wei, Z., & Zhang, M. (2017). Phenolic content and antioxidant activity of eight representative sweet corn varieties grown in South China. *International Journal of Food Properties*, *20*(12), 3043–3055.

Zhou, K., & Yu, L. (2004). Effects of extraction solvent on wheat bran antioxidant activity estimation. *LWT-Food Science and Technology*, *37*(7), 717–721.

Zhou, Z., Ren, X., Wang, F., Li, J., Si, X., Cao, R., Yang, R., Strappe, P., & Blanchard, C. (2015). High pressure processing manipulated buckwheat antioxidant activity, anti-adipogenic properties and starch digestibility. *Journal of Cereal Science, 66*, 31–36. doi:10.1016/j.jcs.2015.09.002.

Zhu, F. (2019). Proanthocyanidins in cereals and pseudocereals. *Critical Reviews in Food Science and Nutrition, 59*(10), 1521–1533.

Zilic, S., Serpen, A., Akıllıoğlu, G., Gökmen, V., & Vančetović, J. (2012). Phenolic compounds, carotenoids, anthocyanins, and antioxidant capacity of colored maize (*Zea mays* L.) kernels. *Journal of Agricultural and Food Chemistry, 60*(5), 1224–1231.

Chapter 4

Applications of Cold Plasma Technology in Grain Processing

Kirty Pant, Mamta Thakur and Vikas Nanda

CONTENTS

4.1 INTRODUCTION

Cold plasma (CP) technology is arising as a non-thermal food processing technique that has drawn the interest of numerous scientists across the world. Cold plasma technology was initially employed for improving the adhesion and printing characteristics of polymers, thereby raising the surface energy of material and the multiplicity of their potential applications in electronics, textiles, paper, glasses and other products (Ekezie et al., 2017). The implementation of CP was gradually expanded into the food industries, and it was introduced as a powerful non-thermal processing tool with versatile forms of utilization. The non-thermal cold plasma technique is extremely beneficial for microbial decontamination of food material, including sporulating and spoilage/pathogenic microbes, owing to the sufficient quantity of reactive oxygen species (ROS) confined in the quasi-neutral plasma gas (Jung et al., 2017; Min et al., 2017).

There are three conventional states of matter – solid, liquid and gas – whereas plasma is counted as the fourth state of matter. The application of energy plays a significant role in the conversion of matter from one state to another, for example, the transformation of matter from solid to liquid, then to gas, and finally, to ionized gas or to plasma (Thirumdas et al., 2015; Niemira, 2012). An ionized gas comprises a wide range of active species, such as electrons, ions, free radicals, etc. (Ekezie et al., 2017; Dasan et al., 2017). Plasma remains in either its elevated or ground state, which obtains a net neutral charge, hence it can be manipulated under different temperature and pressure conditions via stimulating a neutral gas.

Plasma can be categorized into two parts, i.e., thermal and non-thermal plasma. Thermal or hot plasma is produced by applying extreme pressure ($\geq$ 105 Pa up to 50 MW power) and temperature (20,000 K) to attain proper ionization and propagation, in which all the particles are existing in thermodynamic equilibrium, which also helps to distinguish between the electrons and heavier particles because of uniform gas temperature (Ekezie et al., 2017; Scholtz et al., 2015). Meanwhile, non-thermal or cold plasma can be sub-categorized into quasi-equilibrium plasma (temperature range from 100 to 150°C) and nonequilibrium plasma (temperature < 60°C) (Mandal et al., 2018). Cold plasma is generated at low pressure and power, whereas energy is required to facilitate elastic collision among atoms, electrons and gas particles. The particles in cold plasma are not in confined thermodynamic equilibrium and the electrons remain in constant collision with other ions and molecules, thus this phase is designated as non-equilibrium plasma (Ekezie et al., 2017; Muhammad et al., 2018). Generally, energy is applied to dissociate the gas into various reactive species and leads to other subsequent reactions such as ionization, excitation and de-excitation. In the food sector, the application of cold plasma affected by electrical discharges is gaining significant attention because of its potential significance in food processing at low temperatures (Ekezie et al., 2017). Cold plasma technology used to conduct sterilization processes offers a low-cost option as well as enacting minimal changes to food products, hence becoming an alternative to the traditional heat-based techniques. Cold plasma technology in food processing includes upgrading the performance of seed germination, modifying the functionality of food components, enhancing the physicochemical characteristics of grains and decreasing agrochemical residues (Mir et al., 2016; Sarangapani et al., 2017; Sivachandiran et al., 2017). Additionally, it is also employed for the development of packaging materials to improve protective properties and confer antimicrobial activity (Oh et al., 2016; Puligundla et al., 2016). In this chapter, we will discuss the application of cold plasma technology in grain processing, such as the effects of cold plasma treatment on various food components, the modification of physicochemical properties, safety aspects and future scope.

4.2 COLD PLASMA TECHNOLOGY: AN OVERVIEW

Throughout the universe – in the sun and other stars as well as in cometary and planetary atmosphere – plasma is found in most of matter. Plasma exists naturally in lightning and fire on Earth (Foest et al., 2006; Moreau et al., 2008). Plasma, referred to as the 'fourth state of matter,' is produced with a gas energy application that causes the ionization of the light-transmitting gas consisting of charged particles, free radicals and UV rays. Plasma is often distinguished by the presence of positive (and sometimes negative) ions and negative electrons. Different gases, depending on their enthalpy, reactiveness and costs, are used to create cold plasma, such as oxygen, argon, nitrogen or helium (Hoffmann et al., 2013).

As discussed, the plasma treatment may be classified according to its temperature and electron density in (i) thermal (hot) and (ii) non-thermal (cold) plasma. High electron density ranges from 1,021–1,026 m^{-3} are found in thermal plasma. The electron temperature of non-thermal plasma is slightly higher than that of heavy particles and the density of electron is < 1,019 m^{-3} (Tendero, 2006; Moreau et al., 2008). Plasmas are manufactured by applying energy to a known quantity of neutral gas, which leads to ionization and the generation of fraction of free electrons and ions. In addition, a variety of electricity discharges or electron beams may produce non-thermal plasma, and input energy comes as electric energy, which is primarily converting electrical energy into energy electrons and does not heat the gas as a whole. The formation, excitation and ionization of background gas molecules contribute to the formation of excited species, free radicals and ions (Fridman et al., 2005; Foest et al., 2006). Plasma can be formed with all sorts of electrical, thermal, optical (UV) and electromagnetic radiating energy that may ionize the gases. However, electric or electromagnetic fields are widely used for CP generation (Hoffmann et al., 2013).

In terms of food processing, plasma source, electrode design, pressure, voltage, time to process, distance of electrodes and reactivity of gas all have significant functions in determining the speciation of gases, the concentration of reactive species, features of discharge and overall process efficiency. Therefore, the food industry's cold plasma device is held at a relatively low temperature, which is of primary importance to the food industry. This technology is very useful for heat-sensitive materials, as it affects qualitative characteristics minimally. Moreover, the typical electrical discharges used to produce non-thermal plasmas or cold plasma include corona discharge, glow discharge, radiofrequency (RF) discharge, dielectrical barrier discharge (DBD), pulsed corona discharge, microwave discharge and plasma jet (Shrestha et al., 2007; Hoffmann et al., 2013). The two most frequently employed forms of cold plasma processing include DBD and plasma jet due to their simple design and reconfiguration options for many types of goals and requirements (Bourke et al., 2017). Plasma jet is more acceptable and appealing in some biomedical applications than in solid food manufacturing, which has important considerations in limited areas of surface treatment and efficiency (Nishime et al., 2017). For DBD plasma, this mode is a normal disintegration with streamers, and a circuit that is capable of exciting the plasma with an oscillation current is the fundamental concept. At present, the interest in these pressure plasmas in practical food applications has been highlighted because they reduce the need for a vacuum system and permit continuing material processing (Turner, 2016). Moreover, DBD plasma is especially suited to in-package inactivation of fresh products – both reactive oxygen species (ROS) and reactive nitrogen species (RNS), which may be produced directly within the package of sealed products (Misra et al., 2013).

The excited charged species form a uniform plasma discharge inside the package and inactivate microbes such as bacteria (Ziuzina et al., 2014), yeasts and molds (Misra et al., 2013), etc. The reactive species in plasma cause oxidative effects on the outer surface of microbial cells. The formation of cold plasma using humid air leads to the production of active species such as OH and NO radicals (Benstaali et al., 1998; Laroussi et al., 2003). These radicals cause intense bombardment around the microorganisms, causing surface lesions that cannot be repaired, and ultimately lead to the destruction of cells. This whole process is known as etching (Pelletier, 1992). Nitrogen and oxygen gas plasma results in the formation of reactive oxygen-based and nitrogen-based species such as O, O_2, O_3, OH, NO and NO_2. These species can disrupt the movement of biomolecules by acting on the double bond of the unsaturated fatty acid of the membrane cell (Critzer et al., 2007). The activity of the active species in polyunsaturated fatty acid (PUFA) induces a radical loss for a hydrogen atom. In addition, $O2$ oxidizes the radical fatty acid, leading to lipid hydroperoxide formation (Joshi et al., 2011). The operation of the active species results in microbial injury or sometimes death; amino acids and nucleic acids are oxidized (Critzer et al., 2007). Furthermore, the active reactive plasma species, UV photons, can modify microorganisms' DNA and trigger cell replication disruption. In addition to reactive species, UV photons can change the microorganisms' DNA and therefore interfere with cell replication (Boudam et al., 2006).

4.3 PLASMA AND THE BIOCHEMICAL COMPOSITION OF GRAINS

4.3.1 Starch Modification

Starch is one of the significant biopolymers that is extensively used in various fields such as food, pharmaceutical, textile and paper industries, etc. (BeMiller, 1997). According to the industrial point of view, grains (like maize, corn, wheat and rice) are the major source of starch extraction followed by other plant-based food products (such as potato, cassava, etc.) where native starch is being subjected to various processes (physical, chemical and enzymatic) in order to achieve modification and eventually enhancement of the starch properties. Cold plasma is a novel and modern physical modification method that is eco-friendly, fast and does not produce toxic substances, which is why it is also termed green technology (BeMiller et al., 2015; Wongsagonsup et al., 2014). Starch modification during cold plasma treatment is generally influenced by various factors such as voltage applied, feed gas and treatment time (Thirumdas et al., 2017).

There are mainly three different modification mechanisms, named cross-linking, depolymerization and plasma etching, that are responsible for various surface alterations of different biodegradable polymers (Morent et al., 2011). The modification of the starch properties is primarily due to cross-linking of amylose and amylopectin side chains and depolymerization, which is briefly explained in Table 4.1. Subsequently, the plasma treatment leads to a drop in viscosity, molecular weight and gelatinization temperatures. However, plasma etching enhances the surface energy and improves the hydrophilicity of the starch granules (Thirumdas et al., 2017). However, apart from the above-discussed mechanisms, functional groups have also been proposed as a new phenomenon that can lead to starch alteration. Morent et al. (2011) have mentioned that polar hydrophilic groups are incorporated on the biopolymer due to interlinking among biopolymers and the chemically active species developed in the gas discharge plasma.

4.3.2 Effect of Cold Plasma on Physicochemical Properties of Starch

4.3.2.1 Molecular Weight

During cold plasma treatment of starch, factors such as feed gas and plasma reactor are majorly responsible for the increase or decrease in the molecular weight of the treated starch polymer

TABLE 4.1 BRIEF EXPLANATION OF DIFFERENT MAIN MECHANISMS TAKING PLACE DURING COLD PLASMA TREATMENT OF FOOD MATERIAL

Mechanism	Phenomena	Results	Diagrammatic View
Cross-linking or grafting	Takes place between the polymeric chains of starch molecules, induced by free radicals and energetic electrons formed during plasma generation. In this mechanism, cleavage occurs between the reducing ends of two polymeric chains (C-OH) and a new C-O-C linkage is formed between these two chains due to cross-linking	• Removal of water molecule • Hydroxyl radicals (*OH) formed by the oxygen gas • Increase in glycosidic linkage	Cross-linking of starch chains with ionized water ions (Thirumdas et al., 2017)
Depolymerization	Ionization plays the most significant role in this mechanism, which leads to starch modification. The bombardment of high-energetic ions of plasma causes depolymerization of amylose and amylopectin side chains of starch molecules, resulting in smaller fragments. It depends on the active plasma species, the free radicals formed, the concentration of ions and the nature and morphology of the starch	• Endpoint is the abundant production of maltose, maltotriose and maltotetrose • Radiolysis of carbohydrates results in the production of formic acid, acetaldehyde and formaldehyde	Depolymerization of starch branched chains (Thirumdas et al., 2017)

(Continued)

TABLE 4.1 (CONTINUED)   BRIEF EXPLANATION OF DIFFERENT MAIN MECHANISMS TAKING PLACE DURING COLD PLASMA TREATMENT OF FOOD MATERIAL

Mechanism	Phenomena	Results	Diagrammatic View
Plasma etching	The surface material is selectively etched by physical sputtering or by chemical reaction. The effectiveness of etching depends on the plasma density and the energy with which ions impact the substrate surface	• Increased surface energy of substrate • Increase in hydrophilic nature	Etching of starch granule by reactive species (Thirumdas et al., 2017)

(Thirumdas et al., 2017). Plasma treatment of starch could lead to a decline in molecular weight due to the depletion caused by the plasma species. Several researchers have stated that there is a reduction in the molecular weight of starch after it is subjected to plasma treatment because of treatment time and plasma power (Wongsagonsup et al., 2014; Lii et al., 2002b). The reduction in molecular weight indicates the breakdown or depolymerization of the side polymeric chains. Lii et al. (2002a) have described that the corn starch with high amylose content possessed high resistance potential toward the activity of plasma due to the development of high molecular weight moieties of oxidation. The cereal or grain starches show great resistance when subjected to plasma processing as compared to tuber starches, but diminution in mean square radius gyration was extreme in cereal starches (Thirumdas et al., 2017).

4.3.2.2 Crystallinity

The modification of starch performed by physical techniques primarily affects the amorphous zone of the starch polymer (Han et al., 2009; Bao et al., 2005). The starch granules with amorphous regions show the highest associated destruction as compared to crystalline regions due to the sensitivity toward the active spices formed during plasma processing (Laovachirasuwan et al., 2010). The moisture content of the starch molecule also plays a significant role in crystallinity behavior because during plasma processing there is a formation of secondary electrons that are responsible for water molecular degradation. During plasma treatment, activated hydroxyl radicals are formed that cause further reduction in crystallinity due to the breakdown of water molecules. It has been reported that the rate of decrease in crystallinity percentage is mainly influenced by treatment time and the type of feed gas used for plasma generation (Thirumdas et al., 2017).

4.3.2.3 Morphology

Starch molecules subjected to the application of non-thermal cold plasma technology resulted in holes, cavities or fissures on the surface of the polymer. The plasma etching mechanism causes significant changes in the morphology of starch. Lii et al. (2002a) have stated that starch molecules subjected to corona discharges lead to macro-scale deterioration detected under SEM, where the SEM micrographs exposed that the starch variety with smaller-sized particles is less affected than the starch variety with bigger-sized particles. Many researchers have described that oxidation of particles existing on the starch polymer induced by the plasma species caused unevenly scattered depositions on the surface, whereas pores are observed in the case of waxy corn starch (Thirumdas et al., 2017).

4.3.2.4 Rheological Properties

Starch modification by cold plasma treatment leads to depolymerization of amylose and amylopectin chains, which significantly alters the viscosity. The cross-linking mechanism hinders the swelling of starch granules, which is ultimately responsible for the reduction in viscosity of highly cross-linked starches (Ai & Jane, 2015). Li et al. (2011) have stated that starches with low breakdown viscosity typically have a lower tendency toward retrogradation and high thermal stability. In contrast, Michel et al. (1980) described that intrinsic viscosity is directly inversely proportional to the degree of depolymerization and the number of reducing groups. Similar to other physical methods applied to achieve starch modification, cold plasma treatment leads to depolymerization or fragmentation of the starch molecule, which eventually causes a reduction in viscosity (Thirumdas et al., 2017). It can be inferred that cold plasma processing of grains or starch can be used to achieve a reduction in starch paste viscosity, tends toward stronger gel formation and possesses a lower tendency toward retrogradation of starch for a shorter period of treatment time, which is briefly discussed in Table 4.2.

TABLE 4.2   BRIEF DISCUSSION OF COLD PLASMA PROCESSING CONDITIONS AND THEIR EFFECT ON VARIOUS GRAINS AND PRODUCTS

Varieties	Plasma treatment	Quality observations	References
Cereals: wheat, barley, oat, rye and corn Legumes: bean, chickpea, soybean and lentil	Cold plasma (low pressure), 5–20 min, 300 W, 1 kHz, 20 kV, 500 mTorr	• No change in soaking or cooking time or yield of legumes • Minor variation in moisture content of wheat and legumes • No difference in wet gluten content, gluten index or sedimentation in wheat	Selcuk et al. (2008)
Cereal: brown rice	Radiofrequency plasma, air (0.15 mbar), 13.56 MHz, 40–50 W, 5–10 min	• Reduction in cooking time, hardness, chewiness and moisture content • Increase in gelatinization and water uptake ratio • Increase in L value and whiteness index	Thirumdas et al. (2016)
	Dielectric barrier discharge, 15 kHz, 250 W, air, 5–20 min	• Reduction in pH and hardness • Decrease in a* and b* values and increase in L* value	Lee et al. (2016)
Legume: black gram	Radiofrequency plasma, 2 Pa, air (0.15 mbar), 13.56 MHz, 30–50 W, 5–15 min	• Hydrophilization and surface etching • Reduction in cooking time and hardness • Decrease in ash and moisture content	Sarangapani et al. (2017)
Unpeeled almonds	Diffuse coplanar surface barrier discharge, 20 kV, 15 kHz, Air, O2, N2, CO2 and 90% CO2 + 10% Ar, 15 min	• Surface discoloration • Browning occurred due to air and N2 plasma treatment	Hertwig et al. (2017)
Wheat flour	Dielectric barrier discharge, 60–70 kV, 5–10 min, air	• Increase in peak time, peak integral, elastic modulus, viscous modulus, dough strength and optimum mixing time • No significant color variation in observed	Misra et al. (2016)
Refined wheat flour	Dielectric barrier discharge 1–2.5 kV, 50 Hz, 1–5 min	No significant color alteration was observed	Mahendran (2016)

(Continued)

TABLE 4.2 (CONTINUED)   BRIEF DISCUSSION OF COLD PLASMA PROCESSING CONDITIONS AND THEIR EFFECT ON VARIOUS GRAINS AND PRODUCTS

Varieties	Plasma treatment	Quality observations	References
Rice starches	Radiofrequency plasma, 13.56 MHz, 40–60 W, 0.15 mbar, air, 5–10 min	• Reduction in amylose content, gelatinization temperature, retrogradation tendency, turbidity, degree of starch hydrolysis and pasting temperature • Increase in leaching of amylose, pasting, final viscosities, water absorption index, solubility, swelling power and syneresis	Thirumdas et al. (2017)
Wheat starch	Ethylene gas, 65 kV	Reducing sugar increased	Lii et al. (2002b)
	Oxygen gas, 800 V	Polarity decreased	
	Oxygen gas, 65 kV	pH decreased	Lii et al. (2003); Thirumdas et al. (2017)
Corn starch	Air, 1–2 kV	Water content and iodine complex decreased	Lii et al. (2002c); Lii et al. (2002a)
	Ethylene gas, 65 kV	Reducing sugar increased	Lii et al. (2002b)

4.3.2.5 Gelatinization Temperatures

Singh et al. (2003) stated that gelatinization primarily occurs in an amorphous region, as a crystalline region resists it due to strong hydrogen bonding. Gelatinization temperature is directly proportional to the water content, amylose-amylopectin ratio and different arrangement of starch crystallinity. Cold plasma mechanisms such as cross-linking, depolymerization and oxidization show a high impact on the gelatinization temperature of starches. However, the degree of cross-linking can also influence the gelatinization temperatures; a higher degree of cross-linked starch possesses greater gelatinization temperature as compared to lower cross-linked starch. Wongsagonsups et al. (2014) also observed similar results in regard to cold plasma–treated starches. The reduction in crystallinity of plasma-exposed starch resulted in a decline in gelatinization temperature. Another cause of declining gelatinization temperature might be hydrophilicity caused by the formation of active species and a boost in the surface energy of the starch molecule. Plasma etching has reduced the resistance between the surrounding medium and the starch membrane. Starches treated by plasma etching are more prone to gelatinize rapidly than native starch due to enhancement in the water absorption tendency of the starch along with the temperature. Meanwhile, Bie et al. (2016) reported enhancement of the gelatinization temperature as well as a reduction in the enthalpy of starch after cold plasma exposure, which is briefly discussed in Table 4.2. They have described that reduction in the enthalpy of starch is caused by the disfigurement of supramolecular lamellar structural and background regions on the nano level.

4.3.2.6 pH Alteration

Subjecting a starch solution to cold plasma leads to minor alteration in the pH of the solution (Thirumdas et al., 2017; Lii et al., 2002a). The decline in the pH of starch determines the development of chemical groups with acidic nature such as peroxide, carboxyl and carbonyl groups; this indicates the oxidization of starch polymer when being subjected to plasma treatment. The intensification of the acidic character of starch can also be described by the rise in peak intensity at 1,710 cm –1 in plasma-processed starch samples as compared to native starch samples.

4.3.3 Polyphenols and Antioxidant Properties

Antioxidants are significant compounds known for their defensive properties against free radicals. Therefore, this specific area should be well explored in order to elucidate and create a better understanding of the basic interactions among plasma species and bioactive compounds, in a manner that evades nutritional deterioration or any other unwanted results in future applications. Antioxidants guard vegetative cells against the detrimental effects of ROS, such as superoxide, hydroxyl ions, singlet oxygen, peroxyl radicals and peroxynitrite, etc. (Percival, 1998).

However, in the food industries, antioxidant activity is not only used as a direct quality attribute but also referred to as the close indicator of numerous compounds, such as polyphenols, flavonoids and flavanols, existing in the food products. The antioxidant activities possessed by phenolic compounds could be the result of their redox properties (Shan et al., 2005). Several researchers reported that the impact of cold plasma treatment on the phenolic compounds of the food products has an extensive range of variation (Pankaj et al., 2018).

During cold plasma treatment, various types of reactive species (ROS and/or RNS) can occur and interrelate with the surfaces of food material, and this condition will lead to the occurrence of physiological response. A complex antioxidant enzymatic system will lower the detrimental effects of reactive species and convert them into less harmful compounds for vegetable cells (Andre et al., 2010). Harborne and Williams (2000) observed that various plant species can bear UV radiations and lead to the accumulation of flavonoid compounds in epidermal cells. The formation of UV radiations during plasma generation may be responsible for the development of phenolic compounds, which are obtained from the cells of the upper epidermis of leaves. Thirumdas et al. (2016) reported that non-thermal cold plasma treatment improved the phenolic compound content in basmati rice flour. In this study, the researcher observed that the maximum phenolic content was attained by dropping the power from 40 to 30 W and the processing time from ten to five minutes. Furthermore, all plasma-processed flour displayed higher phenolic contents than unprocessed basmati rice flour samples (0.48–0.53 vs. 0.44 mg GAE/100 g, respectively).

Since the application of cold plasma technology will lead to the enhancement of phenolic compound content in edible plant tissues, the favorable features of this technology will be emphasized for implementation in the food industry. However, the deprivation of polyphenolic compounds must be taken into consideration, though the information regarding mechanisms responsible for the loss of anthocyanins is not completely understood yet (Liu et al., 2018). Several enzymes are naturally found in vegetable food, such as polyphenol oxidases, peroxidases, lipoxygenase and laccases, which have been considered responsible for the oxidative degradation of polyphenols in vegetable food products (López-Nicolás & García-Carmona, 2009). The application of plasma treatments can lead to inactivate food enzymes by causing an alteration in the secondary structures (reducing the α-helix structures and increasing the β-sheet) of polyphenol oxidases and peroxidases (Han et al., 2019). Several new studies have been conducted that explore the utilization of plasma treatments on phenolic compounds and show contrasting outcomes regarding the accountability of these enzymes for the deterioration of phenolic compounds present in foodstuffs (Munekata et al., 2020).

4.4 PLASMA APPLICATIONS IN GRAIN PROCESSING

4.4.1 Seed Germination and Initial Plant Growth

The germination of seeds and the primary growth of plants are significantly influenced by the type of treatment given to the seeds (Scholtz et al., 2019). Numerous authors have observed that non-thermal plasma treatment had positive effects on treated seeds that eventually improved their germination potency within several parameters such as germination rate, speed of germination, number of germinated grains, grain vitality, length and weight of seedlings, germination index and root/shoot (R/S) ratio, etc. (Scholtz et al., 2019).

Many studies have shown that the early germination of seeds after treating them with plasma technology is due to the penetration of active plasma entities through seed coat and lead to sudden effect over the interior mechanism system of the cells. The intensification in the rate of germination is because of the alteration occurring on the surface of the seed, resulting in ablation, which further improves the transmission of oxygen and moisture via the seed coat to the embryo, influencing its germination rate (Thirumdas et al., 2015). According to Fridman (2008), the interaction between cells and plasma may lead to cell wall rupture, DNA damage or alteration of protein structure, and can stimulate natural signals like production of growth factors, influencing enzymatic activity, which is responsible for the interruption of the seeds' dormant phase which then leads to an upsurge in the germination rate of seeds.

Sera et al. (2012) observed that plasma-treated wheat grains showed a quick germination rate as compared to untreated wheat seeds. When grain or seeds are treated by air plasma (exposure to oxygen radicals and low-energy bombardment) it causes seed coat ablation and perhaps significantly promotes germination improvement. Filatova et al. (2011) carried out germination studies on legumes where the results displayed an increase in the germination rate by 10–20% in both the laboratory and the field and also observed a 3–15% reduction in fungicidal effect; they inferred that OH radicals and atomic oxygen produced during plasma treatment are the most likely sterilizing agents. Dhayal et al. (2006) conducted a study on the germination rate of safflower seeds where they have proposed that cold plasma is an appropriate technique for the surface modification of seeds due to the generation of high ion particles, initiating the etching of the seed coat and resulting in a 50% increase in the rate of germination. Jiafeng et al. (2014) observed that the germination rate of wheat grains increased by 6.7% when subjected to helium plasma. Similarly, Meng et al. (2017) reported that the application of DBD plasma along with the gases oxygen, argon and nitrogen can significantly enhance the germination potency by 28%, 24% and 35.5% respectively after four minutes of contact. The progressive effect of non-thermal plasma treatment was also observed on barley (Braşoveanu et al., 2015), maize (Henselová et al., 2012), oat (Dubinov et al., 2000), rice (Jo et al., 2014) and mung bean (Zhou et al., 2016).

4.4.2 Surface Decontamination

Microorganisms and various insect pests are mainly responsible for the contamination of the grain's surface (Laca et al., 2006). This chapter is mainly focused on the significance of cold plasma processing in order to ensure the effectiveness of grain surface decontamination. It extends to numerous applications such as the inactivation of microbes and insect pests present on the surface of cereal grains, principally epiphytic bacteria. Applying NTP to winter wheat grains for a time period of ten seconds lessens the fungal colonies by one order of magnitude, as reported by Kordas et al. (2015). The decrease is measured to be by two orders of magnitude as reported by Los et al. (2018). Zahoranová et al. (2016) focused also on microorganisms rather than only fungi and saw a notable reduction in the population of epiphytic bacteria and phyto-pathogenic and toxinogenic filamentous fungi by two orders after the application of NTP. A significant reduction was observed in microbial contamination on wheat and barley grain's surface; this was achieved after cold plasma exposure where argon atmospheric plasma was used for the inactivation of *Geobacillus stearothermophilus* endospores on artificially contaminated wheat grains and polypropylene model substrates. Although decontamination of the smooth surfaces was done efficaciously, showing a reduction of up to four orders of magnitude, the reduction was of only one order of magnitude on the surface of grains.

The author believed that the rough surface of grain (such as randomly arranged bran and front creases of wheat grains) provides protection from the plasma-generated volatile species. Insects are considered significant pests; Shahrzad et al. (2015) accounted for their decontamination by treating two to three instar larvae amid wheat grains. Mean death percentage hit 100% after 20 seconds for the confused flour beetle (*Tribolium confusum*, Coleoptera: *Tenebrionidae*) and the Mediterranean flour moth (*Ephestia kuehniella*, Lepidoptera: *Pyralidae*). Rehman et al. (2018) reported a study in which cold plasma was applied to the egg, larval and grown-up phases of the red flour beetle (*Tribolium castaneum*, Coleoptera: *Tenebrionidae*). One hundred percent death rate can be obtained in the case of all the flour beetle stages depending on plasma subjection time and potency of plasma. Favorable results were noticed on all stages of *T. castaneum* after application of plasma for 15 minutes. It can be concluded that non-thermal plasma is a propitious tool for efficacious decontamination. The deactivation of insect pests and microorganisms present on the surface of cereal grains, along with various other promising applications, is offered by non-thermal plasma.

4.4.3 Destruction of Microorganisms

Cold plasma is an aggregate of excited particles, charged particles, photons and reactive neutrals. Due to these particles, the inactivation of different microorganisms occurs in various types of food. Generally, due to the accumulation and depositions of these particles and charges, the sample ruptures the cell walls of microorganisms. Moreover, due to electrostatic forces, different types of ions and electrons initiate the growth of reactive species, such as reactive oxygen species and reactive nitrogen species.

The collision of charged particles with a reactant, gases and a microorganism leads to the formation of reactive species like O, OH, O2, O3, NO and NO2. These short-lived chemical species rupture the outer membranes of bacteria and cause a defect generally known as chemical sputtering. This mechanism is valid for bacteria and spores. The impact of cold plasma on microorganisms is summed up in Table 4.3 (see also Figure 4.1).

4.4.4 Destruction of Mycotoxins

Nowadays, the rate of prevalence of mycotoxins in food contamination is growing up to 25% for grains, i.e., 25% of cereals globally become inapt for consumption (Misra et al., 2016).

TABLE 4.3   RECENT STUDIES ON THE EFFECT OF COLD PLASMA TREATMENT ON MICROBIAL INACTIVATION TECHNOLOGY AND THEIR OUTCOMES

Food Sample	Exposure Time	Process Gas	Target MOs	Outcomes	References
Wheat	1 hour	Argon	Endospores	5 log reduction	Butscher et al. (2016)
Corn	7 min	Air	Fungi	2 log reduction	Zahoranová et al. (2014)
Brown rice	5–20 min	Air	Aerobic microorganism	0.61, 0.91, 1.4 log reduction according to the processing time	Lee et al. (2016)
	20 min	Air	*Bacillus cereus* KCTC3624 and *Bacillus subtilis*	2.99 and 2.94 log reduction, respectively	
			E. coli O157:H7	2.3 log reduction	
Rapeseed seed	3 min	Dry air	Aerobic microorganism	2.2 log reduction	Puligundla et al. (2017)
			Bacillus cereus	1.2 log reduction	
			E. coli and *Salmonella spp.*	2.0 and 1.8 log reduction, respectively	
Chickpea	5 min	Air	Microflora	2 log reduction Improved textural properties	Mitra et al. (2013)
Unpeeled almonds	15 min	Air and oxygen-based mixtures	*Salmonella enteritidis* PT30	> 5 and 4.8 log reduction, respectively	Hertwig et al. (2017)
		N_2 and air		2.0–6.0 log CFU/g reduction	
Peanuts	20 min	SF_6	Spores	3 log reduction No bad effect on food quality	Selcuk et al. (2008)
Nuts	30 sec	Air	*E. coli*	4 log reduction	Deng et al. (2007)

Inappropriate agricultural and harvesting practices and low efficiency of prevention methods are major reasons behind the undesirable fungal and mycotoxin growth (Misra, 2015).

Cereal grains are the main staple of a nutritive, healthy diet. However, after harvesting, cereal grains may be loaded with microbial contamination including families such as *Pseudomonadaceae, Micrococcaceae, Lactobacillaceae* and *Bacillaceae*, and molds, namely *Alternaria, Fusarium, Aspergillus, Helminthosporium* and *Cladosporium*. In these cases, mycotoxins' increased resistance to decontamination has become one of the most important topics of health safety concerns.

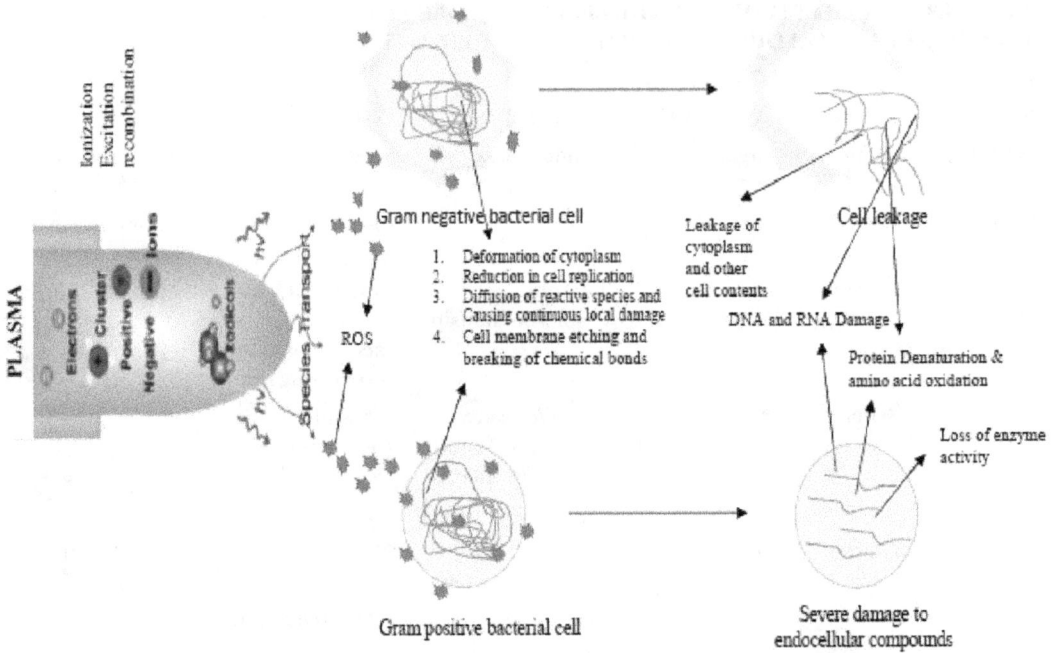

Figure 4.1   Effect of cold plasma processing on a microbial cell.

AflatoxinB1 (AFB1) is a specific carcer causing mycotoxin produced by the *Aspergillus* species AFB1 remains stable in food and cannot be easily destroyed by applying traditional food processing operations (Pleadin et al., 2015), as we briefly discuss in Table 4.4. Numerous reports have been focused on a similar issue related to the *Fusarium* species (fungi), and the mycotoxins associated with it are mainly responsible for grain and fruit contamination. In short, there is an extensive need for a technology that can effectively arrest fungal growth, mycotoxin contamination and pesticide residues in harvested grains, as well as reduce postharvest losses and increase germination potency and safety. In order to eliminate mycotoxins, cold plasma technology is considered a promising technique that is mostly being applied to treat grains, seeds and cereal crops and fresh products (Basaran et al., 2008; Selcuk et al., 2008; Ito et al., 2012; Zahoranová et al., 2016), as briefly discussed in Table 4.4. Cold plasma technology possesses numerous benefits for agricultural processing, such as low-temperature operations, time-efficient processing, being economical and incurring negligible damage to seeds, grains, foods, humans and the environment. The reactive species generated during the plasma processing mainly influence seed germination capacity, microbial cells, enzymes in plant growth and the overall quality of the agricultural commodities. Generally, the foodborne pathogenic organisms existing in low–water activity foods are more likely to exhibit high resistance against heat and other processing treatments that are fatal to living microbial cells in high–water activity surroundings (Beuchat et al., 2013).

Mitra et al. (2013) have reported results regarding chickpea seed decontamination by applying cold atmospheric plasma treatment. They have observed a substantial decrease in the natural microbial load (2 log) associated with the chickpea surface after prolonged exposure to plasma treatments of up to five minutes. Butscher et al. (2015) observed a reduction by 2.15 log

TABLE 4.4   SUMMARY OF APPLICATION OF VARIOUS COLD PLASMA PROCESSING METHODS FOR INACTIVATION OF FUNGAL SPECIES IN DIFFERENT FOOD PRODUCTS

Food Product	Microorganism	Mycotoxin	Plasma Source and Parameter	Outcome	References
Maize grains: peanuts, nuts, maize, cottonseed, wheat, barley, cocoa beans, rice, copra, dried fruits, spices	*A. flavus* *A. parasiticus*	Aflatoxins FB1, FB2, FG1, FG2	Atmospheric pressure fluidized bed plasma (APFBP), 1 atm, dry air; N2, 5–10 kV, 18–25 kHz 665 W, 1–5 min	Dry air reduction: *A. flavus*: 5.48 log and *A. parasiticus*: 5.20 log N2 reduction: *A. flavus*: 5.08 log and *A. parasiticus*: 4.99 log Air was more effective compared to N2	Dasan et al. (2016)
Brown rice cereals	*A. flavus*	Aflatoxins FB1, FB2	Radiofrequency atmospheric cold plasma jet, 1 atm, argon, 40 W, 50–600 kHz, 10 kV (max)	Plasma power of 40 W for 20 min was effective in preventing *A. flavus* growth for 20 days under storage conditions of 25°C and 100% RH	Suhem et al. (2013)
Wheat and barley	*B. atrophaeus* *P. verrucosum*		Dielectric barrier discharge, 1 atm, air, 80 kV, 50 Hz, 5–10 min	Fungi population decreased on barley surface by 2.1 and 1.5 log 10 and wheat surface by 2.5 and 1.7	Los et al. (2018a)
Wheat, bean, chickpea, soybean, barley, oat, rye, lentil, corn	*Aspergillus spp.* *Penicillium spp.*	Aflatoxins FB1, FB2, FG1, FG2, Ochratoxin A (OTA), Patulin (PAT)	Inductively coupled plasma (ICP), 500 mTorr, air, SF6, 1 kHz, 20 kV, 300 W, interval: 5, 10, 20 min	SF6 plasma for 15 min allowed reduction in both species by 3 log 10 Seed germination is retained after plasma treatment	Selcuk et al. (2008)
Hazelnuts, peanuts, pistachio nuts	*A. parasiticus*	Aflatoxins FG1, FG2	Inductively coupled plasma (ICP), 500 mTorr, air, SF6, 1 kHz, 20 kV, 300 W, 5–20 min	SF6 plasma was more effective as compare to air plasma with a 5 log 10 reduction in fungal population No significant organoleptic changes	Basaran et al. (2008)

by 30 seconds in the concentration of *B. amyloliquefaciens* endospores deposited on the upper layer of wheat grains when subjected to a low-pressure fluidized-bed plasma reactor at high power inputs of 900 W. They have found effective destruction of the microbial population during an experimental duration of 60 minutes at a grain surface temperature of 90°C. They have also researched the impact of DBD-generated pulsed plasma processing of inoculated wheat grains with polypropylene model substrates for the inactivation of *Geobacillus stearothermophilus*. In this experiment, pulsed argon plasma discharge treatment was applied with different combinations of treatment time, pulse voltage and frequency. Whereas a 5 log reduction was attained on polypropylene granules within a ten-minute time interval, the maximum reduction of around 3 log was observed to inactivate endospores presented on wheat grain after an hour. In addition, there was no effect found on the nutrition quality parameters of wheat grains (Butscher et al., 2016).

Furthermore, in regards to products with a high surface-to-volume ratio, the intensity and concentration of reactive species (plasma-generated) tend to reduce during treatment due to interfacing with the food surface itself, despite microorganisms deposited on that surface (Hertwig et al., 2015). The above condition suggests that an extensive and thorough study is required for further optimization of the plasma processing parameters in order to accomplish effective and reliable destruction of pathogens associated with different foods with varied surface characteristics.

4.4.5 Degradation of Pesticide Residues

Modern agriculture predominantly relies on pesticide use, and recently, yearly pesticide usage has been evaluated at approximately 2.5 million tons (Gavrilescu, 2005). Pesticides enhance the production of crops by reducing extrinsic hurdles (such as keeping a check on the presence of insects, fungi, viruses, etc.), arresting the growth of weeds and working as growth regulators.

According to data from the European Union's Pesticide Action Network, 2019 reported the presence of various pesticides in food products manufactured in the EU. Thus, there is a necessity to discover efficacious methods to lessen or eradicate pesticide residues from food. Non-thermal and advanced oxidation technologies facilitate the reduction and removal of pesticide residues, as accounted by Misra (2015). Plasma species with positive oxidation potential, like ozone (O_3), hydrogen peroxide, hydroxyl radicals, etc., facilitate pesticide dissipation, as reported by various studies (Sarangapani et al., 2016b).

Limited studies have evaluated pesticide degradation on food with the application of non-thermal plasma. According to one study, in-package atmospheric air plasma treatment resulted in the chemical breakdown/degradation of agrochemicals on blueberries, as evaluated by Sarangapani et al. (2017b). During the above treatment, more than 70% decomposition of the targeted pesticides occurred in a time span of five minutes while keeping the quality specifications of the fruit intact (negligible changes). Hero et al. (2014) reported pesticide dissipation with the application of atmospheric plasma treatment remotely on apples. Moreover, stored food such as cereal crops can be protected from pests with the application of cold plasma. The presence of insects is taken care of with elementary processing and operation without any residues. An innocuous compound is formed by the reaction of free radicals coming from the cold plasma treatment with the chemical constituents of pesticides in comparison to the toxic pesticide residues. Sarangapani et al. (2018) reported that 2,2-dichlorovinyl dimethyl phosphate (or dichlorvos), omethoate, diazinon or paraoxon are various pesticides that have undergone decomposition after the application of cold plasma treatment. O_2-plasma-induced degradation of dichlorvos and omethoate on maize samples was reported by Bai et al. (2009); they also suggested that the chemical structure of pesticide and operational conditions play a significant role in the efficiency of degradation. The intermediates were harmless in comparison to the parent pesticide, as evaluated by the authors.

4.4.6 As Pre-Treatment for Grain Drying

Cold plasma has a salubrious effect in controlling the growth of microorganisms in various dried food products such as cereal, legumes, oilseeds, etc. When cold plasma is used in combination with controlled parameters, energy input and process gases such as oxygen, nitrogen and argon, the energy input can be used to provide an antimicrobial impact on dry foods. The composition of plasma, its discharge properties and its penetration of depth further signify its microbial inactivation capability (Hertwig et al., 2018). A limited penetration depth is the most preferable in the case of food with large volume-to-surface ratios. The reactivity of cold plasma with grains totally depends upon the composition and water activity. Cold plasma is a promising technique in which complete surface decontamination takes place without disturbing the characteristic nutrients of the dried food.

Cold plasma treatment can be used in direct, indirect or semi-direct modes with dried foods. Many seeds, nuts and legumes are temperature sensitive. To overcome this, cold plasma treatment provides a valuable solution for surface decontamination. Lee et al. (2016), Dassan et al. (2017) and Kim et al. (2016) performed different cold plasma experiments on various dry foods, for example, brown rice, maize, hazelnuts and broccoli seeds, to successfully hinder the growth of the spore and mold microorganisms *Bacillus spp.*, *E. coli spp.*, *Aspergillus spp.* and *Salmonella spp.* respectively. It was also concluded from these experiments that cold plasma treatment does not affect the sensorial attributes and physicochemical properties of dried samples. Cold plasma treatment can be effectively used within the range of 0–70°C.

4.5 SAFETY REGULATIONS

As an alternative solution for surface and shelf-life improvement, CP technology has proven very reliable. For its acceptance as an alternative processing technology, the effect of CP on grain quality is very vital (Sarangapani et al., 2018). CP techniques have little or negligible effect on the physical, chemical, nutritional and sensory characteristics of different products due to their non-thermal approach. Cold plasma is accessible, waterless, waste-free and leaves no chemical residue (Pankaj et al., 2018). Cold plasma can produce reactive species like molecular oxygen and ozone, which are common and universal oxidizing agents used to treat wheat flour (Joye et al., 2009). Active species practically vanish automatically when plasma power is switched off; plasma processing is thus environmentally sound and can follow all environmental requirements (Misra et al., 2011). However, there are limited research opportunities on the interaction between CP species and food components on the molecular level. Chaple et al (2020), after application of CP, reported improvements in the hydration properties of wheat flour and its crystallinity. Functional plasma flour could be used to manufacture baked goods and frozen products in different formulations of wheat products. Another challenge is to obtain a measurable dose of CP for grains and refined goods (Cullen and Milosavljevic, 2015). CP technology would need a precise understanding of the mechanisms and quality control in order to realize its full potential on a market scale (Pankaj et al., 2018). But CP processing remains in its early stages and needs more study in order to achieve its potential.

The regulatory approval process is another significant impediment to the introduction of plasma technologies to the food industry. From a regulatory point of view, it takes a tremendous amount of data collection, data reviews and time for approval of a new plasma process, whether it be for direct or indirect food treatment, food packaging treatment or food packaging. Owing to the complexities of plasma chemistry, many potential chemical effects need to be studied, prioritized and evaluated. In addition, each country has its own new technology review and approval process. For example, a plasma system may be equipped for the creation of ozone. In the US, the Food and Drug Administration (FDA) has designated ozone as GRAS and no FDA regulatory review or approval is needed for any device claiming that ozone generation is required. But if reactive gas species other than ozone are generated, it depends on the potential demand and willingness of the company to accept regulatory risks which path they will take to regulatory review and final approval. The plasma community is urged to play an active role in the regulatory review phase for accelerating plasma technology commercialization (Keener and Misra, 2016). In order to provide regulatory consent to atmospheric plasma technology, the community (fabricants, manufacturers, science staff, engineers, consumer groups) should come together to create guidance documents. A technical survey of what atmospheric cold plasma is and how plasma is produced; the recommendation of methods for plasma calculation; data collection and data analysis; regulatory roadmaps for obtaining approval for a food plasma process or system; and consumer concerns about the application of plasma may be some possible guidance documents. The main technologies predicted to grow worldwide in the next five years are high-pressure processing (HPP), microwave heating (MWH) and ultraviolet light (UV). The European market is projected to be more important for cold plasma and pulsed electrical field (PEF) within ten years, while HPP, microwave and UV remain more relevant in North America (Jermann et al., 2015).

4.5.1 Effect on Packing Material

Food materials are guarded against the external environment by food packaging materials while being handled, transported and distributed. The application of cold plasma for the decontamination of packaging materials from the outside is viable, as the possibility of shadow effect is insignificant since plasma glides throughout the surface. Surface modifications generally take place in instances of packaging materials during plasma processing, where it undergoes various surface treatments like cleaning, coating, printing, painting and adhesive bonding (Pankaja et al., 2014). Packaging materials like plastic bottles, lids and films can be quickly and safely sterilized without excessively affecting the material properties or leaving any residues via low-temperature gas plasma sterilization. Sterilization of heat-susceptible packaging materials such as polyethylene, ethylene and polycarbonate can be carried out by cold plasma as it possesses low-temperature conditions. The same goes for surfaces of polymers specifically designed for edible packaging films, having a hydrophobic nature with lower surface energies (Vesel and Mozetic, 2012). Deposition of one material over the surface of another material (in a fine layer), thus conveying some of its attributes, can be attained by the use of plasma as the transport mechanism and catalyst. Hedenqvist and Johansson reported the oxygen transport characteristic of the SiOx coating on polypropylene [PP] films, polyethylene terephthalate [PET] and low- as well as high-density polyethylene [LDPE, HDPE] attained by using cold plasma technology; they compared the experimental data with a computer model and found that the diffusivity was lower as compared to that of normal material (2003). Plasma deposition of heat-liable substances like vitamins, antioxidants and antimicrobials into packaging matter is pursued as a prospective substitute in the emergent field of antimicrobial and active packaging. However, the barrier properties of packaging material can be improved with the application of nanotechnology, which can be accomplished via cold plasma processing.

4.6 FUTURE PERSPECTIVE, CHALLENGES AND LIMITATIONS

Cold plasma is an excellent technique which efficiently used for the purpose of decontamination and the modification of packaging polymers, but there is also huge scope for application in the food processing sector. Its stability and amount of energy consumption based on the type of discharges used during treatment can be utilized to achieve maximum efficiency rates at minimum costs of operation by optimizing these parameters. Various studies have resulted in the effective application of cold plasma on different foods for microbial destruction but they did not describe its effects on the nutritional values and toxicology of processed foods. There is a need to develop standards, and plasma-treated food should be recognized as GRAS after deep research (in vitro and in vivo) in this field.

Upcoming research should be conducted on the applications of cold plasma on food surfaces to modify its physical, chemical and functional characteristics in a cost-effective way. Cold plasma's application in grain processing and its effect on the physicochemical and functional characteristics of grain in real food systems must be exploited well. Future studies should be focused on the reproducibility of outcomes on a large scale and their parallel effects on the quality and nutritional value of grains during storage.

From the industrial point of view, the technology transfer, cost and maintenance of a cold plasma processing unit are very expensive. That is why cold plasma treatment is still not viable from the commercial perspective in the food sector, since the current chapter is mainly concerned with the plasma's properties for the processing of grains. Any technology or processing condition applied to food material should not influence the allergenicity of food components. Numerous experiments have been carried out on cytotoxicity and the mutagenic potential of food when exposed to cold plasma. Many researchers reported that there is no formation of any toxic or mutagenic products after cold plasma processing, but extensive studies must be conducted in this area to ensure the GRAS limit for the consumption of processed food.

According to the economic aspects, the mechanism of inactivation of microbial species during the application of cold plasma is still not well explored and clear because it results in the generation of some 75 species and more than 1,000 chemical reactions occurring at nano-, micro-, milli- and second timescales. Furthermore, the regulatory aspects of plasma treatment should be covered in order to fully ensure the health and quality of food supplies. The effect of cold plasma on nutritional and sensory attributes has to be addressed in order to commercialize this technology. In spite of the need for more research on microbial destruction and decontamination, if the above-discussed points are studied well and clarified, cold plasma might become a revolutionary technology for future food preservation (Jermann et al., 2015).

4.7 CONCLUSION

The cold plasma technique is arising as a novel and green technology of the present era. This technology is gaining recognition due to its unique features, like low or ambient temperature processing for a very short period of time, which facilitates the preservation of the freshness and quality of food products. The application of cold plasma is significantly used in grain surface decontamination and modification, microbial destruction, other post-harvest processing activities and the packaging of food products. The results gained by cold plasma treatment are better than those of conventional techniques, which possess some unfavorable effects on processed food's nutritional quality. Plasma sterilization delivers high-efficacy preservation and does not introduce any toxicity to the medium. Selection of the specific gas medium for plasma treatment is the key step because these gases possess specific germicidal characteristics which enhance plasma sterilization. Preservation is the core benefit achieved by cold plasma treatment via having minimal effects on the sensory and nutrition quality parameters of the food and leaving no chemical residue. Several authors also illustrated additional advantages such as enhancement of germination rate, plant growth, increase in polyphenolic and antioxidant properties, reduction of pesticide residues and many more. Numerous studies reported that food substances subjected to cold plasma treatment can effectively retain their native texture, sensory attributes and functional characteristics. But commercialization of cold plasma technology is not done yet. However, research is being conducted to determine the cost of processing large quantities and different varieties of grain (such as cereal, legumes, oilseed, etc.) with cold plasma at the industrial level and also to ensure safety, quality and wholesomeness. Hence, cold plasma processing can be viewed as a forthcoming technology in grain processing in the near future.

REFERENCES

Ai, Y., & Jane, J. L. (2015). Gelatinization and rheological properties of starch. *Starch-Starke*, *67*(3–4), 213–224.

Andre, C. M., Larondelle, Y., & Evers, D. (2010). Dietary antioxidants and oxidative stress from a human and plant perspective: A review. *Current Nutrition and Food Science*, *6*(1), 2–12.

Bai, Y., Chen, J., Mu, H., Zhang, C., & Li, B. (2009). Reduction of dichlorvos and omethoate residues by O2 plasma treatment. *Journal of Agricultural and Food Chemistry*, *57*(14), 6238–6245.

Bao, J., Ao, Z., & Jane, J. L. (2005). Characterization of physical properties of flour and starch obtained from gamma-irradiated white rice. *Starch-Stärke*, *57*(10), 480–487.

Basaran, P., Basaran-Akgul, N., & Oksuz, L. (2008). Elimination of *Aspergillus parasiticus* from nut surface with low pressure cold plasma (LPCP) treatment. *Food Microbiology*, *25*(4), 626–632.

BeMiller, J. N. (1997). Starch modification: Challenges and prospects. *Starch-Stärke*, *49*(4), 127–131.

BeMiller, J. N., & Huber, K. C. (2015). Physical modification of food starch functionalities. *Annual Review of Food Science and Technology*, *6*, 19–69.

Benstaali, B., Moussa, D., Addou, A., & Brisset, J. L. (1998). Plasma treatment of aqueous solutes: Some chemical properties of a gliding arc in humid air. *The European Physical Journal-Applied Physics*, *4*(2), 171–179.

Beuchat, L. R., Komitopoulou, E., Beckers, H., Betts, R. P., Bourdichon, F., Fanning, S., … & Ter Kuile, B. H. (2013). Low–water activity foods: Increased concern as vehicles of foodborne pathogens. *Journal of Food Protection*, *76*(1), 150–172.

Bie, P., Li, X., Xie, F., Chen, L., Zhang, B., & Li, L. (2016). Supramolecular structure and thermal behavior of cassava starch treated by oxygen and helium glow-plasmas. *Innovative Food Science and Emerging Technologies*, *34*, 336–343.

Boudam, M. K., Moisan, M., Saoudi, B., Popovici, C., Gherardi, N., & Massines, F. (2006). Bacterial spore inactivation by atmospheric-pressure plasmas in the presence or absence of UV photons as obtained with the same gas mixture. *Journal of Physics. Part D: Applied Physics*, *39*(16), 3494.

Bourke, P., Ziuzina, D., Han, L., Cullen, P. J., & Gilmore, B. F. (2017). Microbiological interactions with cold plasma. *Journal of Applied Microbiology*, *123*(2), 308–324.

Braşoveanu, M., Nemţanu, M., Surdu-Bob, C., Karaca, G., & Erper, I. (2015). Effect of glow discharge plasma on germination and fungal load of some cereal seeds. *Romanian Reports in Physics*, *67*(2), 617–624.

Butscher, D., Schlup, T., Roth, C., Müller-Fischer, N., Gantenbein-Demarchi, C., & Rudolf von Rohr, P. (2015). Inactivation of microorganisms on granular materials: Reduction of *Bacillus amyloliquefaciens* endospores on wheat grains in a low pressure plasma circulating fluidized bed reactor. *Journal of Food Engineering*, *159*, 48–56.

Butscher, D., Zimmermann, D., Schuppler, M., & von Rohr, P. R. (2016). Plasma inactivation of bacterial endospores on wheat grains and polymeric model substrates in a dielectric barrier discharge. *Food Control*, *60*, 636–645.

Chaple, S., Sarangapani, C., Jones, J., Carey, E., Causeret, L., Genson, A., … & Bourke, P. (2020). Effect of atmospheric cold plasma on the functional properties of whole wheat (*Triticum aestivum* L.) grain and wheat flour. *Innovative Food Science and Emerging Technologies*, *66*, 102529.

Critzer, F. J., Kelly-Wintenberg, K., South, S. L., & Golden, D. A. (2007). Atmospheric plasma inactivation of foodborne pathogens on fresh produce surfaces. *Journal of Food Protection, 70*(10), 2290–2296.

Cullen, P. J., & Milosavljević, V. (2015). Spectroscopic characterization of a radio-frequency argon plasma jet discharge in ambient air. *Progress of Theoretical and Experimental Physics, 2015*(6).

Dasan, B. G., Mutlu, M., & Boyaci, I. H. (2016). Decontamination of *Aspergillus flavus* and *Aspergillus parasiticus* spores on hazelnuts via atmospheric pressure fluidized bed plasma reactor. *International Journal of Food Microbiology, 216*, 50–59.

Dasan, B. G., Onal-Ulusoy, B., Pawlat, J., Diatczyk, J., Sen, Y., & Mutlu, M. (2017). A new and simple approach for decontamination of food contact surfaces with gliding arc discharge atmospheric non-thermal plasma. *Food and Bioprocess Technology, 10*(4), 650–661.

Deng, S., Ruan, R., Mok, C. K., Huang, G., Lin, X., & Chen, P. (2007). Inactivation of *Escherichia coli* on almonds using nonthermal plasma. *Journal of Food Science, 72*(2), M62–M66.

Dhayal, M., Lee, S. Y., & Park, S. U. (2006). Using low-pressure plasma for *Carthamus tinctorium L.* seed surface modification. *Vacuum, 80*(5), 499–506.

Dubinov, A. E., Lazarenko, E. R., & Selemir, V. D. (2000). Effect of glow discharge air plasma on grain crops seed. *IEEE Transactions on Plasma Science, 28*(1), 180–183.

Ekezie, F. G. C., Sun, D. W., & Cheng, J. H. (2017). A review on recent advances in cold plasma technology for the food industry: Current applications and future trends. *Trends in Food Science and Technology, 69*, 46–58.

Filatova, I., Azharonok, V., Kadyrov, M., Beljavsky, V., Gvozdov, A., Shik, A., & Antonuk, A. (2011). The effect of plasma treatment of seeds of some grain and legumes on their sowing quality and productivity. *Romanian Journal of Physics, 56*, 139–143.

Foest, R., Schmidt, M., & Becker, K. (2006). Microplasmas, an emerging field of low-temperature plasma science and technology. *International Journal of Mass Spectrometry, 248*(3), 87–102.

Fridman, A. (2008). *Plasma Chemistry*. Cambridge University Press.

Fridman, A., Chirokov, A., & Gutsol, A. (2005). Non-thermal atmospheric pressure discharges. *Journal of Physics. Part D: Applied Physics, 38*(2), R1.

Gavrilescu, M. (2005). Fate of pesticides in the environment and its bioremediation. *Engineering in Life Sciences, 5*(6), 497–526.

Han, Y., Cheng, J. H., & Sun, D. W. (2019). Activities and conformation changes of food enzymes induced by cold plasma: A review. *Critical Reviews in Food Science and Nutrition, 59*(5), 794–811.

Han, Z., Zeng, X. A., Yu, S. J., Zhang, B. S., & Chen, X. D. (2009). Effects of pulsed electric fields (PEF) treatment on physicochemical properties of potato starch. *Innovative Food Science and Emerging Technologies, 10*(4), 481–485.

Harborne, J. B., & Williams, C. A. (2000). Advances in flavonoid research since 1992. *Phytochemistry, 55*(6), 481–504.

Hedenqvist, M. S., & Johansson, K. S. (2003). Barrier properties of SiOx-coated polymers: Multi-layer modelling and effects of mechanical folding. *Surface and Coatings Technology, 172*(1), 7–12.

Henselová, M., Slováková, Ľ., Martinka, M., & Zahoranová, A. (2012). Growth, anatomy and enzyme activity changes in maize roots induced by treatment of seeds with low-temperature plasma. *Biologia, 67*(3), 490–497.

Heo, N. S., Lee, M. K., Kim, G. W., Lee, S. J., Park, J. Y., & Park, T. J. (2014). Microbial inactivation and pesticide removal by remote exposure of atmospheric air plasma in confined environments. *Journal of Bioscience and Bioengineering, 117*(1), 81–85.

Hertwig, C., Leslie, A., Meneses, N., Reineke, K., Rauh, C., & Schlüter, O. (2017). Inactivation of *Salmonella enteritidis* PT30 on the surface of unpeeled almonds by cold plasma. *Innovative Food Science and Emerging Technologies, 44*, 242–248.

Hertwig, C., Meneses, N., & Mathys, A. (2018). Cold atmospheric pressure plasma and low energy electron beam as alternative nonthermal decontamination technologies for dry food surfaces: A review. *Trends in Food Science and Technology, 77*, 131–142.

Hertwig, C., Reineke, K., Ehlbeck, J., Knorr, D., & Schlüter, O. (2015). Decontamination of whole black pepper using different cold atmospheric pressure plasma applications. *Food Control, 55*, 221–229.

Hoffmann, C., Berganza, C., & Zhang, J. (2013). Cold Atmospheric plasma: Methods of production and application in dentistry and oncology. *Medical Gas Research*, *3*(1), 21.

Ito, M., Ohta, T., & Hori, M. (2012). Plasma agriculture. *Journal of the Korean Physical Society*, *60*(6), 937–943.

Jermann, C., Koutchma, T., Margas, E., Leadley, C., & Ros-Polski, V. (2015). Mapping trends in novel and emerging food processing technologies around the world. *Innovative Food Science and Emerging Technologies*, *31*, 14–27.

Jiafeng, J., Xin, H., Ling, L. I., Jiangang, L., Hanliang, S., Qilai, X., ... & Yuanhua, D. (2014). Effect of cold plasma treatment on seed germination and growth of wheat. *Plasma Science and Technology*, *16*(1), 54.

Jo, Y. K., Cho, J., Tsai, T. C., Staack, D., Kang, M. H., Roh, J. H., ... & Gross, D. (2014). A non-thermal plasma seed treatment method for management of a seedborne fungal pathogen on rice seed. *Crop Science*, *54*(2), 796–803.

Joshi, S. G., Cooper, M., Yost, A., Paff, M., Ercan, U. K., Fridman, G., & Brooks, A. D. (2011). Nonthermal dielectric-barrier discharge plasma-induced inactivation involves oxidative DNA damage and membrane lipid peroxidation in Escherichia coli. *Antimicrobial Agents and Chemotherapy*, *55*(3), 1053–1062.

Joye, I. J., Lagrain, B., & Delcour, J. A. (2009). Use of chemical redox agents and exogenous enzymes to modify the protein network during breadmaking – A review. *Journal of Cereal Science*, *50*(1), 11–21.

Jung, S., Lee, J., Lim, Y., Choe, W., Yong, H. I., & Jo, C. (2017). Direct infusion of nitrite into meat batter by atmospheric pressure plasma treatment. *Innovative Food Science and Emerging Technologies*, *39*, 113–118.

Keener, K. M., & Misra, N. N. (2016). Future of cold plasma in food processing. In: N. N. Misra, O. Suchluter, & P. J. Cullen (Eds.), *Cold Plasma in Food and Agriculture* (pp. 343–360). Academic Press.

Kim, J. W., Puligundla, P., & Mok, C. (2016). Effect of corona discharge plasma jet on surface-borne microorganisms and sprouting of broccoli seeds. *Journal of the Science of Food and Agriculture*, *97*(1), 128–134.

Kordas, L., Pusz, W., Czapka, T., & Kacprzyk, R. (2015). The effect of low-temperature plasma on fungus colonization of winter wheat grain and seed quality. *Polish Journal of Environmental Studies*, *24*(1), 433–438.

Laca, A., Mousia, Z., Díaz, M., Webb, C., & Pandiella, S. S. (2006). Distribution of microbial contamination within cereal grains. *Journal of Food Engineering*, *72*(4), 332–338.

Laovachirasuwan, P., Peerapattana, J., Srijesdaruk, V., Chitropas, P., & Otsuka, M. (2010). The physicochemical properties of a spray dried glutinous rice starch biopolymer. *Colloids and Surfaces, Part B: Biointerfaces*, *78*(1), 30–35.

Laroussi, M., Mendis, D. A., & Rosenberg, M. (2003). Plasma interaction with microbes. *New Journal of Physics*, *5*(1), 41.

Lee, K. H., Kim, H. J., Woo, K. S., Jo, C., Kim, J. K., Kim, S. H., Park, H. Y., Oh, S. K., & Kim, W. H. (2016). Evaluation of cold plasma treatments for improved microbial and physicochemical qualities of brown rice. *LWT*, *73*, 442–447.

Li, W., Shu, C., Zhang, P., & Shen, Q. (2011). Properties of starch separated from ten mung bean varieties and seeds processing characteristics. *Food and Bioprocess Technology*, *4*(5), 814–821.

Lii, C. Y., Liao, C. D., Stobinski, L., & Tomasik, P. (2002a). Behaviour of granular starches in low-pressure glow plasma. *Carbohydrate Polymers*, *49*(4), 499–507.

Lii, C. Y., Liao, C. D., Stobinski, L., & Tomasik, P. (2002b). Effects of hydrogen, oxygen, and ammonia low-pressure glow plasma on granular starches. *Carbohydrate Polymers*, *49*(4), 449–456.

Lii, C. Y., Liao, C. D., Stobinski, L., & Tomasik, P. (2002c). Exposure of granular starches to low-pressure glow ethylene plasma. *European Polymer Journal*, *38*(8), 1601–1606.

Lii, C. Y., Liao, C. D., Stobinski, L., & Tomasik, P. (2003). Effect of corona discharges on granular starches. *Journal of Food, Agriculture and Environment*, *1*(2), 143–149.

Liu, Y., Tikunov, Y., Schouten, R. E., Marcelis, L. F., Visser, R. G., & Bovy, A. (2018). Anthocyanin biosynthesis and degradation mechanisms in Solanaceous vegetables: A review. *Frontiers in Chemistry*, *6*, 52.

López-Nicolás, J. M., & García-Carmona, F. (2009). Enzymatic and nonenzymatic degradation of polyphenols. In: Laura A. de la Rosa, Emilio Alvarez-Parrilla, & Gustavo A. González-Aguilar (Eds.), *Fruit and Vegetable Phytochemicals: Chemistry, Nutritional Value, and Stability* (pp. 101–129). Wiley-Blackwell.

Los, A., Ziuzina, D., Akkermans, S., Boehm, D., Cullen, P. J., Van Impe, J., & Bourke, P. (2018). Improving microbiological safety and quality characteristics of wheat and barley by high voltage atmospheric cold plasma closed processing. *Food Research International, 106*, 509–521.

Mahendran, R. (2016, November). Effect of cold plasma on mortality of Tribolium castaneum on refined wheat flour. In: *Proceedings of the 10th International Conference on Controlled Atmosphere and Fumigation in Stored Products (CAF 2016)* (pp. 7–11).

Mandal, R., Singh, A., & Singh, A. P. (2018). Recent developments in cold plasma decontamination technology in the food industry. *Trends in Food Science and Technology, 80*, 93–103.

Meng, Y., Qu, G., Wang, T., Sun, Q., Liang, D., & Hu, S. (2017). Enhancement of germination and seedling growth of wheat seed using dielectric barrier discharge plasma with various gas sources. *Plasma Chemistry and Plasma Processing, 37*(4), 1105–1119.

Michel, J. P., Raffi, J., & Saint-Lebe, L. (1980). Experimental study of the radiodepolymerization of starch. *Starch: Stärke, 32*(9), 295–298.

Min, S. C., Roh, S. H., Niemira, B. A., Boyd, G., Sites, J. E., Uknalis, J., & Fan, X. (2017). In-package inhibition of E. coli O157: H7 on bulk Romaine lettuce using cold plasma. *Food Microbiology, 65*, 1–6.

Mir, S. A., Shah, M. A., & Mir, M. M. (2016). Understanding the role of plasma technology in food industry. *Food and Bioprocess Technology, 9*(5), 734–750.

Misra, N. N. (2015). The contribution of non-thermal and advanced oxidation technologies towards dissipation of pesticide residues. *Trends in Food Science and Technology, 45*(2), 229–244.

Misra, N. N., Schlüter, O., & Cullen, P. J. (Eds.) (2016). *Cold Plasma in Food and Agriculture: Fundamentals and Applications*. Academic Press.

Misra, N. N., Tiwari, B. K., Raghavarao, K. S. M. S., & Cullen, P. J. (2011). Nonthermal plasma inactivation of food-borne pathogens. *Food Engineering Reviews, 3*(3–4), 159–170.

Mitra, A., Li, Y.-F., Klämpfl, T. G., Shimizu, T., Jeon, J., Morfill, G. E., & Zimmermann, J. L. (2013). Inactivation of surface-borne microorganisms and increased germination of seed specimen by cold atmospheric plasma. *Food and Bioprocess Technology, 7*(3), 645–653.

Moreau, M., Orange, N., & Feuilloley, M. G. J. (2008). Non-thermal plasma technologies: New tools for bio-decontamination. *Biotechnology Advances, 26*(6), 610–617.

Morent, R., De Geyter, N., Desmet, T., Dubruel, P., & Leys, C. (2011). Plasma surface modification of biodegradable polymers: A review. *Plasma Processes and Polymers, 8*(3), 171–190.

Muhammad, A. I., Liao, X., Cullen, P. J., Liu, D., Xiang, Q., Wang, J., ... & Ding, T. (2018). Effects of nonthermal plasma technology on functional food components. *Comprehensive Reviews in Food Science and Food Safety, 17*(5), 1379–1394.

Munekata, P. E., Domínguez, R., Pateiro, M., & Lorenzo, J. M. (2020). Influence of plasma treatment on the polyphenols of food products—A review. *Foods, 9*(7), 929.

Niemira, B. A. (2012). Cold plasma reduction of Salmonella and Escherichia coli O157: H7 on almonds using ambient pressure gases. *Journal of Food Science, 77*(3), M171–M175.

Nishime, T. M. C., Borges, A. C., Koga-Ito, C. Y., Machida, M., Hein, L. R. O., & Kostov, K. G. (2017). Non-thermal atmospheric pressure plasma jet applied to inactivation of different microorganisms. *Surface and Coatings Technology, 312*, 19–24.

Oh, Y. A., Roh, S. H., & Min, S. C. (2016). Cold plasma treatments for improvement of the applicability of defatted soybean meal-based edible film in food packaging. *Food Hydrocolloids, 58*, 150–159.

Pankaj, S. K., Bueno-Ferrer, C., Misra, N. N., Milosavljević, V., O'donnell, C. P., Bourke, P., ... & Cullen, P. J. (2014). Applications of cold plasma technology in food packaging. *Trends in Food Science and Technology, 35*(1), 5–17.

Pankaj, S. K., Wan, Z., & Keener, K. M. (2018). Effects of cold plasma on food quality: A review. *Foods, 7*(1), 4.

Pelletier, J. (1992). Sterilization by plasma processing. *Aggressologie, 33*, 457–477.

Percival, D. M. (1998). Antioxidants—NUT031. CLINICAL NUTRITION INSIGHTS. Copyright© 1996 *Advanced Nutrition Publications. Inc.*, Revised, 1–4.

Pleadin, J., Vulić, A., Perši, N., Škrivanko, M., Capek, B., & Cvetnić, Ž. (2015). Annual and regional variations of aflatoxin b1 levels seen in grains and feed coming from Croatian dairy farms over a 5-year period. *Food Control, 47*, 221–225.

Puligundla, P., Kim, J. W., & Mok, C. (2017). Effect of corona discharge plasma jet treatment on decontamination and sprouting of rapeseed (*Brassica napus L.*) seeds. *Food Control, 71*, 376–382.

Puligundla, P., Lee, T., & Mok, C. (2016). Inactivation effect of dielectric barrier discharge plasma against foodborne pathogens on the surfaces of different packaging materials. *Innovative Food Science and Emerging Technologies, 36*, 221–227.

Rahman, M. M., Sajib, S. A., Rahi, M. S., Tahura, S., Roy, N. C., Parvez, S., ... & Kabir, A. H. (2018). Mechanisms and signaling associated with LPDBD plasma mediated growth improvement in wheat. *Scientific Reports, 8*(1), 1–11.

Sarangapani, C., Devi, R. Y., Thirumdas, R., Trimukhe, A. M., Deshmukh, R. R., & Annapure, U. S. (2017). Physico-chemical properties of low-pressure plasma treated black gram. *LWT-Food Science and Technology, 79*, 102–110.

Sarangapani, C., Patange, A., Bourke, P., Keener, K., & Cullen, P. J. (2018). Recent advances in the application of cold plasma technology in foods. *Annual Review of Food Science and Technology, 9*, 609–629.

Sarangapani, C., Thirumdas, R., Devi, Y., Trimukhe, A., Deshmukh, R. R., & Annapure, U. S. (2016). Effect of low-pressure plasma on physico–chemical and functional properties of parboiled rice flour. *LWT-Food Science and Technology, 69*, 482–489.

Scholtz, V., Pazlarova, J., Souskova, H., Khun, J., & Julak, J. (2015). Nonthermal plasma—A tool for decontamination and disinfection. *Biotechnology Advances, 33*(6), 1108–1119.

Scholtz, V., Šerá, B., Khun, J., Šerý, M., & Julák, J. (2019). Effects of nonthermal plasma on wheat grains and products. *Journal of Food Quality*, 1–10.

Selcuk, M., Oksuz, L., & Basaran, P. (2008). Decontamination of grains and legumes infected with Aspergillus spp. and Penicillum spp. by cold plasma treatment. *Bioresource Technology, 99*(11), 5104–5109.

Sera, B., Gajdova, I., Cernak, M., et al. (2012). Seed germination and early growth. *Plasma Science, 238*, 1365.

Shahrzad Mohammadi, S., Dorranian, D., Tirgari, S., & Shojaee, M. (2015). The effect of non-thermal plasma to control of stored product pests and changes in some characters of wheat materials. *Journal of Biodiversity and Environmental Sciences, 7*, 150–156.

Shan, B., Cai, Y. Z., Sun, M., & Corke, H. (2005). Antioxidant capacity of 26 spice extracts and characterization of their phenolic constituents. *Journal of Agricultural and Food Chemistry, 53*(20), 7749–7759.

Shrestha, G., Freere, P., Basnet, S. M. S., Jewell, W. T., & Subedi, D. P. (2007, April). Development of a cold plasma generator for atmospheric pressure dielectric barrier discharge. In: *2007 IEEE Region 5 Technical Conference* (pp. 432–435). IEEE.

Singh, N., Singh, J., Kaur, L., Sodhi, N. S., & Gill, B. S. (2003). Morphological, thermal and rheological properties of starches from different botanical sources. *Food Chemistry, 81*(2), 219–231.

Sivachandiran, L., & Khacef, A. (2017). Enhanced seed germination and plant growth by atmospheric pressure cold air plasma: Combined effect of seed and water treatment. *RSC Advances, 7*(4), 1822–1832.

Suhem, K., Matan, N., Nisoa, M., & Matan, N. (2013). Inhibition of *Aspergillus flavus* on agar media and brown rice cereal bars using cold atmospheric plasma treatment. *International Journal of Food Microbiology, 161*(2), 107–111.

Tendero, C., Tixier, C., Tristant, P., Desmaison, J., & Leprince, P. (2006). Atmospheric pressure plasmas: A review. *Spectrochimica Acta, Part B: Atomic Spectroscopy, 61*(1), 2–30.

Thirumdas, R., Deshmukh, R. R., & Annapure, U. S. (2016). Effect of low temperature plasma on the functional properties of basmati rice flour. *Journal of Food Science and Technology, 53*(6), 2742–2751.

Thirumdas, R., Kadam, D., & Annapure, U. S. (2017). Cold plasma: An alternative technology for the starch modification. *Food Biophysics, 12*(1), 129–139.

Thirumdas, R., Saragapani, C., Ajinkya, M. T., Deshmukh, R. R., & Annapure, U. S. (2016). Influence of low pressure cold plasma on cooking and textural properties of brown rice. *Innovative Food Science and Emerging Technologies, 37*, 53–60.

Thirumdas, R., Sarangapani, C., & Annapure, U. S. (2015). Cold plasma: A novel non-thermal technology for food processing. *Food Biophysics, 10*(1), 1–11.

Thirumdas, R., Trimukhe, A., Deshmukh, R. R., & Annapure, U. S. (2017). Functional and rheological properties of cold plasma treated rice starch. *Carbohydrate Polymers, 157*, 1723–1731.

Turner, M. (2016). Chapter 2. Physics of cold plasma. In: N. N. Misra, O. Schl€uter, & P. J. Cullen (Eds.), *Cold Plasma Applications in Food Packaging* (pp. 293–307). Academic Press.

Vesel, A., & Mozetic, M. (2012). Surface modification and ageing of PMMA polymer by oxygen plasma treatment. *Vacuum, 86*(6), 634–637.

Wongsagonsup, R., Deeyai, P., Chaiwat, W., Horrungsiwat, S., Leejariensuk, K., Suphantharika, M., ... & Dangtip, S. (2014). Modification of tapioca starch by non-chemical route using jet atmospheric argon plasma. *Carbohydrate Polymers, 102*, 790–798.

Zaharanova, A., Hensselova, M., Hudecova, D., Kalinakova, B., Kovacik, D., Medvecka, V., & Cernak, M. (2014). Study of low-temperature plasma treatment of plant seeds. In: *14th International Symposium on High Pressure Low Temperature Plasma Chemistry- Book of Contributions* (pp. 563–567).

Zahoranová, A., Henselová, M., Hudecová, D., Kaliňáková, B., Kováčik, D., Medvecká, V., & Černák, M. (2016). Effect of cold atmospheric pressure plasma on the wheat seedlings vigor and on the inactivation of microorganisms on the seeds surface. *Plasma Chemistry and Plasma Processing, 36*(2), 397–414.

Zhou, R., Zhou, R., Zhang, X., Zhuang, J., Yang, S., Bazaka, K., & Ostrikov, K. K. (2016). Effects of atmospheric-pressure N_2, He, air, and O_2 microplasmas on mung bean seed germination and seedling growth. *Scientific Reports, 6*, 32603.

Ziuzina, D., Patil, S., Cullen, P. J., Keener, K. M., & Bourke, P. (2014). Atmospheric cold plasma inactivation of *Escherichia coli, Salmonella enterica serovar Typhimurium* and *Listeria monocytogenes* inoculated on fresh produce. *Food Microbiology, 42*, 109–116.

Chapter 5

Ultrasonic Applications in Bakery and Snack Food Processing Industries

Esra Dogu-Baykut, Celale Kirkin and Meral Kilic-Akyilmaz

CONTENTS

5.1 INTRODUCTION

Ultrasound is sound waves with a higher frequency than the upper limit of the human hearing range (> 16 kHz). It can also be described as pressure waves with a frequency of 20 kHz or more (Dolatowski, Stadnik, & Stasiak, 2007; Piyasena, Mohareb, & McKellar, 2003). Ultrasound technology can be divided into two categories according to frequency ranges: low and high energy. Low-energy (low-power, low-intensity) ultrasound applications have frequencies higher than 100 kHz at intensities lower than 1 Wcm^{-2}. High-energy (high-power, high-intensity) ultrasound, also known as power ultrasound, applications have frequencies between 20 and 100 kHz at intensities higher than 1 Wcm^{-2} (Knorr, Zenker, Heinz, & Lee, 2004). In the food industry, low-energy/high-frequency ultrasound, which is non-destructive to materials, is applied to provide information about physicochemical properties of foods, while high-energy/low-frequency ultrasound is used to alter the physical and chemical properties of foods (Feng & Yang, 2011; Majid, Nayik, & Nanda, 2015).

Food processes are constantly being updated over time with the development of technology. The food industry aims to develop processes that require shorter processing time and less energy and ensure safe and quality food production while preserving the natural properties of food. Over the past few decades, ultrasound has drawn considerable interest due to its advantages over traditional processes in the food industry. As ultrasound offers a cheap, fast, safe, and efficient technology and can be applied at low temperatures, there are many studies investigating its potential use in the food industry (Demirdöven & Baysal, 2009; Kentish & Feng, 2014). This technique can be used in developing new processes or improving existing processes to enhance the quality and safety of processed foods.

Using ultrasonic waves alone as well as combined with temperature and pressure gives effective results in various food processing operations. The number of studies on this technology is increasing day by day as it provides advantages in many processes in the food industry such as crystallization, emulsification, extraction, separation, viscosity alteration, defoaming, extrusion, inactivation of enzymes and microorganisms, fermentation, and heat transfer (Patist & Bates, 2008). In some processes such as cutting, emulsification/homogenization, sterilization/pasteurization, meat tenderization, and degassing, ultrasound is used instead of traditional methods, while in some processes it is used as an aid to traditional methods. Extraction, filtration, drying/dehydration, freezing, thawing, brining, and oxidation are the main food processes in which ultrasound is used to increase the efficiency of traditional processes (Tao & Sun, 2015).

Ultrasound-assisted food processes applied in cereal food production include the processes applied as a pretreatment to grains, extraction, and the processes used for the production or control of bakery foods and snacks. In bakery and snack food processing, the characterization of flours, doughs, and batters is key to having a good final product. Several techniques such as mechanical, chemical, or spectroscopic tests are used to measure the functionality of flours and how it is affected by other ingredients (Álava et al., 2007). Ultrasound can also be applied for analysis and processing in bakery and snack food processing. In this chapter, ultrasound systems, the benefits of ultrasound in the food industry, and the use of ultrasound and combined technologies in cereal, bakery, and snack food processing will be reviewed.

5.2 ULTRASOUND SYSTEMS

5.2.1 Generation of Ultrasound

Ultrasound covers the inaudible sound waves at the frequency range of 20 kHz–10 MHz that can propagate in solid, liquid, and gas media. While ultrasound propagates in a medium, it can generate compression or high pressure and rarefaction (decompression or low pressure) cycles on particles. Particles get closer to and further away from each other and oscillate, causing an increase and decrease in density and pressure in the medium (Kentish & Ashokkumar, 2011). The magnitude of the pressure variations is called amplitude and is directly proportional to the amount of energy applied to the medium (Kentish & Ashokkumar, 2011). The amount of energy emitted by ultrasound in a medium depends on the frequency, power, duration, geometry, and surface area of the ultrasound emitter. The size of the application chamber also influences the overall effect of ultrasound on a medium. These factors determine power intensity, which is the applied power per unit area or volume for a specific ultrasound system. The impact of ultrasound on a medium can be increased when combined with heat (thermosonication), pressure (manosonication), and both heat and pressure (manothermosonication).

Ultrasound applications are categorized into two groups according to intensity or power. Applications at a frequency above 100 kHz and a level of intensity below 1 W/cm^2 are classified as low-intensity ultrasound (LIUS). High-intensity ultrasound (HIUS) is applied at a frequency in the range of 20–100 kHz and intensity between 10 and 1000 W/cm^2. LIUS does not induce changes in the physical and chemical properties of materials, therefore, it has been used for fast, cheap, and non-destructive analysis of food materials. Analysis of composition and quality, monitoring of physical changes in a process, and determination of size, shape, and dimension are possible by the use of LIUS. HIUS has the ability to change the physical and chemical properties of materials and is applied to inactivate microorganisms and enzymes, improve food processing operations, and modify the structure or functional properties of food components.

Ultrasound waves have cavitational and non-cavitational effects in a medium. Noncavitational effects are mechanical vibrations and acoustic streaming (Kentish & Ashokkumar, 2011). Cavitational effects are created by the formation, growth, and collapse of bubbles that cause shear forces, shock waves, and microjets in a medium. As a consequence of these mechanisms, ultrasonic energy induces turbulence, agitation, heat, diffusion, interface instabilities, friction, mechanical rupture, and chemical reactions that are beneficial in various process operations.

Cavitation is the major mechanism of action of HIUS. Cavitation bubbles are formed in a liquid medium due to tensile forces created by propagating ultrasound waves. These and gas microbubbles present in a liquid medium are expanded and compressed by ultrasound waves and they grow. Cavitation bubbles contain mostly water vapor in aqueous systems; nitrogen and oxygen also permeate into bubbles during oscillation (Kentish & Ashokkumar, 2011). The surface area of gas bubbles increases and decreases, hence they oscillate until a critical threshold value at which point the bubbles collapse. Oscillating bubbles will cause fluctuations in velocity and pressure in the surrounding liquid that result in microstreaming and turbulence within the liquid on a microscale (Kentish & Ashokkumar, 2011). The collapse of the bubbles causes a local rise in temperature (5000 K) and pressure (10–100 MPa), shockwaves, turbulence, and shear forces in the medium that can break down chemical bonds and damage cell walls (Ashokkumar et al., 2010; Condón, Raso, & Pagán, 2004; Feng & Yang, 2011). If the bubble is close to a solid surface, an asymmetrical explosion of the bubble occurs, resulting in a microjet of fluid. The flow of the bubbles through the solid surface can erode the surface and dislocate particles from the surface (Kentish & Ashokkumar, 2011). Acoustic standing waves can result from the reflection of the sound wave from a solid surface or an air–liquid interface back into the solution at the same time that a wave is generated, forming filamentary structures on the solid surface. These physical effects of ultrasound are strongest close to system boundaries and fluid–solid and fluid–fluid interfaces, which renders this technology efficient in improving heat and mass transfer in various unit operations (Kentish & Ashokkumar, 2011). Ultrasonic energy is ultimately converted to heat that causes an increase in the temperature of the medium if there is no cooling.

Ultrasound has different physical and chemical effects on a medium depending on its frequency. Transient cavitation occurs at low frequencies around 20–100 kHz and high acoustic intensity. Cavitation bubbles can grow in a few cycles, collapse instantly to smaller bubbles, and these smaller bubbles also collapse rapidly. Stable cavitation bubbles are formed at frequencies above 100 kHz and at a low acoustic intensity where bubbles aggregate in many acoustic cycles. The collapse of bubbles is less violent in this case and growth and collapse events occur continuously (Ashokkumar et al., 2010; Torley & Bhandari, 2007). The chemical effects of ultrasound become dominant at intermediate frequencies (200–500 kHz) as the number of active bubbles is higher than that in transient cavitation (Kentish & Ashokkumar, 2011). As the frequency increases, cavitation will be less intense, and cavitation does not occur at or above a 2.5 MHz frequency (Condón et al., 2004). Noncavitational effects become dominant at frequencies above 1 MHz (Kentish & Ashokkumar, 2011). The frequency of the ultrasound system is chosen according to the physical or chemical effects desired for a particular operation. For instance, the physical effects are important in cleaning and extraction while the chemical effects are preferred in polymerization reactions.

The cavitation threshold also depends on the temperature and pressure of the medium. Larger bubbles will be formed at high temperatures and pressures in the cavitation zone (Condón et al., 2004). The increase in medium pressure also increases pressure inside the bubbles and reduces the number of cavitation bubbles; as a result, more intense and rapid collapse occurs. This phenomenon allows the enhancement of process efficiency without an increase in amplitude. The viscosity of the liquid medium, vapor pressure, and surface tension are dependent mostly on temperature. An increase in the temperature of the medium increases water vapor pressure inside the cavitating bubbles. Water vapor has a stabilizing effect on the bubbles that reduces the intensity of their collapse at temperatures above ambient (Kentish & Ashokkumar, 2011). Cavitation bubbles cannot form efficiently in a medium with high viscosity. Temperature increase also causes a reduction in viscosity that allows the formation of more bubbles (Earnshaw, 1998; Patist & Bates, 2008).

The chemical effects of ultrasound have been used in the modification and synthesis of molecules; however, they could be detrimental to quality in food processing. Local heating and pressure increases create hot points inside the bubbles and nearby fluid that are important in chemical effects on the medium (Knorr et al., 2004). On the other hand, heat is transient and local, therefore a temperature change of 109°C amounts to a total increase in temperature by 5°C (Knorr et al., 2004; Salazar, Chávez, Turó, & García-Hernández, 2009). Heat, pressure, and turbulence increase the rate of mass transfer, new reactions can take place, particles or cells are disrupted, and new particles are produced (Patist & Bates, 2008; Salazar et al., 2009). Water molecules are hydrolyzed to hydrogen and hydroxyl radicals and these react with other components present in a liquid medium. Hydrogen peroxide is also produced in the presence of oxygen. Pyrolysis reactions and oxidation by free radicals need to be considered in regard to the application of ultrasound in food processing. Rubbery off-odor was noted in sonicated milk by Riener et al. (2009) that was explained by a pyrolysis reaction in the bubble and the decomposition of unsaturated fatty acid hydroperoxides. Low-frequency ultrasound and the use of an antioxidant such as ascorbic acid could be a solution in some applications (Ashokkumar et al., 2008).

5.2.2 Ultrasound Systems and Parameters

An ultrasound system consists of three main parts including a power generator, a transducer, and an emitter. Generators convert electricity into a high-frequency alternating current and transmit it to the transducer. Mechanical vibrations are produced by the transducer from the alternating current and the emitter amplifies and transfers these vibrations as sound waves to the medium (Awad, Moharram, Shaltout, Asker, & Youssef, 2012; Mason, 1998). The electrical generator and transducer are placed in the same unit and emitters of different shapes and sizes can be attached to this unit.

The transducer is the principal component, supplying the required ultrasound intensity to the system. Mechanical and electromechanical transducers are used for producing ultrasound (de Castro & Capote, 2007; Mason, 1998). Mechanical types produce ultrasound by the flow of liquid or gas, like whistles. Electromechanical transducers, including piezoelectric and magnetostrictive types, are commonly used in ultrasound processors. Magnetostrictive transducers are based on an alteration in the length of magnetostrictive materials, such as nickel or iron, in an alternating magnetic field. Magnetostrictive types work only with frequencies lower than 100 kHz at 60% efficiency (Mason, 1998; Muthukumarappan, Tiwari, O'Donnell, & Cullen, 2009; Torley & Bhandari, 2007). Piezoelectric types are made of piezoelectric materials that expand and contract in an oscillating electric field, such as barium titanate, quartz, or ceramic, and that convert electrical energy to acoustic energy (Bhargava, Mor, Kumar, & Sharanagat, 2021; de Castro & Capote, 2007; Feng & Yang, 2011). Piezoelectric transducers can work in the whole frequency range at more than 95% efficiency, therefore, they are commonly used in commercial systems.

Ultrasounds can be emitted by a bath or a probe system depending on the target operation. Ultrasonic baths are simple, cheap, and versatile systems that contain one or more transducers placed under a tank that directly or indirectly emit ultrasound to a medium. A processing tank can be placed in an ultrasonic bath with coupling fluid, or feed fluid can be placed directly into the ultrasonic bath (Gogate & Pandit, 2015). Direct application is more efficient compared to indirect application, but corrosion of the bath walls can occur due to contact with the feed fluid. Temperature control is also difficult in an indirect application as the temperature of the coupling fluid increases with the application of ultrasound (Gogate & Pandit, 2015). Maximum intensity is limited to $1-5$ W/cm^2 in these systems to prevent cavitational damage on the walls of the tank (Mason, 1998).

Probe or horn systems use sonotrodes to emit the ultrasound directly into a medium. Sonotrodes, with attachment tips in some designs, are made of metal that amplifies and emits vibrations transmitted from the transducer. The intensity of ultrasound can be controlled by the shape and dimensions of the sonotrode and the amplitude supplied to the transducer. The vibrational amplitude provided by the vibrating tip of the sonotrode is measured as the maximum displacement in micrometers (μm) as it emits the acoustic waves into the medium. Each sonotrode has a maximum amplitude specified by the manufacturer. Probes should be made of mechanically resistant materials; titanium and aluminum alloys are commonly used for this purpose. Probe systems have some disadvantages over bath systems. Application tips can be worn out by cavitation, especially at high power levels with extended use, which reduces the efficiency of the horn. In addition, metal contamination can occur from the tip (Mason, 1998; Torley & Bhandari, 2007). Pulsed application in some systems allows the intermittent application of ultrasound that increases the durability of the system (Gogate & Pandit, 2015). Temperature increases are another disadvantage of probe systems that occurs proportionally to the power intensity applied. A temperature increase could be up to 25°C in probe systems and around 1°C for bath systems (Salazar et al., 2009). These limitations were resolved by the development of continuous units including flow cells and tube reactors. In these units, the probe is attached to a tube or placed in a chamber where feed liquid is pumped continuously. The flow rate is adjusted for controlling the duration of the treatment and the temperature is controlled by a cooling jacket surrounding the application zone (Torley & Bhandari, 2007). Flow cells or bath systems can be connected in series to treat large volumes of feed.

Ultrasound systems have varying frequencies in an application field due to their mechanism of formation. As a result, power intensity is variable according to location and it may be zero at some points, which can cause process variability (Feng & Yang, 2011). Probe systems are very efficient in locations close to the probe surface; however, the effect decreases exponentially and becomes zero at 2–5 cm away from the horn surface (Gogate & Pandit, 2015). The solution to this problem is to use multiple transducers working at the same or a different frequency in a proper geometrical arrangement (Feng & Yang, 2011; Gogate & Pandit, 2015). Transducers can also be placed against a reflector in a parallel plate arrangement to enhance efficiency (Gogate & Pandit, 2015). Mechanical agitation is also required for homogeneous application of ultrasound in large volumes of fluid feed. Stepped-plate transducers with extended surfaces were also developed that allow uniform application of ultrasound on a large volume of food (Gallego-Juárez, 2017).

5.2.3 Benefits and Use of Ultrasound in the Food Industry

LIUS application is based on interactions of ultrasound waves with materials. Attenuation, absorption, or scattering of acoustic waves are measured and quantified in determining composition, structure, and dimensions (Bhargava et al., 2021). Ultrasonic sensors at varying frequencies have been developed for measuring distance, level, flow, temperature, pressure, and concentration of components in gas and/or liquid mixtures. These sensors can be used for nondestructive evaluation of materials, imaging, and process control in the food industry (Gallego-Juárez, 2017).

HIUS application can improve the efficiency of food processing operations by increasing heat and mass transfer. Many potential areas of usage for ultrasound exist in the food industry, including homogenization, emulsification, defoaming/degassing, crystallization, freezing/thawing, drying, fermentation, microbial/enzymatic inactivation, extraction, cutting/slicing, and filtration (Table 5.1). In addition, high-power ultrasonic waves can be applied for polymerization, depolymerization, oxidation, and synthesis of nanoparticles in food and other industries (Boufi et al., 2018; Chemat, Zill-e-Huma, & Khan, 2011; Knorr et al., 2004; Mason, 1998; Patist & Bates, 2008; Salazar et al., 2009; Torley & Bhandari, 2007).

HIUS has been used for the inactivation of microbial cells and enzymes, which is important for the safety and quality of food products. Microbial inactivation by ultrasound is achieved by shear disruption, heating, and free-radical production (Bermudez-Aguirre, Mobbs, & Barbosa-Canovas, 2011; Feng & Yang, 2011; Piyasena et al., 2003). Cell membranes are thinned and disrupted by the cavitational effects of ultrasound depending on the resistance of the microorganism (Bermudez-Aguirre et al., 2011; Corbo et al., 2009; Knorr et al., 2004). Spores, Gram(+), aerobic, and cocci bacteria have been found to be more stable against ultrasound (Chandan, 2006; Condón et al., 2004; Feng & Yang, 2011).

Enzymes can also be inactivated by depolymerization caused by shear forces, denaturation, and the action of free radicals (Feng & Yang, 2011; Mason, Paniwnyk, & Chemat, 2003; O'Donnell, Tiwari, Bourke, & Cullen, 2010). Ultrasound has been used in the inactivation of alkaline phosphatase in milk and of pectin methylesterase, peroxidase, and lipoxygenase in fruit and vegetables (Chandrapala, Oliver, Kentish, & Ashokkumar, 2012; Feng & Yang, 2011; O'Donnell et al., 2010). The food matrix is heterogeneous with many components that can reduce the effect of ultrasound required for complete inactivation of microorganisms and enzymes. The combination of

TABLE 5.1 POTENTIAL APPLICATIONS AND BENEFITS OF ULTRASOUND IN THE FOOD INDUSTRY

Applications	Advantages	Products
Cutting	Reduction in total cutting force Reduction in product sticking to the cutting tools Reduction in cracking, crumbling, and product losses Enhancement of cutting quality and consistency Reduction in cleaning time	Bakery products (breads, cakes, pizzas, sandwiches) Granola bars and snack bars Hard and soft cheeses Semi-frozen meat and fish Fresh or frozen vegetables
Emulsification/ homogenization	Reduction in surface tension Faster dissolution, erosion, and breakage of bubbles Increase in emulsion stability and shelf life High energy efficiency Easy to operate and clean	Milk, fruit juice, oil Emulsions (ketchup, mayonnaise)
Sterilization/ pasteurization/enzyme inactivation	Minimization of nutritional and flavor losses Increase in homogeneity of application reduction in energy requirement	Milk and dairy products Fruit and vegetable juices
Meat tenderization	Shortening of the aging period Reduction in shear force Increase in myofibrillar protein Extraction and water holding capacity Improvement of juiciness and tenderness	Meat and meat products (chicken, beef, pork, surimi)
Degassing/defoaming	Reduction in the use of anti-foam chemical additives Reduction in defoaming heating/cooling, and its related energy costs Improvement of microbiological safety	Fermented products (beer) Carbonated drinks Milk

Applications	Advantages	Products
Extraction	Increase in extraction yield and rate Reduction in extraction temperatures Protection of the degradation of compounds affected by temperature Reduction in solvent consumption Reduction in energy consumption and costs	Fruits and vegetables Meat and meat products (beef, chicken, fish) Herbs and spices
Filtration	Reduction in flow resistance Reduction in membrane-fouling rates Increase in permeate flux Increase in filter life	Fruit and vegetable juices Water
Drying/dehydration	Reduction in drying time Enhancement of organoleptic quality Prevention of nutritional loss Reduction in energy consumption and costs	Fruits and vegetables Herbs and spices Meat and meat products (beef, chicken, fish)
Freezing/thawing	Reduction in ice crystal size Reduction in thawing loss Enhancement of heat and mass transfer rate Enhancement of organoleptic quality	Fruits and vegetables Meat (beef, pork, cod) Milk and dairy products
Brining	Reduction in NaCl consumption Enhancement of mass transfer rate Enhancement of organoleptic quality and product stability	Meat (beef, chicken, fish) Vegetables Cheese
Oxidation	Enhancement of the aging rate Reduction in cost and space for the aging	Alcoholic beverages (wine)

ultrasound with heat and pressure has been applied to intensify the effect of ultrasound on the inactivation of microorganisms and enzymes (Chemat et al., 2011; O'Donnell et al., 2010).

HIUS has been used for increasing the rate and efficiency of crystallization and freezing operations. Fast cooling and the formation of many crystal seeds are possible by way of an increase in heat and mass transfer by ultrasound. The application of HIUS can shorten the time of operation and improve the quality and stability of food products by enhancing the rate of crystallization. Meat, fruit/vegetables, fish, and cheese have been effectively frozen by ultrasound (Chemat et al., 2011). Likewise, ultrasound-assisted thawing has been found to be faster than conventional thawing and has prevented slow-thawing-induced microbial growth and quality losses by oxidation (Bhargava et al., 2021; Li et al., 2020).

In membrane filtrations, the application of ultrasound prevents fouling and concentration polarization on the membrane surface by applying vibrations to solid particles and opening channels for the flow of feed. Ultrasound is also effective in the disruption of cellular membranes, allowing penetration of solvent and extraction of intracellular materials such as sugar, oil, protein, and bioactive compounds (Chemat et al., 2011; Salazar et al., 2009). Ultrasound-assisted extraction can be conducted at low temperatures that can improve the recovery and activity of extracted components. In addition, less energy, solvent, and time are used by ultrasound-assisted extraction, which reduces the cost of operation (Zhang, Wang, Zhang, Zhou, & Guo, 2014). However, the efficiency of an extraction process depends on the power intensity, frequency, solvent, and matrix, which need to be investigated for each application. Brining can be fastened by ultrasound, reducing the loss of quality due to bloating, structural damage, and the enzymatic softening of brined foods (Bhargava et al., 2021). Fermentation can also be accelerated by improved cellular diffusion of materials in yogurt, wine, and beer processing.

Ultrasound can enhance the efficiency of mixing and size reduction operations in food processing. The collapse of bubbles can break down the interface between two phases while high pressure and temperature can improve mixing (Feng & Yang, 2011). Fine and stable emulsions can be formed by HIUS treatment. Ultrasound is also effective as a homogenization method and creates a narrow particle size distribution. In cutting/slicing operations, ultrasound allows precise portioning and reduces the amount of wasted food (Bhargava et al., 2021).

Ultrasound can be used in drying to accelerate the movement of water, creating air turbulence and forming new microchannels by cavitation in fruit and vegetable tissues (Bhargava et al., 2021; Chemat et al., 2011; Salazar et al., 2009). Ultrasound can be applied before or during drying operations and in osmotic pretreatments. In either case, it can reduce drying time and improve the texture and color of dried foods (Bhargava et al., 2021). High-intensity airborne ultrasound has improved the rate of forced-air drying (Gallego-Juarez & Riera, 2011). Stepped-plate transducers have also been used in direct contact with solid food samples, which allowed drying at room temperature (Gallego-Juarez & Riera, 2011; Musielak, Mierzwa, & Kroehnke, 2016).

Ultrasound is mainly a physical operation that has an outstanding homogenization effect on liquid–solid dispersions. The physical effect of ultrasound is utilized in extractions that are principally mass transfer operations. Shear forces created by ultrasonic waves cause size reduction and cell rupture. In addition, propagating ultrasound waves in a medium induces hydration and swelling of the matrix, resulting in the enlargement of pores and contact surface between solvent and solutes. These changes in the matrix enhance the penetration of solvent into the cells and the diffusion of molecules into the solvent, increasing the rate of mass transfer (Chemat et al., 2017; Bhargava et al., 2021).

Ultrasound can also modify the structure of macromolecules present in foods such as polysaccharides and proteins and change the structural-functional properties of these components.

HIUS can induce changes in the molecular structure by cavitation-induced shear forces, free radicals, and heat. Protein structure and conformation can be changed by the breakage of peptide bonds, unfolding, and denaturation depending on the intensity of ultrasound, resulting in changes in solubility and functionality such as gelation and emulsification (Arzeni et al., 2012). Depolymerization of polysaccharides can occur by scission of the backbone, or functional side groups can be modified by HIUS, which leads to changes in viscosity, gelation, and water-holding capacity in aqueous systems (Weiss, Kristbergsson, & Kjartansson, 2011). Starch granules of rice were damaged by ultrasound application, but their fine structure was not affected (Yang et al., 2019). This caused changes in the pasting and retrogradation behavior of rice starch. The effects of HIUS on macromolecules depend on power intensity, duration, and temperature as well as the structure and concentration of the macromolecule and the conditions of the medium (Yu et al., 2013). Ultrasound can also improve yield and efficiency in the modification of starch by enzymes or chemical reactions (Yujie Zhang et al., 2020). Crystallization behavior, hydrogenation, and interesterification of lipids can be modified by HIUS (Weiss et al., 2011). On the other hand, free radicals created by ultrasound can be detrimental by the induction of oxidation in lipid-containing foods, especially by intensive treatment.

5.3 ULTRASOUND-ASSISTED PROCESSES APPLIED TO CEREALS AND GRAINS

Grains can be treated with ultrasound or ultrasound-assisted methods in order to improve their functional and quality properties or to replace the traditional and conventional processing methods with a more efficient alternative. The ultrasound-assisted processing of grains includes germination, hydration, fortification, hydrolysis, and nixtamalization (Table 5.2). Besides the processing or pretreatment of grains, ultrasound technology can also be utilized in the extraction of proteins, starch, dietary fiber, oil, and antioxidants from grains as it can improve efficiency. Applications of ultrasound in cereal food processing are summarized in Figure 5.1.

5.3.1 Processing of Grains Using Ultrasound

5.3.1.1 Germination

Germination enhances the nutritional, functional, sensory, and physicochemical properties of the grain (Charoenthaikij et al., 2009; Singh, Rehal, Kaur, & Jyot, 2015). Sprouted grains have already been used to produce malt and food products, such as sprouted rice used in Asia (Ding & Feng, 2018). However, the germination process needs to be controlled to improve the quality properties of the grain and to prevent undesired changes and damages due to uncontrolled and excessive sprouting.

Several methods, such as temperature control during steeping, oxygen control, and non-thermal methods including ultrasonication, can be utilized in germination control (Ding & Feng, 2018; Rifna, Ratish Ramanan, & Mahendran, 2019; Jianfei Wang, Ma, & Wang, 2019). Ultrasound-controlled germination can improve the contents of gamma-aminobutyric acid (GABA) and antioxidants, reducing the sugar and bioaccessibility of minerals (Ding, Hou, Dong, et al., 2018; Ding, Hou, Nemzer, et al., 2018; Ding, Johnson, Chu, & Feng, 2019; Xia et al., 2020). Ultrasound treatment can also accelerate the germination process, improve water uptake, and decrease cooking time (Cui, Pan, Yue, Atungulu, & Berrios, 2010; Xia et al., 2020; Yaldagard, Mortazavi, & Tabatabaie, 2008a). Starch hydrolysis is improved and retrogradation is decreased by ultrasound-controlled germination (Ding, Hou, Dong, et al., 2018; Ding, Hou, Nemzer, et al., 2018).

TABLE 5.2   STUDIES EVALUATING THE USE OF ULTRASOUND IN THE PROCESSING OF GRAINS

Process	Sample	Treatment Conditions	References
Germination	Barley	20 kHz 20, 60, 100% 5, 10, 15 min	(Yaldagard, Mortazavi, & Tabatabaie, 2007; Yaldagard et al., 2008a)
		20 kHz 20, 60, 100% 5, 10, 15 min 30, 50, 70°C	(Yaldagard, Mortazavi, & Tabatabaie, 2008b)
	Brown rice	28 kHz 5, 10, 15, 30 min	(Xia et al., 2020)
		16 kHz 30 min 25, 40, 55°C	(Cui et al., 2010)
	Red rice	25 kHz 5 min 23–24°C	(Ding et al., 2018)
	Brown and red rice	25 kHz 5 min 23–24°C	(Ding, Hou, Dong, et al., 2018)
	Oats	25 kHz 5 min 23–24°C	(Ding et al., 2019)
	Sorghum	20 kHz 40%, 60% 5, 10 min	(Hassan et al., 2020, 2017)
	Wheat	25 kHz 5, 30 min	(Ding, Hou, Nemzer, et al., 2018)
Hydration	Barley	20 kHz 10, 15, 20, 25°C 750, 1,500 W	(de Carvalho et al., 2018)
		25 kHz 20, 30, 40, 50, 60°C 300 W	(Shafaei, Nourmohamadi-Moghadami, & Kamgar, 2019)
		37 kHz 35°C 154 W	(Borsato, Jorge, Mathias, & Jorge, 2019)
	Corn	25 kHz 41 W/L	(Miano et al., 2017)
	Sorghum	40 kHz Room temperature, 53°C	(Patero & Augusto, 2015)
	Wheat	25 kHz, 360 W 40 kHz, 480 W 30, 40, 50, 60, 70°C	(Shafaei, Nourmohamadi-Moghadami, & Kamgar, 2018)

(Continued)

TABLE 5.2 (CONTINUED)   STUDIES EVALUATING THE USE OF ULTRASOUND IN THE PROCESSING OF GRAINS

Process	Sample	Treatment Conditions	References
		20 kHz 300–1,500 W	(Guimaraes, Polachini, Augusto, & Telis-Romero, 2020)
Fortification	Rice	40 kHz 130 W 5 min	(Bonto et al., 2020)
		53 kHz 180 W 15°C 1–5 min	(Tiozon Jr. et al., 2021)
Hydrolysis	Corn gluten meal	28 kHz 45–65 W/L 15–25 min	(Liang et al., 2017)
		20, 28, 40 kHz 90 W/L 40 min 30°C	(L. Luo et al., 2017)
		20/28/40 kHz, 20/35/50 kHz, 20/40/60 kHz (triple-frequency mode) 30, 60, 90, 120, 150 W/L 60, 90, 120, 150, 180 min	(Qu et al., 2018)
		40 kHz 40 min 20°C	(Zhou et al., 2017)
	Wheat gluten	20/35 kHz (dual-frequency mode) 150 W/L 10 min 30°C	(Yanyan Zhang, Li, Li, Ma, & Zhang, 2018)
		25, 47, 69 kHz 0.707 W/cm^2 30 min 50°C	(Zhu et al., 2011)
Nixtamalization	Corn	20 kHz 1.85 W/g Cooking for 1 h at boiling temperature/1 h steeping at 25°C	(Janve et al., 2013)
		1.25 W/cm 20 kHz 1.25 W/cm^3 Cooking for 1 h at 84°C followed by steeping for 1.5 h	(Janve et al., 2015)

(Continued)

TABLE 5.2 (CONTINUED)   STUDIES EVALUATING THE USE OF ULTRASOUND IN THE PROCESSING OF GRAINS

Process	Sample	Treatment Conditions	References
		25 kHz 843 W/m² Cooking for 60 min at 85 and 95°C/steeping for 8 h	(Moreno-Castro et al., 2015)
		42 kHz Cooking at 80°C for 30, 45, 60 min followed by steeping for 8, 16 h	(Robles-Ozuna et al., 2016)

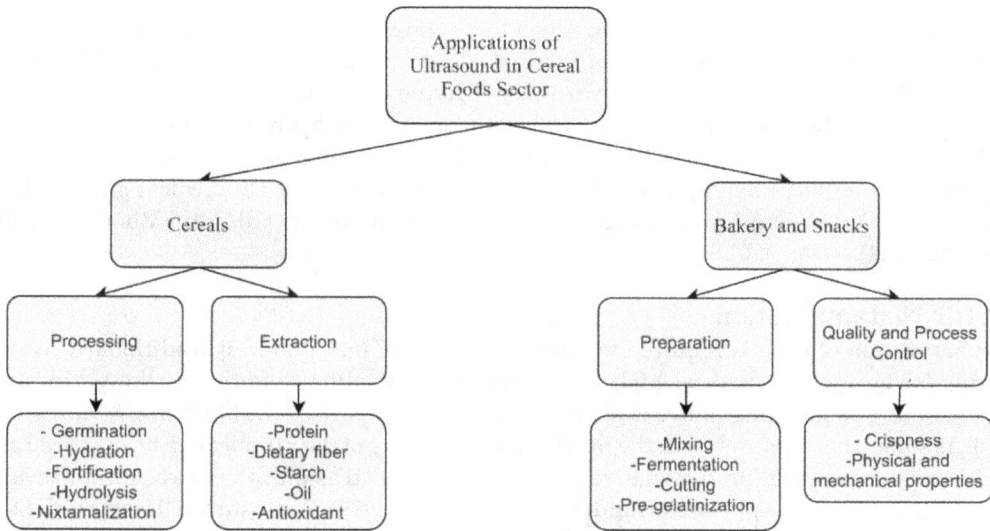

Figure 5.1   Applications of ultrasound in the cereal processing.

Changes in the fatty acid profile due to ultrasound-aided germination have also been reported by Hassan, Imran, Ahmad, and Khan (2017).

5.3.1.2 Hydration

Hydration is one of the main and crucial steps of the malting of barley as it determines the quality of the malt (de Carvalho, Polachini, Darros-Barbosa, Bon, & Telis-Romero, 2018). Hydration is applied in the processing of other grain products as well, such as corn and sorghum. It is required for wet milling of the grain. During the hydration process, the grain is soaked in water to facilitate water uptake. Thus, water absorption during hydration needs to be controlled to gain a standard final product quality (Gruwel, Chatson, Yin, & Abrams, 2001). Conventionally, the hydration process can take quite a long time, up to 36 hours for corn, for instance (Singh & Eckhoff, 1996). Ultrasound treatment has been reported to be able to successfully replace conventional hydration for several grain products. It significantly reduces the time required for hydration without negative effects on product properties (Miano et al., 2015; Miano, Ibarz, & Augusto, 2017).

5.3.1.3 Fortification

Ultrasound treatment can be used in the fortification of cereals with nutrients. Alberto Claudio Miano and Augusto (2018) proposed an ultrasound-assisted hydration process for the fortification of carioca beans with iron. Similarly, Bonto et al. (2020) successfully applied ultrasound in the iron fortification of rice by sonicating the samples in an ultrasonic bath and gained a 28-fold increase in the initial iron content. Moreover, the folate content of rice was increased by 4,054-fold by ultrasonication in a study by Tiozon Jr. et al. (2021). The authors reported that there was no negative effect on textural properties; however, thermal and pasting properties and color values (lightness and yellowness) were changed.

5.3.1.4 Hydrolysis

Gluten is a valuable byproduct obtained by the wet milling of gluten-containing grains, such as corn and wheat. It can be added in bread production to improve the protein content, yet its utilization in the production of other foods is limited due to insufficient functional properties (Elmalimadi et al., 2017). For instance, it has low solubility in water and its emulsifying and foaming abilities are also weak (Elmalimadi et al., 2017; Yanyan Zhang et al., 2015; Zhu, Su, Guo, Peng, & Zhou, 2011). Controlled hydrolysis of gluten can improve its functional properties. Ultrasound is applied as a pretreatment before the enzymatic hydrolysis of gluten. It was also reported by several researchers that ultrasound pretreatment increased angiotensin-converting enzyme (ACE) activity and improved the antioxidant activity of corn and wheat glutens (Liang et al., 2017; L. Luo, Zhang, Wang, Ma, & Dong, 2017; Yanyan Zhang et al., 2015; Zhou et al., 2013; Zhu et al., 2011).

5.3.1.5 Nixtamalization

Nixtamalization of corn is required for the production of masa flour. It traditionally involves cooking for around one hour at high temperature and alkaline conditions followed by steeping, washing, and grinding (Janve et al., 2013; Ramírez-Araujo, Gaytán-Martínez, & Reyes-Vega, 2019). As some undesired losses in the nutritional properties of the product can be observed after traditional nixtamalization, alternative methods, such as extrusion, microwave, ohmic heating, and ultrasound have been tested (Ramírez-Araujo et al., 2019). For instance, Janve et al. (2013) suggested HIUS decreased the time required for steeping during the nixtamalization process from 20 hours to one hour. Furthermore, the nixtamalization time was reduced to 30 min by the aid of ultrasound in the study of Moreno-Castro et al. (2015). Janve, Yang, and Sims (2015) also reported that some changes in the instrumental color values and textural properties of masa, baked chips, and tortilla chips were observed when power ultrasound-assisted nixtamalization of corn was applied, but none of these changes influenced the acceptability for consumers.

5.3.2 Extraction

5.3.2.1 Extraction of Proteins

Ultrasound technology was utilized by Drochioiu et al. (2016) in the improvement of extraction of proteins from maize seeds. Bedin et al. (2020) compared ultrasound-assisted extraction, microwave-assisted extraction, and alkaline extraction for isolating rice bran proteins. They reported that ultrasound-assisted extraction and microwave-assisted extraction were more efficient compared to alkaline extraction. The optimum parameters for ultrasound for rice bran protein extraction were suggested as the ratio of defatted rice bran to water 0.5:10 at a frequency of 20 kHz with a pulsed application of 30 s for 10 min.

Ultrasonication was found to improve the extraction yield and antioxidant properties of rice bran protein isolates (Iscimen & Hayta, 2018). Ly et al. (2018) also reported that both the protein extraction yield from defatted rice bran and its gelation capacity were increased and the extraction time was decreased by ultrasound-assisted extraction. However, the foaming capacity and stability of protein concentrate were lowered by ultrasound.

Li et al. (2017) utilized counter-current ultrasound-assisted alkali extraction for isolating rice dreg protein from rice dreg flour. They used an ultrasonic reactor operating at 20 kHz with power up to 448 W at different temperatures. The authors reported that the counter-current ultrasound-assisted alkali extraction significantly decreased the extraction time and enhanced the extraction yield and efficiency. Zhao, Bean, and Wang (2008) used sonication to improve protein extraction from mashed and unmashed sorghum meal. They observed that sonication resulted in a higher amount of protein extracted from mashed sorghum compared to nonsonicated control, but the extraction yield from unmashed sorghum was lower when sonication was applied. The authors noted that the amount of polymeric proteins was higher in sonicated samples and a correlation existed between the amount of extractable proteins and fermentation properties (ethanol yield and conversion efficiency).

In addition to studies showing the positive effects of ultrasound technology on protein extraction from grains, its combination with other methods is also promising. For instance, multiple-mode ultrasound treatment in combination with α-amylase degradation improved the extraction rate, purity, and functional properties of proteins extracted from rice bran (Yang et al., 2018). In another study, Zhu, Sun, and Zhou (2009) used ultrasound-assisted extraction in combination with reverse micelles to improve the extraction yield of defatted wheat germ proteins. They observed that the treatment decreased the extraction time and increased the extraction efficiency.

5.3.2.2 Extraction of Dietary Fiber

Bran is one of the main byproducts of grain processing that is rich in dietary fiber content. The extraction of dietary fiber from wheat bran generally requires alkaline reagents that can alter functional properties. Ultrasound-assisted extraction techniques can offer a good alternative to conventional methods (Jing Wang, Sun, Liu, & Zhang, 2014). For instance, the extraction time and alkali usage were significantly decreased during ultrasound-assisted extraction of hemicelluloses from wheat bran (Hromádková, Košt'álová, & Ebringerová, 2008). Moreover, the combination of ultrasound technology with enzymatic extraction used for the extraction of arabinoxylan from wheat bran increased extraction efficiency and yield, while the extraction time was reduced (Jing Wang et al., 2014). In another study, ultrasound alone was not sufficient for the complete extraction of arabinoxylans from brewer's spent grain but its combination with alkaline extraction improved purity (less starch) and reduced time, amount of alkali reagent, and energy (Reis, Coelho, Coimbra, & Abu-Ghannam, 2015). However, the authors noted that ultrasound with alkali caused some degradation of polysaccharides which resulted in losses of arabinoxylan. Nevertheless, ultrasound allowed extraction at room temperature for 25 min compared to 7 h at 45°C (Reis et al., 2015).

5.3.2.3 Extraction of Starch

The isolation of starch from rice is conventionally achieved using alkali conditions and consequently produces high amounts of alkali and salt wastes (L. F. Wang & Wang, 2004b, 2004a). The combination of a neutral protease digestion method with HIUS was suggested that resulted in high extraction yield and short extraction time without producing salt wastes and compromising the physicochemical properties of the rice starch (L. F. Wang & Wang, 2004b). Cameron and

Wang (2006) reported that the use of ultrasound extraction in combination with neutral protease shortened the processing time and could eliminate the SO_2 addition during the process of isolating corn starch. Similarly, ultrasound extraction with or without the use of sodium dodecyl sulfate as a surfactant gave promising results when applied for the isolation of rice starch (L. F. Wang & Wang, 2004a).

5.3.2.4 Extraction of Oil

Han et al. (2018) utilized an ultrasound-assisted aqueous enzymatic extraction method for isolating corn germ oil. Moreover, Gordon et al. (2018) used ultrasound-assisted extraction of fatty acids from barley, corn, and sorghum as a pre-processing method before gas chromatography analysis. They reported that the extraction yield of oil was increased in corn and barley, yet an opposite trend was observed for sorghum.

5.3.2.5 Extraction of Antioxidants

Natural antioxidants with health benefits have been extracted from foods and food wastes, including grains, for potential use in functional foods and nutraceuticals. Various antioxidants are present in grains, including phenolic acid, alkylresorcinols, carotenoids, tocopherols, anthocyanins, flavonoids, and proanthocyanidins, and the majority are present in bran and germ (Luthria, Lu, & John, 2015). Grains contain mostly insoluble bound phenolics esterified to cell wall polysaccharides compared to the free forms present in fruits and vegetables (Acosta-Estrada, Gutiérrez-Uribe, & Serna-Saldívar, 2014; Luthria et al., 2015).

Ultrasound application with a probe unit working at 700–900 W power and 20 kHz frequency for 12 min increased the yield in the extraction of carotenoids from maize meal by 3.6-fold compared to conventional extraction (Ye, Feng, Xiong, & Xiong, 2011). The authors recommended temperatures lower than 40°C for preventing structural damage to carotenoids. Viell et al. (2020) extracted soluble phenolics from whole brown teff grains by ultrasound application with a probe system at a power of 120 W and frequency of 20 kHz for 5 min. They found that the efficiency of the extraction for phenolics was similar to that obtained by using a homogenizer at 12,000 rpm for 10 min at 30°C; however, total flavonoid content and antioxidant capacity were higher in the case of ultrasound-assisted extraction (Viell et al., 2020). On the other hand, ultrasound application in a bath at 30.8 kHz 100 W for 60 min at temperatures between 20–60°C did not improve the yield of phenolics and antioxidant activity from quinoa seed flour, which was attributed to the nature of the plant matrix, the degradation of compounds at high temperature (> 40°C) and long time (> 20 min), the type of compounds to be extracted, and the pretreatment of the seeds (milling to very fine particles) (Carciochi & Dimitrov, 2014).

Extraction time was found to be the most significant factor in the extraction of phenolics from wheat bran by ultrasound application in a bath system working at 40 kHz and 250 W where optimum conditions were 60°C and 25 min (Wang, Sun, Cao, Tian, & Li, 2008). Cengiz, Babacan, Akinci, Tuncer Kesci, and Kaba (2020) determined optimum temperature and time in an ultrasonic bath at a frequency of 37 kHz and a power of 120 W for the extraction of each phenolic acid present in wheat bran from ancient cultivars, which varied between 21.71–74.85°C and 19.01–34.14 min. The authors determined an optimum combination of 44.57°C and 23 min by multiple response prediction. In another study with a new black wheat cultivar, optimum conditions for extraction of phenolic compounds were found to be 50°C and 40 min for an ultrasound system working at a frequency of 35 kHz and a power of 300 W preceded by extrusion (Chen et al., 2020).

Jha, Das, and Deka (2017) found that ultrasound extraction in a bath unit with optimum conditions of 35 kHz, 160 W, 10 min, and 49.46°C followed by microwave treatment (2,450 MHz, 31.11

s) yielded more and different types of phenolics than conventional solvent extraction from black rice husk. Tabaraki and Nateghi (2011) also reported optimum conditions for the extraction of phenolics from rice bran as 51–54°C and 40–45 min for an ultrasonic bath working at 35 kHz and 140 W. Turrini, Boggia, Leardi, Borriello, and Zunin (2018) reported that phenolics in flour, caryopses, and leaves of a new rice cultivar called 'Violet Nori' can be extracted in a shorter time and with less solvent by an ultrasound probe system compared with conventional solvent extraction. Optimum conditions were found as 5 min at 30% amplitude (200 W maximum) and 80% pulse with a temperature lower than 60°C. In another study by Setyaningsih, Saputro, Carrera, and Palma (2019), phenolic compounds from whole rice grains were recovered at pH 4.25 and 45°C for 25 min by a probe ultrasound system with 24 kHz frequency and 200 W power working at an amplitude of 47%. The authors stated that a rice cell is solid, compact, and complex with fibrous layers and proteins compared to softer fruit matrices, therefore the extraction time is longer. In addition, the importance of extraction kinetics was emphasized as there was no change in concentration after 25 min of extraction. Wang et al. (2020) also found that the yield of extraction of anthocyanins from red rice bran was maximum at a solid–liquid ratio of 1:17.46 (w/v) with an ethanol concentration of 78.37% by application of ultrasound with a device working at 24 kHz and 400 W.

The extraction of antioxidants from food wastes such as spent grains from alcohol or beer production can be improved by ultrasound application. Izadifar (2013) found that ultrasound treatment damaged cell walls and specifically increased the surface area, pore volume, and pore size of spent wheat distillers dried grain. This led to an increase in the extraction yield and rate compared to those of conventional extraction. The author recommended 87 W and 30s to 5 min for treatment with a probe ultrasound system. The recovery of phenolics from brewers' spent grain was increased by 57.9% by ultrasound applied at 400 W for 55 min with a probe system in a study by Martín-García et al. (2019).

Ultrasound provides advantages over conventional solvent extraction by reducing the time, heat, and amount of solvent and reagent needed, which has a potential for reduction in equipment and operational costs (Chemat et al., 2017). Optimum conditions for extraction vary by the antioxidant, grain matrix, and equipment used. The parameters of ultrasound extraction are power intensity, amplitude, frequency, number of cycles, and duration, in addition to the parameters of extraction including solvent composition, solid-to-solvent ratio, particle size of solids, duration, and temperature. In general, extraction efficiency is improved by an increase in power intensity and a decrease in frequency at moderate temperatures; however, applied power is limited by the degradation of the molecule of interest. In addition, the chemical effects of ultrasound can induce a reduction in antioxidant activity due to degradation and oxidation reactions (Barba, Brianceau, Turk, Boussetta, & Vorobiev, 2015; Moses, Rajauria, & Tiwari, 2017). Although heat treatment improves the efficiency of extraction, elevation in temperature by ultrasound can enhance the degradation of solutes, especially above 40°C. Therefore, optimization of parameters of extraction and ultrasound is required for the extraction of an antioxidant from a particular matrix.

5.4 APPLICATIONS OF ULTRASOUND IN BAKERY AND SNACK FOODS

Ultrasound technology can be used to improve the quality of bakery and snack foods in mixing, fermentation, and pre-gelatinization processes. Cutting and portioning operations can also be achieved precisely and with less energy by the application of ultrasound. In addition, ultrasound can be used to modify functional properties of gluten, such as foaming capacity, foam stability, emulsion stability, and rheological properties (H. Zhang, Claver, Zhu, & Zhou, 2011). Aside from the applications used in processing, ultrasound technology can also be helpful in process control. Ultrasound devices can be used in quality control for rapid online monitoring of product crispness and rheological properties.

5.4.1 Ultrasound-assisted Preparation of Bakery and Snacks Foods

5.4.1.1 Mixing

Ultrasound treatment can be applied using a bath or probe equipment during the preparation of cake batters. Esmaeilzadeh Kenari and Nemati (2020) reported that both probe and bath ultrasound treatments improved the emulsification performance in cupcakes. The sensory properties were improved and staling was delayed in ultrasound-treated samples, but the treatment with probe equipment was more effective.

Ultrasound processing has been successfully utilized in the preparation step of foam-type products, such as chocolate mousse, chocolate Genoise, and sponge cake (Pingret, Fabiano-Tixier, Petitcolas, Canselier, & Chemat, 2011), as it provided a more homogeneous product. Tan, Chin, and Yusof (2011) also used the aid of ultrasound treatment at 2.5 kW for the mixing of sponge cake batter; they observed that the density and rheology of the batter and the cake quality were improved. Similarly, up to 19% volume increase was detected in bread samples that were made with dough mixed with the aid of ultrasound treatment (Pa, Chin, Yusof, & Aziz, 2014).

Ultrasound improved properties of the cake and batter in eggless cake, such as batter density and viscosity, cake volume, and texture (Movahhed, Mohebbi, Koocheki, Milani, & Ansarifar, 2020). Tan et al. (2015) applied ultrasound treatment on whey proteins to replace egg proteins in an eggless cake. They reported that the cake batter properties and the cake quality were good and similar to those of a cake baked with egg. Moreover, the ultrasound treatment increased the number of microbubbles in the foam of cakes prepared with ultrasound-treated whey proteins (Tan, Chin, Yusof, & Abdullah, 2016). The distribution of the microbubbles in the foam was also more even. In another study by Chan, Tan, and Chin (2018), 20% of sugar replacement with ultrasound-treated pectin resulted in better aeration and textural properties in the cake batter.

5.4.1.2 Fermentation

Fermentation of dough with the assistance of ultrasound has the potential to improve bread quality. For instance, D. Luo et al. (2018) reported a reduction in hardness and an improvement in the specific volume and shelf life of steamed bread.

5.4.1.3 Cutting

Ultrasound treatment can also be used in the cutting of several bakery products, including bread, cake, and dumplings (Liu, Jia, Xu, & Li, 2015; Zahn, Schneider, Zucker, & Rohm, 2005). Zahn et al. (2005) reported the effects of amplitude and frequency on the ultrasonic cutting process of bakery products. In addition, Liu et al. (2015) claimed that ultrasound enabled the cutting of thinner slices of bread, rye bread, and cake compared to conventional methods.

5.4.1.4 Pre-gelatinization

Jalali et al. (2020) investigated the effects of ultrasound treatment (with 0, 30%, and 70% intensity) and pre-gelatinization of cornflour on some quality properties of gluten-free bread. They observed that using pre-gelatinized cornflour with ultrasound treatment (30% intensity) provided higher sensory acceptability and better technological and visual properties compared to the control (no treatment) and other samples.

5.4.2 Quality Control of Bakery and Snack Foods Using Ultrasound

5.4.2.1 Control of Crispness

Crispy texture is one of the important quality features of biscuits, crackers, and some other bakery products. Crispness can be measured by sensory, instrumental, or morphological methods. Instrumental methods are based on mechanical and acoustic measurements (Saeleaw & Schleining, 2011). The use of acoustic measurement to determine the crispness of foods began after the 1990s. From this date to the present, some studies have been conducted investigating the combination of acoustic detection of the crispness in relation to mechanical (force/displacement) measurements in biscuits and other bakery products. Being non-destructive and practical, this mechanical–acoustic combining strategy can be recommended for both laboratory and industrial applications for evaluating the food texture.

A low frequency (3–30 kHz) acoustic spectrometer was developed and used to assess the texture of some products such as wafer sheets, crispbread, ring-shaped rolls, and crackers according to the amplitude of the penetrated acoustic signal. The results of the acoustic method gave a high correlation with the results from the texture analyzer (traditional method) for hardness measurements (Juodeikiene & Basinskiene, 2004). Chen, Karlsson, and Povey (2005) designed an acoustic envelope detector that was attached to the texture analyzer to evaluate the crispness of six kinds of biscuits based on the force/displacement behavior of food materials and their acoustic nature. The acoustic ranking of the biscuits obtained from the instrumental evaluation was in good agreement with that obtained from the sensory panel tests. Arimi et al. (2010) developed an acoustic system that can be used as an attachment with any texture-analyzing instrument and can also be used during sensory evaluation. In this system, the sound emitted during the fracture of crispy food was used as an indicator of instrumental crispness. The same group used this acoustic system with an Instron Universal Testing Machine to monitor changes in the crispiness of biscuits at different water activities. The developed acoustic system was successfully applied to monitor changes in crispiness due to hydration (Arimi, Duggan, O'Sullivan, Lyng, & O'Riordan, 2010).

In a study examining the effect of different formulations on the crispness of ciabatta bread crust, the correlation between acoustic parameters related to crispness and instrumental texture evaluation was investigated. The pulsed acoustic spectrometer was developed, which works in a signal frequency range of 5–120 kHz and measures the amplitude of the acoustic signal reflected from the surface of the tested sample. This nondestructive acoustic system was found to be in correlation with instrumental texture assessment and recommended for monitoring the texture of bakery products (Zadeike, Jukonyte, Juodeikiene, Bartkiene, & Valatkeviciene, 2018).

5.4.2.2 Control of Physical and Mechanical Properties

Batters are produced by the incorporation of air phase into a cake mixture by agitation (Gómez, Oliete, García-Álvarez, Ronda, & Salazar, 2008). Thus, physical properties, such as the number of air bubbles and rheology are the most important parameters in the evaluation of batter quality (Gómez et al., 2008). Moreover, the expansion of cake can be affected by the dough viscosity (Gaines & Donelson, 1982; Shrestha, Vali, & Choudhary, 1990). Similar to cake batters, the rheological characteristics of wheat dough are important for the prediction of the final product quality (Ross, Pyrak-Nolte, & Campanella, 2004).

Ultrasound has been recommended as an alternative to rheological measurements of cereal products by researchers as it provides online and more hygienic monitoring of dough quality (Gómez et al., 2008; Elfawakhry, Hussein, & Becker, 2013; Kerhervé et al., 2019). Ultrasound can also replace traditional methods used in the characterization of cake batters and the measurement of their physical and mechanical properties. In addition, it offers more effective monitoring in regard to quality control compared to sensory panels (Gómez et al., 2008). The studies that are evaluating the use of ultrasonic measurements for predicting the physical and mechanical properties of several cereal products are summarized in Table 5.3.

The utilization of ultrasound for the assessment of rheological properties of doughs was comprehensively reviewed by Scanlon and Page (2015). In order to examine the utilization of ultrasound technology through the characterization of physical and mechanical properties of the product, the velocity and attenuation are measured and compared to those obtained by conventional methods. Furthermore, longitudinal mechanical moduli (M' and M") can be calculated using the density and the ultrasonically measured values to predict the rheological properties of the product (Scanlon & Page, 2015). Furthermore, the rheological properties and breadmaking performance of doughs are influenced by the size and the number of bubbles (Scanlon, 2013; Scanlon & Page, 2015). Ultrasonic measurements can also be used in the assessment of bubble sizes.

5.5 CONCLUSIONS

Ultrasound has been shown to positively affect many food processing operations. While LIUS improves quality by non-destructive analysis and process control, HIUS provides the ability to enhance yield and reduce energy and time and hence the cost of processing. HIUS can accelerate many food processing unit operations by improving heat and mass transfer, especially on interfaces. An increase in temperature and pressure by cavitation can also accelerate unit operations and physical and chemical reactions. Ultrasound is a mechanical treatment that can reduce the use of heat, therefore, it can aid in the protection of flavor, color, texture, and nutritional value depending on the used power intensity. However, intensive use can degrade food polymers, and oxidation can be induced by free radicals created by acoustic cavitation. As each food system has a multi-component matrix in different compositions and structures, each application of HIUS should be optimized in terms of frequency, power intensity, and duration to achieve desired yield and quality.

Ultrasound shortens the time and increases the efficiency of treatments and extraction in the processing of grains for food ingredients and products. The bakery and cake industry can benefit from ultrasound in mixing and fermentation as well as in process and quality control systems. High-intensity ultrasound stabilizes dispersed food systems such as emulsions successfully by way of its homogenization effect. Almost all unit operations can be achieved in a short time by ultrasound treatment. The technology is cheap, versatile, and can be combined with existing equipment. The development of efficient large-scale systems is needed before the implementation of this novel technology in various food processes. New transducers and emitting system designs for the production and transmission of acoustic waves are also needed for efficient and uniform treatment of large volumes of liquid and solid foods before engaging in wider industrial application of ultrasound technology in the food industry.

TABLE 5.3   SUMMARY OF STUDIES EVALUATING THE USE OF ULTRASONIC MEASUREMENTS FOR THE EVALUATION OF THE PHYSICAL AND MECHANICAL PROPERTIES DURING THE PRODUCTION OF SEVERAL BAKERY AND SNACK FOODS

Sample	Purpose	Frequency	Ultrasonic Measurement	Conventional Measurement	Result	References
Bread dough	Evaluation of ultrasonic measurements at different frequencies for predicting breadmaking performance	0.3–6 MHz	Longitudinal velocity, attenuation coefficient	Dough: farinograph curves, fracture stress, Hencky strain at fracture stress, energy required for uniaxial extension, uniaxial and biaxial strain hardening, cell elongation, and mean cell area. Bread: crumb image analysis, strain hardening, loaf volume, baking performance	Significant correlation between the measurements at lower frequencies (0.3 and 3 MHz) and conventional measurements was obtained for both the dough and the bread	(Peressini et al., 2017)
	Performance of ultrasound in the assessment of the effects of mixing time and shortening on the dough properties of bread dough prepared with wheat flour and shortening	50 kHz	Velocity, attenuation coefficient	-	Ultrasound measurements were useful in predicting the effects of the mixing time and the addition of ingredients on dough quality	(Mehta, Scanlon, Sapirstein, & Page, 2009)
Cake batter	Utilization of ultrasound in the evaluation of physical properties	1 MHz	Acoustic impedance	Batter: batter density, storage modulus (G'), loss modulus (G''), complex modulus (G*), consistency (K), flow index (n). Cake: volume index, symmetry, cake density	Significant correlation of acoustic impedance with viscosity, G'' and G*	(Gómez et al., 2008)

(Continued)

TABLE 5.3 (CONTINUED)   SUMMARY OF STUDIES EVALUATING THE USE OF ULTRASONIC MEASUREMENTS FOR THE EVALUATION OF THE PHYSICAL AND MECHANICAL PROPERTIES DURING THE PRODUCTION OF SEVERAL BAKERY AND SNACK FOODS

Sample	Purpose	Frequency	Ultrasonic Measurement	Conventional Measurement	Result	References
Gluten-free rice-flour-based dough	Evaluation of the rheological properties by ultrasound and Mixolab device	100 kHz	Velocity, attenuation	Dough: Mixolab measurements	The ultrasound measurements agreed with the findings by the Mixolab device	(Rosell, Marco, García-Alvárez, & Salazar, 2011)
Noodle dough	Feasibility of ultrasound to evaluate rheological properties	1.0 MHz	Phase velocity, attenuation, complex longitudinal modulus	Dough: stress relaxation measurement	The ultrasound technique was useful for evaluating the mechanical properties	(Hatcher et al., 2014)
	Online quality control of the dough	200 kHz	Phase velocity, attenuation, complex longitudinal modulus (laboratory and online)	–	Both laboratory and online measurements were successful in monitoring of quality	(Kerhervé et al., 2019)
	Evaluating the mechanical properties of yellow alkaline noodles	40 kHz	Velocity, attenuation, and storage mechanical modulus	Dough: density, stress relaxation test measurement	The technique was useful for the evaluation of the mechanical properties, and the ultrasonic measurements agreed with the measurements obtained by stress relaxation tests	(Bellido & Hatcher, 2010)
	Utilization of ultrasound to assess the texture of raw noodle	40 kHz	Velocity, attenuation, longitudinal mechanical moduli, and loss tangent	Raw noodle sheet: density Cooked noodle: maximum cutting stress	The discrimination of noodle flours by ultrasonic measurement was possible	(Diep et al., 2014)

(Continued)

TABLE 5.3 (CONTINUED)   SUMMARY OF STUDIES EVALUATING THE USE OF ULTRASONIC MEASUREMENTS FOR THE EVALUATION OF THE PHYSICAL AND MECHANICAL PROPERTIES DURING THE PRODUCTION OF SEVERAL BAKERY AND SNACK FOODS

Sample	Purpose	Frequency	Ultrasonic Measurement	Conventional Measurement	Result	References
Noodle (cooked)	Assessment of rheological properties by ultrasound	11 MHz	Phase velocity, attenuation coefficient, longitudinal mechanical moduli, and loss tangent	Cooked dough: stress relaxation measurement	Detection of the rheological differences was possible by high-frequency ultrasonic measurement	(Daugelaite, Strybulevych, Scanlon, Page, & Hatcher, 2016)
Wheat dough	Discrimination of wheat varieties by ultrasound and selection of the best frequency	0.5–10 MHz	Phase velocity, attenuation coefficient, longitudinal mechanical moduli, and loss tangent	Dough: micro-extension measurements (resistance to extension, extensibility, area)	Correlations between ultrasound measurements (10 MHz) and mixograph, farinograph, and micro-extension measurements	(Salimi Khorshidi, Thandapilly, & Ames, 2018)
	Using ultrasound for discrimination of flours	100 kHz	Velocity, attenuation	Flour: gel protein elastic modulus, farinograph measurement Dough: extensograph measurements, Alveograph measurements	Ultrasound findings correlated with the rheological measurements obtained by an extensograph and an Alveograph	(Álava et al., 2007)
	Explain the effect of mixing on rheological properties by ultrasound measurement	3.5 MHz	Ultrasound velocity, relative attenuation, longitudinal mechanical moduli	Flour: mixograph measurements Dough: density, mixograph measurements, density, G', G''	Ultrasound (velocity, relative attenuation) and conventional rheology measurements (storage modulus and loss modulus) were in agreement	(Ross et al., 2004)

REFERENCES

Acosta-Estrada, B. A., Gutiérrez-Uribe, J. A., & Serna-Saldívar, S. O. (2014). Bound phenolics in foods: A review. *Food Chemistry, 152*, 46–55. doi:10.1016/j.foodchem.2013.11.093

Álava, J. M., Sahi, S. S., García-Álvarez, J., Turó, A., Chávez, J. A., García, M. J., & Salazar, J. (2007). Use of ultrasound for the determination of flour quality. *Ultrasonics, 46*(3), 270–276. doi:10.1016/j.ultras.2007.03.002

Arimi, J. M., Duggan, E., O'sullivan, M., Lyng, J. G., & O'riordan, E. D. (2010). Development of an acoustic measurement system for analyzing crispiness during mechanical and sensory testing. *Journal of Texture Studies, 41*(3), 320–340. doi:10.1111/j.1745-4603.2010.00224.x

Arimi, J. M., Duggan, E., O'Sullivan, M., Lyng, J. G., & O'Riordan, E. D. (2010). Effect of water activity on the crispiness of a biscuit (Crackerbread): Mechanical and acoustic evaluation. *Food Research International, 43*(6), 1650–1655. doi:10.1016/j.foodres.2010.05.004

Arzeni, C., Martinez, K., Zema, P., Arias, A., Perez, O. E., & Pilosof, A. M. R. (2012). Comparative study of high intensity ultrasound effects on food proteins functionality. *Journal of Food Engineering, 108*(3), 463–472. doi:10.1016/j.jfoodeng.2011.08.018

Ashokkumar, M., Bhaskaracharya, R., Kentish, S., Lee, J., Palmer, M., & Zisu, B. (2010). The ultrasonic processing of dairy products – An overview. *Dairy Science and Technology, 90*(2–3), 147–168. doi:10.1051/dst/2009044

Ashokkumar, M., Sunartio, D., Kentish, S., Mawson, R., Simons, L., Vilkhu, K., & Versteeg, C. K. (2008). Modification of food ingredients by ultrasound to improve functionality: A preliminary study on a model system. *Innovative Food Science and Emerging Technologies, 9*(2), 155–160. doi:10.1016/j.ifset.2007.05.005

Awad, T. S., Moharram, H. A., Shaltout, O. E., Asker, D., & Youssef, M. M. (2012). Applications of ultrasound in analysis, processing and quality control of food: A review. *Food Research International, 48*(2), 410–427. doi:10.1016/j.foodres.2012.05.004

Barba, F. J., Brianceau, S., Turk, M., Boussetta, N., & Vorobiev, E. (2015). Effect of alternative physical treatments (ultrasounds, pulsed electric fields, and high-voltage electrical discharges) on selective recovery of bio-compounds from fermented grape pomace. *Food and Bioprocess Technology, 8*(5), 1139–1148. doi:10.1007/s11947-015-1482-3

Bedin, S., Maria Netto, F., Bragagnolo, N., & Taranto, O. P. (2020). Reduction of the process time in the achieve of rice bran protein through ultrasound-assisted extraction and microwave-assisted extraction. *Separation Science and Technology, 55*(2), 300–312. doi:10.1080/01496395.2019.1577449

Bellido, G. G., & Hatcher, D. W. (2010). Ultrasonic characterization of fresh yellow alkaline noodles. *Food Research International, 43*(3), 701–708. doi:10.1016/j.foodres.2009.11.010

Bermudez-Aguirre, D., Mobbs, T., & Barbosa-Canovas, G. V. (2011). Ultrasound applications in food processing. In: H. Feng, G. V. BarbosaCanovas, & J. Weiss (Eds.), *Ultrasound Technologies for Food and Bioprocessing* (pp. 65–105). New York: Springer. doi:10.1007/978-1-4419-7472-3_3

Bhargava, N., Mor, R. S., Kumar, K., & Sharanagat, V. S. (2021). Advances in application of ultrasound in food processing: A review. *Ultrasonics Sonochemistry, 70*, 105293. doi:10.1016/j.ultsonch.2020.105293

Bonto, A. P., Jearanaikoon, N., Sreenivasulu, N., & Camacho, D. H. (2020). High uptake and inward diffusion of iron fortificant in ultrasonicated milled rice. *LWT, 128*, 109459. doi:10.1016/j.lwt.2020.109459

Borsato, V. M., Jorge, L. M. M., Mathias, A. L., & Jorge, R. M. M. (2019). Ultrasound assisted hydration improves the quality of the malt barley. *Journal of Food Process Engineering, 42*(6), e13208. doi:10.1111/jfpe.13208

Boufi, S., Haaj, S. B., Magnin, A., Pignon, F., Imperor-Clerc, M., & Mortha, G. (2018). Ultrasonic assisted production of starch nanoparticles: Structural characterization and mechanism of disintegration. *Ultrasonics Sonochemistry, 41*, 327–336. doi:10.1016/j.ultsonch.2017.09.033

Cameron, D. K., & Wang, Y. J. (2006). Application of protease and high-intensity ultrasound in corn starch isolation from degermed corn flour. *Cereal Chemistry Journal, 83*(5), 505–509. doi:10.1094/CC-83-0505

Carciochi, R. A., & Dimitrov, K. (2014). Optimization of antioxidant phenolic compounds extraction from quinoa (Chenopodium quinoa) seeds. *Journal of Food Science and Technology*, *52*(7), 4396–4404. doi:10.1007/s13197-014-1514-4

Cengiz, M. F., Babacan, U., Akinci, E., Tuncer Kesci, S., & Kaba, A. (2020). Extraction of phenolic acids from ancient wheat bran samples by ultrasound application. *Journal of Chemical Technology and Biotechnology*, *96*, 134–141. doi:10.1002/jctb.6519

Chan, Y. Ten, Tan, M. C., & Chin, N. L. (2018). Effect of partial sugar replacement with ultrasonically treated citrus pectin on aeration and rheological properties of batter. *Journal of Food Processing and Preservation*, *42*(12), e13827. doi:10.1111/jfpp.13827

Chandan, R. C. (2006). Starter cultures for yoghurt and fermented milks. In: R. C. Chandan, C. H. White, A. Kilara, & Y. H. Hui (Eds.), *Manufacturing Yogurt and Fermented Milks* (pp. 89–116). Ames: Blackwell Publishing.

Chandrapala, J., Oliver, C., Kentish, S., & Ashokkumar, M. (2012). Ultrasonics in food processing – Food quality assurance and food safety. *Trends in Food Science and Technology*, *26*(2), 88–98. doi:10.1016/j.tifs.2012.01.010

Charoenthaikij, P., Jangchud, K., Jangchud, A., Piyachomkwan, K., Tungtrakul, P., & Prinyawiwatkul, W. (2009). Germination conditions affect physicochemical properties of germinated brown rice flour. *Journal of Food Science*, *74*(9), C658–C665. doi:10.1111/j.1750-3841.2009.01345.x

Chemat, F., Rombaut, N., Sicaire, A.-G., Meullemiestre, A., Fabiano-Tixier, A.-S., & Abert-Vian, M. (2017). Ultrasound assisted extraction of food and natural products. Mechanisms, techniques, combinations, protocols and applications. A review. *Ultrasonics Sonochemistry*, *34*, 540–560. doi:10.1016/j.ultsonch.2016.06.035

Chemat, F., Zill-e-Huma, & Khan, M. K. (2011). Applications of ultrasound in food technology: Processing, preservation and extraction. *Ultrasonics Sonochemistry*, *18*(4), 813–835. doi:10.1016/j.ultsonch.2010.11.023

Chen, J., Karlsson, C., & Povey, M. (2005). Acoustic envelope detector for crispness assessment of biscuits. *Journal of Texture Studies*, *36*(2), 139–156. doi:10.1111/j.1745-4603.2005.00008.x

Chen, X., Li, X., Zhu, X., Wang, G., Zhuang, K., Wang, Y., & Ding, W. (2020). Optimization of extrusion and ultrasound-assisted extraction of phenolic compounds from jizi439 black wheat bran. *Processes*, *8*(9), 1153. doi:10.3390/PR8091153

Condón, S., Raso, J., & Pagán, R. (2004). Microbial inactivation by ultrasound. In: G. V. Barbosa-Canovas, M. S. Tapia, & M. P. Cano (Eds.), *Novel Food Processing Technologies* (pp. 423–442). Boca Raton: CRC Press. doi:10.1201/9780203997277.ch19

Corbo, M. R., Bevilacqua, A., Campaniello, D., D'Amato, D., Speranza, B., & Sinigaglia, M. (2009). Prolonging microbial shelf life of foods through the use of natural compounds and non-thermal approaches – A review. *International Journal of Food Science and Technology*, *44*(2), 223–241. doi:10.1111/j.1365-2621.2008.01883.x

Cui, L., Pan, Z., Yue, T., Atungulu, G. G., & Berrios, J. (2010). Effect of ultrasonic treatment of brown rice at different temperatures on cooking properties and quality. *Cereal Chemistry Journal*, *87*(5), 403–408. doi:10.1094/CCHEM-02-10-0034

Daugelaite, D., Strybulevych, A., Scanlon, M. G., Page, J. H., & Hatcher, D. W. (2016). Use of ultrasound to investigate glucose oxidase and storage effects on the rheological properties of cooked Asian noodles. *Cereal Chemistry Journal*, *93*(2), 125–129. doi:10.1094/CCHEM-01-15-0006-R

de Carvalho, G. R., Polachini, T. C., Darros-Barbosa, R., Bon, J., & Telis-Romero, J. (2018). Effect of intermittent high-intensity sonication and temperature on barley steeping for malt production. *Journal of Cereal Science*, *82*, 138–145. doi:10.1016/j.jcs.2018.06.005

de Castro, M. D. L., & Capote, F. P. (2007). Introduction: Fundamentals of ultrasound and basis of its analytical uses. In: M. D. L. DeCastro & F. P. Capote (Eds.), *Analytical Applications of Ultrasound* (pp. 1–34). Amsterdam: Elsevier.

Demirdöven, A., & Baysal, T. (2009). The use of ultrasound and combined technologies in food preservation. *Food Reviews International*, *25*(1), 1–11. doi:10.1080/87559120802306157

Diep, S., Daugelaite, D., Strybulevych, A., Scanlon, M., Page, J., & Hatcher, D. (2014). Use of ultrasound to discern differences in Asian noodles prepared across wheat classes and between varieties. *Canadian Journal of Plant Science, 94*(3), 525–534. doi:10.4141/CJPS2013-043

Ding, J., & Feng, H. (2018). Controlled germination for enhancing the nutritional value of sprouted grains. In: Hao Feng, B. Nemzer, & J. W. DeVries (Eds.), *Sprouted Grains: Nutritional Value, Production, and Applications* (pp. 91–112). Washington: AACC International Press. doi:10.1016/B978-0-12-811525-1.00005-1

Ding, J., Hou, G. G., Dong, M., Xiong, S., Zhao, S., & Feng, H. (2018). Physicochemical properties of germinated dehulled rice flour and energy requirement in germination as affected by ultrasound treatment. *Ultrasonics Sonochemistry, 41*, 484–491. doi:10.1016/j.ultsonch.2017.10.010

Ding, J., Hou, G. G., Nemzer, B. V., Xiong, S., Dubat, A., & Feng, H. (2018). Effects of controlled germination on selected physicochemical and functional properties of whole-wheat flour and enhanced gamma-aminobutyric acid accumulation by ultrasonication. *Food Chemistry, 243*, 214–221. doi:10.1016/j.foodchem.2017.09.128

Ding, J., Johnson, J., Chu, Y. F., & Feng, H. (2019). Enhancement of gamma-aminobutyric acid, avenanthramides, and other health-promoting metabolites in germinating oats (*Avena sativa* L.) treated with and without power ultrasound. *Food Chemistry, 283*, 239–247. doi:10.1016/j.foodchem.2018.12.136

Ding, J., Ulanov, A. V., Dong, M., Yang, T., Nemzer, B. V., Xiong, S., … Feng, H. (2018). Enhancement of gamaaminobutyric acid (GABA) and other health-related metabolites in germinated red rice (*Oryza sativa* L.) by ultrasonication. *Ultrasonics Sonochemistry, 40*(A), 791–797. doi:10.1016/j.ultsonch.2017.08.029

Dolatowski, Z. J., Stadnik, J., & Stasiak, D. (2007). Applications of ultrasound in food technology. *Acta Scientiarum Polonorum Technologia Alimentaris, 6*(3), 88–99.

Drochioiu, G., Ciobanu, C. I., Bancila, S., Ion, L., Petre, B. A., Andries, C., … Murariu, M. (2016). Ultrasound-based protein determination in maize seeds. *Ultrasonics Sonochemistry, 29*, 93–103. doi:10.1016/j.ultsonch.2015.09.007

Earnshaw, R. G. (1998). Ultrasound: A new opportunity for food preservation. In: M. J. W. Povey & T. J. Mason (Eds.), *Ultrasound in Food Processing* (pp. 181–191). London: Blackie Academic & Professional.

Elfawakhry, H., Hussein, M. A., & Becker, T. (2013). Investigations on the evaluation of rheological properties of cereal based viscoelastic fluids using ultrasound. *Journal of Food Engineering, 116*, 404–412. doi:10.1016/j.jfoodeng.2012.12.021

Elmalimadi, M. B., Stefanovic, A. B., Sekuljica, N. Z., Zuza, M. G., Lukovic, N. D., Jovanovic, J. R., & Knezevic-Jugovic, Z. D. (2017). The synergistic effect of heat treatment on alcalase-assisted hydrolysis of wheat gluten proteins: Functional and antioxidant properties. *Journal of Food Processing and Preservation, 41*(5), e13207. doi:10.1111/jfpp.13207

Esmaeilzadeh Kenari, R., & Nemati, A. (2020). The effectiveness of ultrasound bath and probe treatments on the quality of baking and shelf life of cupcakes. *Food Science and Nutrition, 8*(6), 2929–2939. doi:10.1002/fsn3.1595

Feng, H., & Yang, W. (2011). Ultrasonic processing. In: H. Q. Zhang, G. V. Barbosa-Cánovas, V. B. Balasubramaniam, C. P. Dunne, D. F. Farkas, & J. T. Yuan (Eds.), *Nonthermal Processing Technologies for Food* (pp. 135–150). West Sussex: John Wiley & Sons.

Gaines, C. S., & Donelson, J. R. (1982). Cake batter viscosity and expansion upon heating. *Cereal Chemistry, 59*(4), 237–240.

Gallego-Juárez, J. A. (2017). Basic principles of ultrasound. In: Mar Villamiel, J. V. García-Pérez, A. Montilla, J. A. Cárcel, & J. Benedito (Eds.), *Ultrasound in Food Processing: Recent Advances* (pp. 4–26). West Sussex: John Wiley & Sons. doi:10.1002/9781118964156.ch1

Gallego-Juarez, J. A., & Riera, E. (2011). Technologies and applications of airborne power ultrasound in food processing. In: H. Feng, G. V. BarbosaCanovas, & J. Weiss (Eds.), *Ultrasound Technologies for Food and Bioprocessing* (pp. 617–641). New York: Springer. doi:10.1007/978-1-4419-7472-3_25

Gogate, P. R., & Pandit, A. B. (2015). Design and scale-up of sonochemical reactors for food processing and other applications. In: J. A. Gallego-Juárez & K. F. Graff (Eds.), *Power Ultrasonics: Applications of High-Intensity Ultrasound* (pp. 726–755). Cambridge: Woodhead Publishing. doi:10.1016/B978-1-78242-028-6.00024-7

Gómez, M., Oliete, B., García-Álvarez, J., Ronda, F., & Salazar, J. (2008). Characterization of cake batters by ultrasound measurements. *Journal of Food Engineering, 89*(4), 408–413. doi:10.1016/j.jfoodeng.2008.05.024

Gordon, R., Chapman, J., Power, A., Chandra, S., Roberts, J., & Cozzolino, D. (2018). Comparison of ultrasound-assisted extraction with static extraction as pre-processing method before gas chromatography analysis of cereal lipids. *Food Analytical Methods, 11*(11), 3276–3281. doi:10.1007/s12161-018-1304-0

Gruwel, M. L. H., Chatson, B., Yin, X. S., & Abrams, S. (2001). A magnetic resonance study of water uptake in whole barley kernels. *International Journal of Food Science and Technology, 36*(2), 161–168. doi:10.1046/j.1365-2621.2001.00445.x

Guimaraes, B., Polachini, T. C., Augusto, P. E. D., & Telis-Romero, J. (2020). Ultrasound-assisted hydration of wheat grains at different temperatures and power applied: Effect on acoustic field, water absorption and germination. *Chemical Engineering and Processing-Process Intensification, 155*, 108045. doi:10.1016/j.cep.2020.108045

Han, C., Liu, Q., Jing, Y., Wang, D., Zhao, Y., Zhang, H., & Jiang, L. (2018). Ultrasound-assisted aqueous enzymatic extraction of corn germ oil: Analysis of quality and antioxidant activity. *Journal of Olea Science, 67*(6), 745–754. doi:10.5650/jos.ess17241

Hassan, S., Imran, M., Ahmad, M. H., Khan, M. I., Xu, C., Khan, M. K., & Muhammad, N. (2020). Phytochemical characterization of ultrasound-processed sorghum sprouts for the use in functional foods. *International Journal of Food Properties, 23*(1), 853–863. doi:10.1080/10942912.2020.1762644

Hassan, S., Imran, M., Ahmad, N., & Khan, M. K. (2017). Lipids characterization of ultrasound and microwave processed germinated sorghum. *Lipids in Health and Disease, 16*(1), 125. doi:10.1186/s12944-017-0516-4

Hatcher, D. W., Salimi, A., Daugelaite, D., Strybulevych, A., Scanlon, M. G., & Page, J. H. (2014). Application of ultrasound to the evaluation of rheological properties of raw Asian noodles fortified with barley β-glucan. *Journal of Texture Studies, 45*(3), 220–225. doi:10.1111/jtxs.12067

Hromádková, Z., Košťálová, Z., & Ebringerová, A. (2008). Comparison of conventional and ultrasound-assisted extraction of phenolics-rich heteroxylans from wheat bran. *Ultrasonics Sonochemistry, 15*(6), 1062–1068. doi:10.1016/j.ultsonch.2008.04.008

Iscimen, E. M., & Hayta, M. (2018). Optimisation of ultrasound assisted extraction of rice bran proteins: Effects on antioxidant and antiproliferative properties. *Quality Assurance and Safety of Crops and Foods, 10*(2), 165–174. doi:10.3920/QAS2017.1186

Izadifar, Z. (2013). Ultrasound pretreatment of wheat dried distiller's grain (DDG) for extraction of phenolic compounds. *Ultrasonics Sonochemistry, 20*(6), 1359–1369. doi:10.1016/j.ultsonch.2013.04.004

Jalali, M., Sheikholeslami, Z., Elhamirad, A. H., Haddad Khodaparast, M. H., & Karimi, M. (2020). The effect of the ultrasound process and pre-gelatinization of the corn flour on the textural, visual, and sensory properties in gluten-free pan bread. *Journal of Food Science and Technology, 57*(3), 993–1002. doi:10.1007/s13197-019-04132-7

Janve, B., Yang, W., Kozman, A., Sims, C., Teixeira, A., Gunderson, M. A., & Rababah, T. M. (2013). Enhancement of corn nixtamalization by power ultrasound. *Food and Bioprocess Technology, 6*(5), 1269–1280. doi:10.1007/s11947-012-0816-7

Janve, B., Yang, W., & Sims, C. (2015). Sensory and quality evaluation of traditional compared with power ultrasound processed corn (*Zea mays*) tortilla chips. *Journal of Food Science, 80*(6), S1368–S1376. doi:10.1111/1750-3841.12892

Jha, P., Das, A. J., & Deka, S. C. (2017). Optimization of ultrasound and microwave assisted extractions of polyphenols from black rice (*Oryza sativa* cv. Poireton) husk. *Journal of Food Science and Technology, 54*(12), 3847–3858. doi:10.1007/s13197-017-2832-0

Juodeikiene, G., & Basinskiene, L. (2004). Non-destructive texture analysis of cereal products. *Food Research International, 37*(6), 603–610. doi:10.1016/j.foodres.2003.12.014

Kentish, S., & Ashokkumar, M. (2011). The physical and chemical effects of ultrasound. In: Hao Feng, G. V. Barbosa-Cánovas, & J. Weiss (Eds.), *Ultrasound Technologies for Food and Bioprocessing* (pp. 1–12). New York: Springer. doi:10.1007/978-1-4419-7472-3_1

Kentish, S., & Feng, H. (2014). Applications of power ultrasound in food processing. *Annual Review of Food Science and Technology, 5*(1), 263–284. doi:10.1146/annurev-food-030212-182537

Kerhervé, S. O., Guillermic, R. M., Strybulevych, A., Hatcher, D. W., Scanlon, M. G., & Page, J. H. (2019). Online non-contact quality control of noodle dough using ultrasound. *Food Control, 104*, 349–357. doi:10.1016/j.foodcont.2019.04.024

Knorr, D., Zenker, M., Heinz, V., & Lee, D. U. (2004). Applications and potential of ultrasonics in food processing. *Trends in Food Science and Technology, 15*(5), 261–266. doi:10.1016/j.tifs.2003.12.001

Li, D., Zhao, H., Muhammad, A. I., Song, L., Guo, M., & Liu, D. (2020). The comparison of ultrasound-assisted thawing, air thawing and water immersion thawing on the quality of slow/fast freezing bighead carp (*Aristichthys nobilis*) fillets. *Food Chemistry, 320*, 126614. doi:10.1016/j.foodchem.2020.126614

Li, K., Ma, H., Li, S., Zhang, C., & Dai, C. (2017). Effect of ultrasound on alkali extraction protein from rice dreg flour. *Journal of Food Process Engineering, 40*(2), e12377. doi:10.1111/jfpe.12377

Liang, Q., Ren, X., Ma, H., Li, S., Xu, K., & Oladejo, A. O. (2017). Effect of low-frequency ultrasonic-assisted enzymolysis on the physicochemical and antioxidant properties of corn protein hydrolysates. *Journal of Food Quality, 2017*, 2784146. doi:10.1155/2017/2784146

Liu, L., Jia, W., Xu, D., & Li, R. (2015). Applications of ultrasonic cutting in food processing. *Journal of Food Processing and Preservation, 39*(6), 1762–1769. doi:10.1111/jfpp.12408

Luo, D., Wu, R., Zhang, J., Zhang, K., Xu, B., Li, P., ... Li, X. (2018). Effects of ultrasound assisted dough fermentation on the quality of steamed bread. *Journal of Cereal Science, 83*, 147–152. doi:10.1016/j.jcs.2018.07.016

Luo, L., Zhang, Y., Wang, K., Ma, H., & Dong, M. (2017). In situ and real-time monitoring of an ultrasonic-assisted enzymatic hydrolysis process of corn gluten meal by a miniature near infrared spectrometer. *Analytical Methods, 9*(25), 3795–3803. doi:10.1039/c7ay00887b

Luthria, D. L., Lu, Y., & John, K. M. M. (2015). Bioactive phytochemicals in wheat: Extraction, analysis, processing, and functional properties. *Journal of Functional Foods, 18*(B), 910–925. doi:10.1016/j.jff.2015.01.001

Ly, H. L., Tran, T. M. C., Tran, T. T. T., Ton, N. M. N., & Le, V. V. M. (2018). Application of ultrasound to protein extraction from defatted rice bran. *International Food Research Journal, 25*(2), 695–701.

Majid, I., Nayik, G. A., & Nanda, V. (2015). Ultrasonication and food technology: A review. *Cogent Food and Agriculture, 1*(1), 1071022. doi:10.1080/23311932.2015.1071022

Martín-García, B., Pasini, F., Verardo, V., Díaz-De-cerio, E., Tylewicz, U., Gómez-Caravaca, A. M., & Caboni, M. F. (2019). Optimization of sonotrode ultrasonic-assisted extraction of proanthocyanidins from brewers' spent grains. *Antioxidants, 8*(8), 282. doi:10.3390/antiox8080282

Mason, T. J. (1998). Power ultrasound in food processing. The way forward. In: M. J. W. Povey & T. J. Mason (Eds.), *Ultrasound in Food Processing* (pp. 105–126). London: Blackie Academic & Professional.

Mason, T. J., Paniwnyk, L., & Chemat, F. (2003). Ultrasound as a preservation technology. In: P. Zeuthen & L. Bøgh-Sørensen (Eds.), *Food Preservation Techniques* (pp. 303–337). Cambridge: Woodhead Publishing. doi:10.1016/B978-1-85573-530-9.50020-6

Mehta, K. L., Scanlon, M. G., Sapirstein, H. D., & Page, J. H. (2009). Ultrasonic investigation of the effect of vegetable shortening and mixing time on the mechanical properties of bread dough. *Journal of Food Science, 74*(9), E455–E461. doi:10.1111/j.1750-3841.2009.01346.x

Miano, A. C., & Augusto, P. E. D. (2018). The ultrasound assisted hydration as an opportunity to incorporate nutrients into grains. *Food Research International, 106*, 928–935. doi:10.1016/j.foodres.2018.02.006

Miano, A. C., Forti, V. A., Abud, H. F., Gomes-Junior, F. G., Cicero, S. M., & Augusto, P. E. D. (2015). Effect of ultrasound technology on barley seed germination and vigour. *Seed Science and Technology, 43*(2), 297–302. doi:10.15258/sst.2015.43.2.10

Miano, A. C., Ibarz, A., & Augusto, P. E. D. (2017). Ultrasound technology enhances the hydration of corn kernels without affecting their starch properties. *Journal of Food Engineering, 197*, 34–43. doi:10.1016/j.jfoodeng.2016.10.024

Moreno-Castro, L. E., Quintero-Ramos, A., Ruiz-Gutierrez, M. G., Sanchez-Madrigal, M. A., Melendez-Pizarro, C. O., Perez-Reyes, I., & Lardizabal-Gutierrez, D. (2015). Nixtamalization assisted with ultrasound: Effect on mass transfer and physicochemical properties of nixtamal, masa and tortilla. *Revista Mexicana de Ingenieria Quimica, 14*(2), 265–279.

Moses, J. A., Rajauria, G., & Tiwari, B. K. (2017). Effect of ultrasound on anthocyanins. In: M. Villamiel, J. V. García-Pérez, A. Montilla, J. A. Cárcel, & J. Benedito (Eds.), *Ultrasound in Food Processing* (1st ed., pp. 485–505). West Sussex: John Wiley & Sons. doi:10.1002/9781118964156.ch19

Movahhed, M. K., Mohebbi, M., Koocheki, A., Milani, E., & Ansarifar, E. (2020). Application of TOPSIS to evaluate the effects of different conditions of sonication on eggless cake properties, structure, and mass transfer. *Journal of Food Science, 85*(5), 1479–1488. doi:10.1111/1750-3841.15117

Musielak, G., Mierzwa, D., & Kroehnke, J. (2016). Food drying enhancement by ultrasound – A review. *Trends in Food Science and Technology, 56*, 126–141. doi:10.1016/j.tifs.2016.08.003

Muthukumarappan, K., Tiwari, B. K., O'Donnell, C. P., & Cullen, P. J. (2009). Ultrasound processing: Rheological and functional properties of food. In: J. Ahmed, H. S. Ramaswamy, S. Kasapis, & J. I. Boye (Eds.), *Novel Food Processing: Effects on Rheological and Functional Properties* (pp. 85–102). Boca Raton: CRC Press.

O'Donnell, C. P., Tiwari, B. K., Bourke, P., & Cullen, P. J. (2010). Effect of ultrasonic processing on food enzymes of industrial importance. *Trends in Food Science and Technology, 21*(7), 358–367. doi:10.1016/j.tifs.2010.04.007

Pa, N. F. C., Chin, N. L., Yusof, Y. A., & Aziz, N. A. (2014). Power ultrasound assisted mixing effects on bread physical properties. In: R. Chin, and N. L. Man and H. C. Talib (Eds.), *2nd International Conference on Agricultural and Food Engineering, CAFEi2014 – New Trends Forward* (pp. 60–66). Amsterdam: Elsevier. doi:10.1016/j.aaspro.2014.11.009

Patero, T., & Augusto, P. E. D. (2015). Ultrasound (US) enhances the hydration of sorghum (*Sorghum bicolor*) grains. *Ultrasonics Sonochemistry, 23*, 11–15. doi:10.1016/j.ultsonch.2014.10.021

Patist, A., & Bates, D. (2008). Ultrasonic innovations in the food industry: From the laboratory to commercial production. *Innovative Food Science and Emerging Technologies, 9*(2), 147–154. doi:10.1016/j.ifset.2007.07.004

Peressini, D., Braunstein, D., Page, J. H., Strybulevych, A., Lagazio, C., & Scanlon, M. G. (2017). Relation between ultrasonic properties, rheology and baking quality for bread doughs of widely differing formulation. *Journal of the Science of Food and Agriculture, 97*(8), 2366–2374. doi:10.1002/jsfa.8048

Pingret, D., Fabiano-Tixier, A. S., Petitcolas, E., Canselier, J. P., & Chemat, F. (2011). First investigation on ultrasound-assisted preparation of food products: Sensory and physicochemical characteristics. *Journal of Food Science, 76*(2), C287–C292. doi:10.1111/j.1750-3841.2010.02019.x

Piyasena, P., Mohareb, E., & McKellar, R. C. (2003). Inactivation of microbes using ultrasound: A review. *International Journal of Food Microbiology, 87*(3), 207–216. doi:10.1016/S0168-1605(03)00075-8

Qu, W., Sehemu, R. M., Zhang, T., Song, B., Yang, L., Ren, X., & Ma, H. (2018). Immobilized enzymolysis of corn gluten meal under triple-frequency ultrasound. *International Journal of Food Engineering, 14*(5–6), 20170347. doi:10.1515/ijfe-2017-0347

Ramírez-Araujo, H., Gaytán-Martínez, M., & Reyes-Vega, M. L. (2019). Alternative technologies to the traditional nixtamalization process: Review. *Trends in Food Science and Technology, 85*, 34–43. doi:10.1016/j.tifs.2018.12.007

Reis, S. F., Coelho, E., Coimbra, M. A., & Abu-Ghannam, N. (2015). Improved efficiency of brewer's spent grain arabinoxylans by ultrasound-assisted extraction. *Ultrasonics Sonochemistry, 24*, 155–164. doi:10.1016/j.ultsonch.2014.10.010

Riener, J., Noci, F., Cronin, D. A., Morgan, D. J., & Lyng, J. G. (2009). Characterisation of volatile compounds generated in milk by high intensity ultrasound. *International Dairy Journal, 19*(4), 269–272. doi:10.1016/j.idairyj.2008.10.017

Rifna, E. J., Ratish Ramanan, K., & Mahendran, R. (2019). Emerging technology applications for improving seed germination. *Trends in Food Science and Technology, 86*, 95–108. doi:10.1016/j.tifs.2019.02.029

Robles-Ozuna, L. E., Ochoa-Martinez, L. A., Morales-Castro, J., Gallegos-Infante, J. A., Quintero-Ramos, A., & Madera-Santana, T. J. (2016). Effect of nixtamalization conditions ultrasound assisted on some physicochemical, structural and quality characteristics in maize used for pozole. *CyTA-Journal of Food, 14*(2), 324–332. doi:10.1080/19476337.2015.1110201

Rosell, C. M., Marco, C., García-Alvárez, J., & Salazar, J. (2011). Rheological properties of rice-soybean protein composite flours assessed by mixolab and ultrasound. *Journal of Food Process Engineering, 34*(6), 1838–1859. doi:10.1111/j.1745-4530.2009.00501.x

Ross, K. A., Pyrak-Nolte, L. J., & Campanella, O. H. (2004). The use of ultrasound and shear oscillatory tests to characterize the effect of mixing time on the rheological properties of dough. *Food Research International, 37*(6), 567–577. doi:10.1016/j.foodres.2004.02.013

Saeleaw, M., & Schleining, G. (2011). A review: Crispness in dry foods and quality measurements based on acoustic-mechanical destructive techniques. *Journal of Food Engineering, 105*(3), 387–399. doi:10.1016/j.jfoodeng.2011.03.012

Salazar, J., Chávez, J. A., Turó, A., & García-Hernández, M. J. (2009). Effect of ultrasound on food processing. In: J. Ahmed, H. S. Ramaswamy, S. Kasapis, & J. I. Boye (Eds.), *Novel Food Processing: Effects on Rheological and Functional Properties* (pp. 65–84). Boca Raton: CRC Press.

Salimi Khorshidi, A., Thandapilly, S. J., & Ames, N. (2018). Application of low-intensity ultrasound as a rapid, cost-effective tool to wheat screening: A systematic frequency selection. *Journal of Cereal Science, 82*, 190–197. doi:10.1016/j.jcs.2018.06.011

Scanlon, M. G. (2013). What has low-intensity ultrasound informed us about wheat flour dough rheology? *Cereal Foods World, 58*(2), 61–65. doi:10.4141/CJPS2013-043

Scanlon, M. G., & Page, J. H. (2015). Probing the properties of dough with low-intensity ultrasound. *Cereal Chemistry Journal, 92*(2), 121–133. doi:10.1094/CCHEM-11-13-0244-IA

Setyaningsih, W., Saputro, I. E., Carrera, C. A., & Palma, M. (2019). Optimisation of an ultrasound-assisted extraction method for the simultaneous determination of phenolics in rice grains. *Food Chemistry, 288*, 221–227. doi:10.1016/j.foodchem.2019.02.107

Shafaei, S. M., Nourmohamadi-Moghadami, A., & Kamgar, S. (2018). An insight into thermodynamic aspects of ultrasonication effect on hydration mechanism of wheat. *Journal of Food Process Engineering, 41*(7), e12862. doi:10.1111/jfpe.12862

Shafaei, S. M., Nourmohamadi-Moghadami, A., & Kamgar, S. (2019). The combined effect of ultrasonication and hydration temperature on water absorption of barley: Analysis, modeling, kinetics, optimization, and thermodynamic parameters of the process. *Journal of Food Processing and Preservation, 43*(4), e13905. doi:10.1111/jfpp.13905

Shrestha, N., Vali, S. A., & Choudhary, P. N. (1990). Quality characteristics of cakes prepared from different fats and oil. *Journal of Food Science and Technology (Mysore), 27*(6), 400–401.

Singh, A. k., Rehal, J., Kaur, A., & Jyot, G. (2015). Enhancement of attributes of cereals by germination and fermentation: A review. *Critical Reviews in Food Science and Nutrition, 55*(11), 1575–1589. doi:10.1080 /10408398.2012.706661

Singh, V., & Eckhoff, S. R. (1996). Effect of soak time, soak temperature, and lactic acid on germ recovery parameters. *Cereal Chemistry, 73*(6), 716–720.

Tabaraki, R., & Nateghi, A. (2011). Optimization of ultrasonic-assisted extraction of natural antioxidants from rice bran using response surface methodology. *Ultrasonics Sonochemistry, 18*(6), 1279–1286. doi:10.1016/j.ultsonch.2011.05.004

Tan, M. C., Chin, N. L., & Yusof, Y. A. (2011). Power ultrasound aided batter mixing for sponge cake batter. *Journal of Food Engineering, 104*(3), 430–437. doi:10.1016/j.jfoodeng.2011.01.006

Tan, M. C., Chin, N. L., Yusof, Y. A., & Abdullah, J. (2016). Novel 2D and 3D imaging of internal aerated structure of ultrasonically treated foams and cakes using X-ray tomography and X-ray microtomography. *Journal of Food Engineering, 183*, 9–15. doi:10.1016/j.jfoodeng.2016.03.008

Tan, M. C., Chin, N. L., Yusof, Y. A., Taip, F. S., & Abdullah, J. (2015). Improvement of eggless cake structure using ultrasonically treated whey protein. *Food and Bioprocess Technology, 8*(3), 605–614. doi:10.1007/ s11947-014-1428-1

Tao, Y., & Sun, D. W. (2015). Enhancement of food processes by ultrasound: A review. *Critical Reviews in Food Science and Nutrition, 55*(4), 570–594. doi:10.1080/10408398.2012.667849

Tiozon Jr., R. N., Camacho, D. H., Bonto, A. P., Oyong, G. G., & Sreenivasulu, N. (2021). Efficient fortification of folic acid in rice through ultrasonic treatment and absorption. *Food Chemistry, 335*, 127629. doi:10.1016/j.foodchem.2020.127629

Torley, P., & Bhandari, B. (2007). Ultrasound in food processing and preservation. In: M. S. Rahman (Ed.), *Handbook of Food Preservation* (2nd ed., pp. 713–732). Boca Raton: CRC Press. doi:10.1201/9781420017373.ch29

Turrini, F., Boggia, R., Leardi, R., Borriello, M., & Zunin, P. (2018). Optimization of the ultrasonic-assisted extraction of phenolic compounds from *Oryza sativa* L. 'Violet Nori' and determination of the antioxidant properties of its caryopses and leaves. *Molecules, 23*(4), 844. doi:10.3390/molecules 23040844

Viell, F. L. G., Madeira, T. B., Nixdorf, S. L., Gomes, S. T. M., Bona, E., & Matsushita, M. (2020). Comparison between ultra-homogenisation and ultrasound for extraction of phenolic compounds from teff (Eragrostis tef (Zucc.)). *International Journal of Food Science and Technology, 55*(7), 2700–2709. doi:10.1111/ijfs.14523

Wang, Jianfei, Ma, H., & Wang, S. (2019). Application of ultrasound, microwaves, and magnetic fields techniques in the germination of cereals. *Food Science and Technology Research, 25*(4), 489–497. doi:10.3136/fstr.25.489

Wang, Jing, Sun, B., Cao, Y., Tian, Y., & Li, X. (2008). Optimisation of ultrasound-assisted extraction of phenolic compounds from wheat bran. *Food Chemistry, 106*(2), 804–810. doi:10.1016/j.foodchem.2007.06.062

Wang, Jing, Sun, B., Liu, Y., & Zhang, H. (2014). Optimisation of ultrasound-assisted enzymatic extraction of arabinoxylan from wheat bran. *Food Chemistry, 150*, 482–488. doi:10.1016/j.foodchem.2013.10.121

Wang, L. F., & Wang, Y. J. (2004a). Application of high-intensity ultrasound and surfactants in rice starch isolation. *Cereal Chemistry Journal, 81*(1), 140–144. doi:10.1094/CCHEM.2004.81.1.140

Wang, L. F., & Wang, Y. J. (2004b). Rice starch isolation by neutral protease and high-intensity ultrasound. *Journal of Cereal Science, 39*(2), 291–296. doi:10.1016/j.jcs.2003.11.002

Wang, Y., Zhao, L., Zhang, R., Yang, X., Sun, Y., Shi, L., & Xue, P. (2020). Optimization of ultrasound-assisted extraction by response surface methodology, antioxidant capacity, and tyrosinase inhibitory activity of anthocyanins from red rice bran. *Food Science and Nutrition, 8*(2), 921–932. doi:10.1002/fsn3.1371

Weiss, J., Kristbergsson, K., & Kjartansson, G. T. (2011). Engineering food ingredients with high-intensity ultrasound. In: H. Feng, G. V. BarbosaCanovas, & J. Weiss (Eds.), *Ultrasound Technologies for Food and Bioprocessing* (pp. 239–285). New York: Springer. doi:10.1007/978-1-4419-7472-3_10

Xia, Q., Tao, H., Li, Y., Pan, D., Cao, J., Liu, L., ... Barba, F. J. (2020). Characterizing physicochemical, nutritional and quality attributes of wholegrain *Oryza sativa* L. subjected to high intensity ultrasound-stimulated pre-germination. *Food Control, 108*, 106827. doi:10.1016/j.foodcont.2019.106827

Yaldagard, M., Mortazavi, S. A., & Tabatabaie, F. (2007). The effectiveness of ultrasound treatment on the germination stimulation of barley seed and its alpha-amylase activity. In: C. Ardil (Ed.), *Proceeding of World Academy of Science, Engineering and Technology* (p. 489+). Canakkale: World Academy of Science, Engineering and Technology – WASET.

Yaldagard, M., Mortazavi, S. A., & Tabatabaie, F. (2008a). Application of ultrasonic waves as a priming technique for accelerating and enhancing the germination of barley seed: Optimization of method by the Taguchi approach. *Journal of the Institute of Brewing, 114*(1), 14–21. doi:10.1002/j.2050-0416.2008.tb00300.x

Yaldagard, M., Mortazavi, S. A., & Tabatabaie, F. (2008b). The effect of ultrasound in combination with thermal treatment on the germinated barley's alpha-amylase activity. *Korean Journal of Chemical Engineering, 25*(3), 517–523. doi:10.1007/s11814-008-0087-1

Yang, W., Kong, X., Zheng, Y., Sun, W., Chen, S., Liu, D., ... Ye, X. (2019). Controlled ultrasound treatments modify the morphology and physical properties of rice starch rather than the fine structure. *Ultrasonics Sonochemistry, 59*, 104709. doi:10.1016/j.ultsonch.2019.104709

Yang, X., Li, Y., Li, S., Oladejo, A. O., Wang, Y., Huang, S., ... Duan, Y. (2018). Effects of ultrasound-assisted α-amylase degradation treatment with multiple modes on the extraction of rice protein. *Ultrasonics Sonochemistry, 40*(A), 890–899. doi:10.1016/j.ultsonch.2017.08.028

Ye, J., Feng, L., Xiong, J., & Xiong, Y. (2011). Ultrasound-assisted extraction of corn carotenoids in ethanol. *International Journal of Food Science and Technology, 46*(10), 2131–2136. doi:10.1111/j.1365-2621.2011.02727.x

Yu, S., Zhang, Y., Ge, Y., Zhang, Y., Sun, T., Jiao, Y., & Zheng, X.-Q. (2013). Effects of ultrasound processing on the thermal and retrogradation properties of nonwaxy rice starch. *Journal of Food Process Engineering, 36*(6), 793–802. doi:10.1111/jfpe.12048

Zadeike, D., Jukonyte, R., Juodeikiene, G., Bartkiene, E., & Valatkeviciene, Z. (2018). Comparative study of ciabatta crust crispness through acoustic and mechanical methods: Effects of wheat malt and protease on dough rheology and crust crispness retention during storage. *LWT – Food Science and Technology, 89*, 110–116. doi:10.1016/j.lwt.2017.10.034

Zahn, S., Schneider, Y., Zucker, G., & Rohm, H. (2005). Impact of excitation and material parameters on the efficiency of ultrasonic cutting of bakery products. *Journal of Food Science, 70*(9), E510–E513.

Zhang, H., Claver, I. P., Zhu, K.-X., & Zhou, H. (2011). The effect of ultrasound on the functional properties of wheat gluten. *Molecules, 16*(5), 4231–4240. doi:10.3390/molecules16054231

Zhang, Y., Dai, Y., Hou, H., Li, X., Dong, H., Wang, W., & Zhang, H. (2020). Ultrasound-assisted preparation of octenyl succinic anhydride modified starch and its influence mechanism on the quality. *Food Chemistry-X, 5*, 100077. doi:10.1016/j.fochx.2020.100077

Zhang, Y., Li, J., Li, S., Ma, H., & Zhang, H. (2018). Mechanism study of multimode ultrasound pretreatment on the enzymolysis of wheat gluten. *Journal of the Science of Food and Agriculture, 98*(4), 1530–1538. doi:10.1002/jsfa.8624

Zhang, Y., Ma, H., Wang, B., Qu, W., Li, Y., He, R., & Wali, A. (2015). Effects of ultrasound pretreatment on the enzymolysis and structural characterization of wheat gluten. *Food Biophysics, 10*(4), 385–395. doi:10.1007/s11483-015-9393-4

Zhang, Y. B., Wang, L. H., Zhang, D. Y., Zhou, L. L., & Guo, Y. X. (2014). Ultrasound-assisted extraction and purification of schisandrin B from *Schisandra chinensis* (Turcz.) Baill seeds: Optimization by response surface methodology. *Ultrasonics Sonochemistry, 21*(2), 461–466. doi:10.1016/j.ultsonch.2013.09.009

Zhao, R., Bean, S. R., & Wang, D. (2008). Sorghum protein extraction by sonication and its relationship to ethanol fermentation. *Cereal Chemistry Journal, 85*(6), 837–842. doi:10.1094/CCHEM-85-6-0837

Zhou, C., Hu, J., Yu, X., Yagoub, A. E. A., Zhang, Y., Ma, H., … Otu, P. N. Y. (2017). Heat and/or ultrasound pretreatments motivated enzymolysis of corn gluten meal: Hydrolysis kinetics and protein structure. *LWT-Food Science and Technology, 77*, 488–496. doi:10.1016/j.lwt.2016.06.048

Zhou, C., Ma, H., Ding, Q., Lin, L., Yu, X., Luo, L., … Yagoub, A. E.-G. A. (2013). Ultrasonic pretreatment of corn gluten meal proteins and neutrase: Effect on protein conformation and preparation of ACE (angiotensin converting enzyme) inhibitory peptides. *Food and Bioproducts Processing, 91*(4), 665–671. doi:10.1016/j.fbp.2013.06.003

Zhu, K.-X., Su, C.-Y., Guo, X.-N., Peng, W., & Zhou, H.-M. (2011). Influence of ultrasound during wheat gluten hydrolysis on the antioxidant activities of the resulting hydrolysate. *International Journal of Food Science and Technology, 46*(5), 1053–1059. doi:10.1111/j.1365-2621.2011.02585.x

Zhu, K.-X., Sun, X.-H., & Zhou, H.-M. (2009). Optimization of ultrasound-assisted extraction of defatted wheat germ proteins by reverse micelles. *Journal of Cereal Science, 50*(2), 266–271. doi:10.1016/j.jcs.2009.06.006

Chapter 6

Application of Ozone Technology for Grain Processing Industries

Gabriela John Swamy and Kasiviswanathan Muthukumarappan

CONTENTS

6.1 INTRODUCTION

Worldwide there are more than 1.1 billion tons of grains stored every year. Insects and fungi create serious quality problems in stored grains and annual storage losses are estimated at more than USD 800 million (Kells, Mason, Maier, & Woloshuk, 2001). The only way to eliminate pests completely from a food grain without leaving pesticide residues is fumigation. Currently, there are only two registered fumigants for stored food, methyl bromide and phosphine. Because of environmental concerns, the US government mandated that methyl bromide be eliminated from use in 2005. Phosphine tablets are used to control infestation by insects of grains stored in airtight structures such as metal silos, barrels, or brick silos, all of which can be sealed so that the gas cannot escape. Unfortunately, in some product storage environments, insects already exhibit some levels of phosphine resistance, and some show resistance to methyl bromide (Chaudhry, 1997). Loss of fumigants, resistance to remaining fumigants, and a trend by consumers to move away from residual chemicals necessitate the development of additional control strategies.

Ozone is a naturally occurring substance found in our atmosphere and it can also be produced synthetically. The characteristic fresh and clean smell in the air following a thunderstorm represents freshly generated ozone in nature. Ozone formation can also be observed in that faint garlicky aroma around the back of televisions, laser printers, or other electronic equipment, which indicates ozone is being produced. Ozone (O_3) can be generated by electrical discharges in the air and is currently used in the medical industry to disinfect against microorganisms and viruses, as a means of reducing odor, and for removing taste, color, and environmental pollutants in industrial applications. The attractive aspect of ozone is that it decomposes rapidly (half-life of 20–50 min) to molecular oxygen without leaving a residue. In 1982, the US Food and Drug Administration (FDA) classified ozone for treating bottled water as "generally recognized as safe" (GRAS) (FDA, 1982). In 2001, the USDA/FSIS approved ozone for use on meat and poultry products, including the treatment of ready-to-eat meat and poultry products just prior to packaging, with no labeling issues in regard to it being declared a treated product. With respect to the use of ozone in the direct-contact disinfection of food, there have been many studies conducted that show benefits ranging from microbial load reduction to increased shelf life of many different types of food products.

Thermal treatment is widely utilized for the processing and preservation of fruits and vegetables due to its effectiveness against microbial activation. But the process of heating tends to alter the organoleptic properties and destroys the thermo-labile nutrients (Choi & Nielsen, 2005). As a result, increasing interest in non-thermal technologies has emerged in order to attain efficient microbial inactivation with maximum retention of the organoleptic and nutritive value of foods (Diels & Michiels, 2006). Among the non-thermal methods, ozonation is an important one that has gained acceptance from consumers and is routinely used for washing fruits and vegetables in industries and as a disinfectant in bottled drinking waters. Ozone as an oxidant has numerous potential applications in the food industry because of its advantages over traditional food preservation techniques. Application of ozone either in gaseous or liquid form in fruit and vegetable processing is often employed for the inactivation of pathogen and spoilage microorganisms (Cullen et al., 2009). Apart from the wide spectrum of microbial inactivation, ozone also has the potential to kill storage pests and degrade mycotoxins. One of the potential advantages of ozone is that excess ozone auto-decomposes rapidly to produce oxygen and thus leaves no residues in food. Its efficacy against a wide range of microorganisms including bacteria, fungi, viruses, protozoa, and bacterial fungal spores has been reported (Cullen et al., 2009).

Effective application of ozone in food grain preservation would address the growing concern over the use of harmful pesticides to kill storage pests. Currently, commonly used pesticides (fumigants) for grain storage include aluminum phosphide, methyl bromide, phosphine, etc. The use of ozone as a food grain fumigant is a viable alternative from both environmental and economical perspectives. This chapter deals with the efficacy of ozone treatment for storing and preserving different types of food grains and its impact on the product quality of various food grains, as well as the current status of ozone application in grain processing industries.

6.2 STRUCTURE AND PROPERTIES OF OZONE

Ozone is derived from the Greek word "*Ozein*", meaning "smell". Ozone is a form of oxygen that contains three atoms (O_3) compared to the standard two atoms (O_2) in a molecule of oxygen. It is the allotropic form of oxygen, arranged as an isosceles triangle with an angle of 116.8 degrees between two oxygen bonds. The bond distance between the two oxygen atoms is 1.27 angstroms. The chemical structure of ozone is shown in Figure 6.1. Ozone as a gas is blue; both liquid (–111.9°C at 1 atmosphere) and solid ozone (–192.7°C) are an opaque blue-black color. The

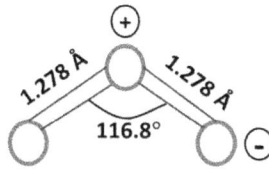

Figure 6.1 Chemical structure of ozone.

relatively high (+2.075 V) electrochemical potential (E^0, Volt) indicates that ozone is a very favorable oxidizing agent (Eqn. 6.1).

$$O_3(g) + 2H^+ + 2e^- \Longleftrightarrow O_2(g) + H_2O \{E^0 = 2.075\ V\} \tag{6.1}$$

Ozone is a bluish gas relatively unstable at normal temperatures and pressures. Ozone is denser than air at 0°C and 1 atmospheric pressure. It is partially soluble in water and the solubility varies with temperature. The solubility of ozone in water decreases with increasing temperature. At 0°C, solubility is 0.640 L ozone/L water, whereas at 60°C it is insoluble in water. The solubility is 13 times more than that of oxygen at 0–30°C and it is progressively more soluble in colder water. Ozone has a characteristic pungent odor and oxidizing properties and is the strongest disinfectant suitable for contact with foods (Muthukumarappan, Halaweish, & Naidu, 2000). Ozone is characterized by high electrochemical potential (+2.075 V) indicating strong oxidizing properties. The oxidation potential conveys bactericidal and virucidal properties and an ability to diffuse through biological membranes (Hunt & Mariñas, 1997). It is a potent antimicrobial agent against bacteria, fungi, viruses, protozoa, and also against bacterial and fungal spores. Ozone inactivates microbes through oxidization, and residual ozone spontaneously decomposes to oxygen, making it an environmentally friendly antimicrobial agent for use in the food industry (Patil & Bourke, 2012). Table 6.1 represents the properties of ozone.

TABLE 6.1   PROPERTIES OF OZONE

Properties	Value
Chemical formula	O_3
Occurrence	Bluish gas, dark blue fluid
Molecular weight	47.99 g/mole
Boiling point	-111.9 ± 0.3°C
Melting point	-192.5 ± 0.4°C
Critical temperature	-12.1°C
Critical pressure	54.6 atm
Density	2.14 kg/m³
Heat of formation	144.7 kJ/mole
Melting point	-192.7°C
Oxidation strength	2.075 V
Solubility in water	3 ppm at 20°C
Specific gravity	1.658

Source: Asokapandian, Periasamy, & Swamy, 2018.

6.3 GENERATION OF OZONE

Ozone exists naturally at low concentrations in the lower atmosphere. Natural ozone is found in the stratosphere at levels up to 6 ppm. The natural production of ozone is by either lightning or UV radiation. Ozone is a highly reactive and unstable gas; hence it is generated at the point of application for commercially demanded treatments. The half-life of ozone is about 20 to 30 min in distilled water at 20°C (Khadre, Yousef, & Kim, 2001). Ozone is generated by the rearrangement of atoms when strong O–O bonds are subjected to breaking by significant energy input with the formation of free oxygen radicals. The generation of ozone for commercial usage is achieved by four recognized methods: the electrical or corona discharge method, the electrochemical method, the ultraviolet method, and the radiochemical method (Muthukumarappan, O'Donnell, & Cullen, 2009; O'Donnell, Tiwari, Cullen, & Rice, 2012).

The electrical or corona discharge method is often called a silent discharge, where the molecular oxygen is ionized by applying a high-power alternating current. The dried air or oxygen is passed through an electric field produced between two high-voltage electrodes separated by a dielectric material, usually glass. Initially, the electrical current causes the "split" in the oxygen molecules (O_2) to form oxygen atoms (O), and later the individual oxygen atom combines with the remaining oxygen molecules to form ozone (O_3). During the ozone generation, around 80% of the applied energy is converted into heat and needs to be removed immediately to avoid the decomposition of ozone into oxygen molecules and atoms. Normally, 3% to 6% ozone is yielded in the gas mixture discharged from the ozonator if high-purity oxygen is used as the feed gas, whereas for dry air only 1% to 3% ozone is obtained (Muthukumarappan, O'Donnell, & Cullen, 2008). The advantages of this method are the effective generation of a high concentration of ozone, the higher durability of corona cells than that of a UV lamp, and better cost-effectiveness than that of a UV ozone generator for a large-scale installation.

In the electrochemical method of ozone generation, an electrical current is applied between an anode and cathode placed in an electrolytic solution containing water and highly electronegative anions. A mixture of oxygen and ozone is produced at the anode. The merits of this method are the use of low-voltage DC current, no preparation of feed gas, compact equipment size, and probably high concentration of ozone generation (Mahapatra, Muthukumarappan, & Julson, 2005).

Ozone generation using ultraviolet light is a photochemical process in which ambient air is passed over an ultraviolet lamp emitting UV light of 140–190 nm wavelength. Through photodissociation, oxygen molecules are split into oxygen atoms, which further combine with other oxygen molecules to form ozone. The disadvantage of this method is that a very low concentration of ozone (0.1% w/w) is produced, which limits the practical application of this method.

High-energy irradiation of oxygen by radioactive substances can also produce ozone. Usually, isotopes such as ^{137}Cs, ^{60}Co, or ^{90}Sr are used for the excitation of circulating air that initiates the dissociation of oxygen molecules to form oxygen atoms, which combine to form ozone. This technique is rarely associated with commercial usage due to its complicated application and the danger of radioactive contamination (Heim & Glas, 2011).

6.4 REGULATIONS FOR USAGE OF OZONE

Ozone was granted GRAS (Generally Recognized as Safe) status by the FDA for use in the treatment of bottled waters. In 2001, ozone was approved as a direct food additive for use as an antimicrobial agent in foods and bottled water. According to the Federal Register for Secondary Food Additives, "Ozone (CAS Reg. No. 10028-15-6) may be safely used in the treatment, storage, and processing of foods, including meat and poultry" (FDA, 2001). As per the regulations of the US Environmental Protection Agency, a residual dissolved ozone concentration of 0.4 mg/L is permitted. In order to achieve pathogen reduction of juice with ozone treatment, HACCP regulations must be carried out as a process that will produce, at a minimum, a 5-log reduction of the most resistant microorganism of public health concern, and the treatment process must be validated. At the same time, ozone at high concentrations is potentially harmful to human beings. Exposure to ozone levels of 1–2 ppm could cause irritation of the throat, dry cough, headache, and chest pain. As per the Occupational Safety and Health Administration, the short-term and long-term limits for ozone exposure in the workplace are 0.3 ppm for 15 min and 0.1 ppm for 8 h d/40 h workweek (Suslow, 2004).

6.5 APPLICATIONS OF OZONE IN GRAINS DURING STORAGE

The patented Oxygreen process was one of the most significant advances for the application of ozone in food grains (Dubois et al., 2006, 2008). This process involves a premoistening of grains in a closed batch reactor followed by ozonation. Similarly, Yvin et al. (2005) patented a process to obtain flour with enhanced microbial safety from ozonated grains. Within the food industry, ozone is employed for fresh fruit and vegetable decontamination. However, a limited number of studies have been reported on the ozone treatment of cereals and cereal-based products as an alternative to chlorine treatment.

Ozone treatment of grain is generally applied in silos or vessels. Prior to ozone application, it is necessary to characterize the dynamics of ozone movement through the various grain types to optimize ozone generators for use on large commercial storage bins (Shunmugam et al., 2005). Ozone moves through grain slowly because the gas reacts with the chemical constituents present in the outer layer of the grain (seed coat). The diffusion of ozone into the grain depends upon the grain characteristics. Movement of ozone within a silo or column filled with grain can be in any of three directions; namely, movement in the transverse direction (x,z) or movement in the vertical direction (y) under the influence of ozone gas velocity (vf) and adsorption of ozone by the grain surface and possible reactions leading to degradation of ozone. Adsorption of ozone and subsequent penetration into the grain depends upon several intrinsic and extrinsic factors (k) such as surface characteristics of the grain, microbial contamination, presence of insects, and moisture content, etc. Penetration and movement of ozone within a grain column can be expressed by the differential kinetic–diffusion equation (Eqn. 6.2) (Raila et al., 2006).

$$\frac{\partial C_0}{dt} = D\left(\frac{1}{r}\frac{\partial}{\partial r}\left[r\frac{\partial C_0}{\partial r}\right] + \frac{\partial^2 C_0}{\partial h^2}\right) - v_a\left(\frac{\partial C_0}{\partial h}\right) - kC_0 \tag{6.2}$$

where: C_0 = ozone concentration; D = diffusivity; r = radius of the bottom of the grain mound; h = grain mound height; k = factor of ozone absorption; v_a = air seepage velocity in the grain layer; and t = time of exposure to ozone.

Ozone movement through the grain layer is restricted by the highly reactive nature of ozone. Kells et al. (2001) divided the movement of ozone into two distinct phases for maize. The first phase is the contact of ozone with grains, during which the concentration of ozone reduces as it moves in the y-direction through the grain due to the interaction with organic materials present on or in the vicinity of the grain surface rapidly degrading the ozone through an oxidation reaction. The movement of ozone in the first phase is restricted due to the ozone demand of the organic matter (Cullen et al., 2009; Kim et al., 1999). The second phase corresponds to the free movement of ozone through grain layers once these reactive sites are eliminated (Kells et al., 2001; Mendez et al., 2003). Ozone adsorption in the grain layer depends on ozone concentration in the feed gas, duration of exposure, gas flow rate, temperature, grain characteristics, the presence of other organic matter such as insects, and the surface microbial status of the grain. The presence of moisture also plays an important role in ozone reactivity with grain because water solubilizes ozone and increases contact between gas and grain. Raila et al. (2006) observed slower ozone penetration between grain layers with higher mycological contamination.

6.6 REGULATION OF GRAIN PESTS USING OZONE

Grains are frequently stored for periods of up to 36 months at ambient temperatures in bulk silos and are often fumigated to prevent infestation and contamination. Within the grain processing industry, ozone is employed as a replacement for the existing fumigants such as methyl bromide and phosphine for the control of storage pests. Ozone as a fumigant is reported to kill stored-grain insects such as *Tribolium castaneum*, *Rhyzopertha dominica*, *Oryzaephilus surinamensis*, *Sitophilus oryzae*, and *Ephestia elutella* (Sousa et al., 2008). The respiratory system is the major entry route for toxic gases into an insect body, determining the lethality of a fumigant as the respiratory system of insects consists of highly branched cuticle-lined tubes extending throughout the body. Insects breathe discontinuously to minimize oxidative damage due to oxygen toxicity. Ozone causes oxidative tissue damage even at low concentrations, resulting in DNA strand breaks, alteration of pulmonary function, bronchial responsiveness, membrane oxidation, or mutations in vivo. Increased respiration rate with increasing temperatures may result in more gas exchange due to the overall increase in metabolic and respiration rate.

Laboratory and field studies report the efficacy of ozone in controlling both phosphine-susceptible and phosphine-resistant strains of *Silophilus zeamais*, *S. oryzae*, *R. dominica*, and *T. castaneum* (Tiwari et al., 2010a). Ozone toxicity for insects varies depending on the stage within its life cycle. For example, larval and pupal stages of *T. castaneum* are ozone sensitive with sensitivity decreasing with age (McDonough, Mason, & Woloshuk, 2011). The mortality rate of two flour beetles (*Ephestia kuehniella* and *Tribolium confusum*) was studied and it was observed that insect mortality during ozonation was not only dependent upon the life stages specific to both the species but was also insect-specific (Isikber & Athanassiou, 2015). The results showed a higher susceptibility and high mortality for all three larvae, pupae, and adult stages of *E. kuehniella* (90–100%) compared to *T. confusum* (1.3–22.7%) for similar experimental conditions. Likewise, an elevated susceptibility rate for insects is stated for Indian meal moth (*Plodia interpunctella*) compared to *T. confusum* (Leesch, 2003). A high mortality rate for maize weevil, red flour beetle, and the Indian meal moth exposed to ozone (50 ppm for 3 days or 25 ppm for 5 days) has been reported (Tiwari et al., 2010a). Table 6.2 lists reported applications of ozone for the control of insects.

6.7 MICROBIAL DECONTAMINATION AND MYCOTOXINS

Ozone in gaseous or aqueous form is reported to reduce levels of the natural microflora, as well as bacterial, fungal, and mold contamination in cereals and cereal products, including spores of

TABLE 6.2   EFFECTS OF OZONE TREATMENT ON GRAIN PESTS

Grain	Pest	Results	References
Wheat	• *Rhyzopertha dominica*	• The optimized treatment conditions for ozone fumigation of stored wheat grain were 12% (w/w) grain moisture, 2.5 g/m³ ozone concentration, and 8h of treatment • The mortality of *R. dominica* adults, pupae, larvae, and eggs was 97, 100, 99, and 100%	Mishra, Palle, Srivastava, & Mishra, 2019
Wheat	• *Plodia interpunctella* • *Sitophilus oryzae* • *Tribolium castaneum* • *Rhyzopertha dominica* • *Cryptolestes ferrugineus* • *Oryzaephilus surinamensis*	• Ozone treatments on eggs and larvae of *Plodia interpunctella* were not effective, but pupae were more susceptible • *Sitophilus oryzae* (L.) adults were the most susceptible species with 100% mortality reached after 2 d in all ozone treatments • *Tribolium castaneum* adults had 100% mortality only after 4 d at 50 or 70 ppmv • For *Rhyzopertha dominica* (F.), *Cryptolestes ferrugineus* (Stephens), and *Oryzaephilus surinamensis* (L.), 100% mortality was never achieved, and progeny were produced at all ozone concentrations	Bonjour et al., 2011
Wheat	• *Ephestia kuehniella* • *Tribolium confusum*	• For *E. kuehniella*, empty space ozone treatment resulted in complete mortality of adults, pupae, and larvae, while only 62.5% of the eggs were killed • For *T. confusum*, ozone treatment resulted in very low mortality of adults, pupae, and eggs, ranging from 4.2 to 14.1% while only larvae had high mortality (74%) • Ozone flush treatment at 30-min intervals for 5 h resulted in almost complete mortality of all life stages of *E. kuehniella* placed in the top position of 2 kg wheat, whereas eggs of *E. kuehniella* placed in the bottom position of 2 kg wheat were hard to kill. • For *T. confusum*, larvae placed in the bottom position of 2 kg wheat were easily killed, whereas eggs, pupae, and adults survived	Işıkber & Öztekin, 2009
Broken kernel mix	*Oryzaephilus mercator*	• Mortality was higher when insects were treated without food • When food was not provided, a minimum of 11,030 ppm for 1 h is required to kill 99% of eggs, the most tolerant stage, whereas 500 ppm for 1 h is required to kill 99% of larvae, the least tolerant • When provided with food, adults were the most tolerant and larvae the least tolerant • Adults require exposure time of 7.7 h of 100 ppm ozone to kill 99% of insects in the absence of food.	Mahroof, Amoah, & Wrighton, 2018

(Continued)

TABLE 6.2 (CONTINUED)   EFFECTS OF OZONE TREATMENT ON GRAIN PESTS

Grain	Pest	Results	References
Whole wheat flour, cornmeal	*Lasioderma serricorne*	• The concentration–mortality estimates suggested that, generally, higher concentrations were required to kill 99% of insects when treated without food compared to when treated with food made of whole wheat flour and brewer's yeast • In the absence of food, larvae were the most tolerant and adults were the least tolerant to ozone treatment and required 15 h, 974 ppm, and 3,769 ppm to kill 99% of the individuals respectively • In the presence of food, eggs and pupae were the most and least tolerant respectively • An exposure time of $7.1 \times 1,030$ h was required to kill 99% of adults treated in the presence of food and 99 h in the absence of food • When adult insects were exposed to 100 ppm ozone for 1–6 h in the absence of food, a weak relationship between survival rate and exposure duration observed initially became stronger in subsequent days	Amoah & Mahroof, 2018
Wheat, rice	• *Sitophilus oryzae* • *Tribolium castaneum* • *Rhyzopertha dominica*	• *Sitophilus oryzae* (L.), *Rhyzopertha dominica* (F.), and *Tribolium castaneum* (Herbst) adults were exposed to atmosphere containing 0.1, 0.2, or 0.4μg/ml initial ozone at 23–25°C and 50% r.h. • The experiments showed that the effects of ozone on respiration had two distinct phases. Phase 1 involved a lower respiration rate of the adult stored-product *Coleoptera* under ozone atmosphere and reflected the need for insects to reduce ozone toxicity • After 1 h, CO_2 production of *S. oryzae* was 3.19, 2.63, 2.27, and 1.99 μl/mg for the ozone concentration of 0, 0.1, 0.2, and 0.4 μg/ml • The results also showed that there were decreases in the rate of respiration in *R. dominica* and *T. castaneum* with an increase in ozone concentration • During phase 2, respiration of *S. oryzae*, *R. dominica*, and *T. castaneum* adults treated with ozone increased as the ozone degraded to oxygen • After 7 h, the effect of ozone on CO_2 production, relative to the control, changed from a decrease to an increase	Lu, Ren, Du, Fu, & Gu, 2009

Bacillus, Coliform bacteria, *Micrococcus, Flavobacterium, Alcaligenes, Serratia, Aspergillus,* and *Penicillium.* Some studies show that up to 3 log reductions of microorganisms in cereal grains can be achieved depending on ozone concentration, temperature, and relative humidity conditions. Transmission electron microscopic micrographs of *Bacillus* spores treated with ozone suggest that ozone inactivates spores by degrading the outer spore component (spore coat layers comprise approximately 50% of the spore volume), thus exposing the cortex and core to the action of ozone (Tiwari et al., 2010a).

Fungal or mold contamination of food grains is one of the most important issues determining grain quality, with both qualitative and quantitative losses reported due to microbes (Figure 6.2). Microorganisms that are present in food grains, either on their surface or internally, deteriorate the nutritional quality of the products and produce metabolites (e.g., mycotoxins) that are dangerous to human and animal health. Ozone has been effectively used to control fungal growth and reduce mycotoxin contamination. Fungal and mold growth lead to the release of secondary metabolites known as mycotoxins. These are known to exhibit carcinogenic, teratogenic, immunosuppressive properties and cause several physiological disorders both in humans and animals. The human health risks from mycotoxins in cereals and cereal-based products (e.g., beer) and the risk to animals from contaminated feed and cereal byproducts is widely reported. Application of ozone in grain handling and storage could reduce or eliminate mycotoxins and undesirable microflora (fungi and molds) from grain and grain products.

Ozone is reported to be effective in the detoxification and degradation of commonly occurring mycotoxins such as aflatoxin, patulin, cyclopiazonic acid, secalonic acid D, ochratoxin A, and ZEN. Like the killing of insects, fungal inactivation and the subsequent decontamination of toxins depend upon several factors including ozone concentration, exposure time, pH, and moisture content of the grain mass. For example, Raila et al. (2006) observed approximately 2.2 times

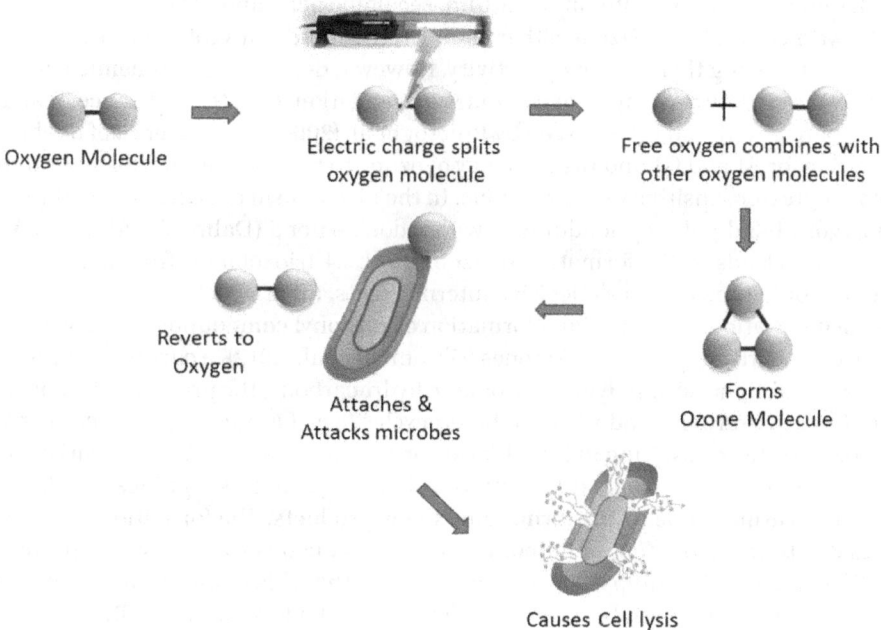

Oxygen Molecule

Electric charge splits
oxygen molecule

Free oxygen combines with
other oxygen molecules

Reverts to
Oxygen

Attaches &
Attacks microbes

Forms
Ozone Molecule

Causes Cell lysis

Figure 6.2   Microbial inactivation by ozone treatment.

higher fungal decontamination of wheat grains at a grain moisture content of 15.2% and 3 times higher decontamination at 22.0% grain moisture compared to dry conditions. This could be due to the greater efficacy of ozone in aqueous media compared to a gaseous phase, higher fungal growth in wet conditions than in dry conditions, and slower movement of ozone within grain layers in moist conditions, thus allowing greater exposure time. Hence, moisture is an important factor influencing the efficacy of ozone. Application of dry ozone gas is reported to be less effective compared to moist conditions. For example, a 90% reduction in deoxynivalenol (also known as vomitoxin) during moist ozone (1.1 mol %) treatment and a 70% reduction after 1 h of ozone treatment under similar experimental conditions in corn was observed by Young, Zhu, and Zhou (2006) and Tiwari et al. (2010a). The efficacy of ozone is also reported to be influenced by the pH of the medium. The researchers also observed a rapid degradation of trichothecene mycotoxins at a low pH (pH 4–6) compared to a higher pH (pH 7–8). However, the effect of pH is unimportant as far as ozonation of food grain is concerned.

Grain temperature also influences the efficacy of ozone in the degradation of mycotoxins. A greater degradation of aflatoxins in peanut kernels at higher temperatures was observed by Proctor, Ahmedna, Kumar, and Goktepe (2004). A reduced treatment time from 15 min to 10 min for a temperature rise from 25°C to 75°C to obtain a degradation of 77% for AFB1 and 80% for AFG1, respectively, was reported. Further, they achieved a greater degradation in peanut kernels compared to flour, most likely due to the larger exposure area and the fact that fungal or mold contamination is principally found on the grain surface. Lower doses of ozone (5 ppm in atmosphere) are reported to inhibit surface growth, sporulation, and mycotoxin production by *Aspergillus flavus* and *Fusarium moniliform* (Mason, Woloshuk, & Maier, 1997). Likewise, an ozone dose of 0.16 mg/g barley is sufficient to achieve 96% inactivation of fungal spores within 5 min as suggested by Allen, Wu, and Doan (2003). Ozone gas was demonstrated to be effective for the degradation and detoxification of common mycotoxins (aflatoxins B1, B2, G1, and G2, cyclopiazonic acid, fumonisin B1, ochratoxin A, patulin, secalonic acid, and zearalenone) in an aqueous solution (Tiwari et al., 2010a). Ozone either completely degrades mycotoxins or causes chemical modifications, reducing their biological activity. However, degradation or chemical modification is specific to the structure of mycotoxins under investigation. Greater resistance of aflatoxin B2 and G2 compared to B1 and G2 was seen by Proctor et al. (2004). The presence of double bonds at C8–C9 position for B1 and G1 and the tendency of ozone to react at olefinic positions indicate the possibility for greater sensitivity toward ozone. In the mechanism of degradation, the ozone molecule undergoes 1–3 dipolar cycloadditions with a double bond (Dalmázio, Almeida, Augusti, & Alves, 2007). This leads to the formation of ozonides (1,2,4-trioxolanes) from alkenes and ozone with aldehyde or ketone oxides as decisive intermediates, all of which have finite lifetimes. The oxidative disintegration of ozonide and formation of carbonyl compounds result, while oxidative work-up leads to carboxylic acids or ketones (O'Donnell et al., 2012). Apart from the degradation of double-bonded aliphatic or polycyclic aromatic hydrocarbons, the presence of chlorinated ring structures (e.g., Ochratoxin) and nitrogen heterocycles (e.g., Fumonisin) in mycotoxins are also subject to ozone attack, resulting in free chlorine or amino acids (Zhu, 2018; Tiwari et al., 2010b).

Studies show that the biological toxicity of certain mycotoxins significantly decreases following ozone treatment due to the formation of new products. The formation of Fumonisin B1 derivatives due to the ozonation of Fumonisin B1 in an aqueous solution was reported (Lemke, Ottinger, Ake, Mayura, & Phillips, 2001). They proposed that this might be due to the reaction of ozone with the primary amine (–NH2) group leading to the formation of nitrogen oxide (–NO$_2$ or –N$_2$O) coupled with the formation of ketone by the Criegee mechanism. However, the authors observed biological toxicity in a mycotoxin-sensitive bioassay, possibly due to the presence of an intact primary amine in the Fumonisin B1 derivative. It has been shown that the primary amine group is necessary for the biological activity of fumonisin-like compounds.

6.8 APPLICATIONS OF OZONE IN GRAIN'S COMPOSITION QUALITY

Starch is a biodegradable biopolymer that acts as a source of energy in humans as well as in plants. The majority of starch is extracted from corn, wheat, potato and cassava, banana, and amaranth. Irrespective of the source, starch granules consist of two different kinds of glucose polymers, namely amylose and amylopectin. There are a few essential factors that affect the properties of starch and its applicability in the food industry: (i) the ratio and chain length between amylose and amylopectin; and (ii) the packing pattern of the amylopectin in a semi-crystalline domain (Pandiselvam et al., 2019). When the native starch molecules do not exhibit the desired functional properties, they can be modified using the physical, chemical, enzymatic, and genetic methods to improve their functional and rheological properties for use in foods. In general, chemical methods (acetylation, cross-linking, hydroxypropylation, oxidation, and etherification) are highly efficient in modifying starch. The starch that is modified using a chemical method possesses several advantages in the food industry due to its low retrogradation and gel synersis, while the texture of gel, paste clarity, and film adhesion are improved. Oxidation using sodium hypochlorite is a commonly employed technique, but this technique is deterred by more time consumption, the formation of chemical residues, and the generation of toxic wastes. To counter these challenges, ozone treatment, a novel green technological approach, is being used because it does not cause any harm to the environment and leaves no residues in the food product.

The modification of starches by using ozone is very efficient, as ozone reacts with starch molecules even at room temperatures and even in the absence of a catalyst, and does not require any controlled or specific conditions for the reaction to progress. The different treatment conditions used for starch modification using ozone are listed in Table 6.3. In addition to their nutritional quality, proteins in food should possess specific functional properties to facilitate processing. The functional properties of proteins, therefore, serve as a basis for product performance. Functional properties of proteins are dictated by their physicochemical properties, which broadly govern the behavior of proteins in foods. These properties can be broadly classified into two different groups: hydrodynamic or hydration-related properties, which include water absorption, solubility, viscosity, gelation, and aroma retention capacity, and surface-active properties, such as emulsification, foaming, and film formation capacity. Treatment of ozone on the deproteinization of starch granules increases the degree of oxidation. Resident starches exhibit higher carboxylic content as compared to deproteinized starch and deproteinization causes reduction in pasting viscosity. These types of changes can prove that the presence of protein, polyphenols, and non-starch polysaccharides reduces the effect of ozone on starch granules; thus a modified starch molecule is not as reactive as unmodified grain molecules. Although a basic scientific explanation is available in relation to the action of ozone on proteins, little is known about the change in functional properties of proteins in food systems after ozonolysis.

6.9 CONCLUSION

Ozone is an effective sanitizing agent with promising applications in the food processing industries. It acts as a potential substitute for chemical sanitizers and disinfectants. Even at low

TABLE 6.3   STARCH MODIFICATION BY OZONATION

Source	Treatment Conditions	Result of Ozone Treatment	References
Potato starch	• Ozone generated by coronal-discharge method from industrial oxygen (95% purity) • Flow rate – 0.5 L/min • Ozone concentration – 47 mg/L • Time – 15, 30, 45, 60 min	• There was a more significant amount of carboxyl groups than carboxyl groups after ozonation in all the processed samples • The relative crystallinity (RC) presented no significant ($p < 0.05$) difference between the native and the ozonated starches • After 30 min of ozonation, the granules with irregular shapes were more frequent and there were some fissures and pores on the granules' surfaces • The relative setback (RSB) of the ozonated potato starches were higher than the native starch, besides having a lower amount of hydroxyl groups if compared to the native sample • Furthermore, the 15- and 30-min samples presented even higher values than the 45- and 60-min samples. This can be explained by the molecular depolymerization since smaller molecular chains present a more suitable size for re-association due to their higher mobility	Castanha, da Matta Junior, & Augusto, 2017; Castanha, Santos, Cunha, & Augusto, 2019
Cassava starch	• Time – 60 min • pH – 3.5, 6.5, and 9.5 • Temperature – 25°C	• The pH 6.5 and 9.5 seemed to favor the cross-linking between the depolymerized starch molecules during ozonation • The pH 3.5 was more effective in reducing the peak viscosity, breakdown, setback, and final viscosity of cassava starch during ozonation in an aqueous solution • No differences in the granule surface morphology were observed in the ozone-treated cassava starches compared to native starch	Klein et al., 2014

(Continued)

TABLE 6.3 (CONTINUED)   STARCH MODIFICATION BY OZONATION

Source	Treatment Conditions	Result of Ozone Treatment	References
Maize starch	• Ozone generated by coronal-discharge method from industrial oxygen (95% purity) • Flow rate – 1 L/min • Ozone concentration – 42 mg/L US conditions were: • Processing time – 8 h • Temperature of 24–26°C • Frequency – 25 kHz • Volumetric power – 72 W/L	• The ultrasound (US) treatment alone did not show influence on the starch's physical characteristics • On the other hand, the O_3 treatment, alone or in combination with US, led to significant changes in starch molecules by increasing carbonyl and carboxyl groups and the apparent amylose content, while decreasing pH and the starch molecular size distribution • The granules' particle size distribution (PSD), morphology, and crystallinity were not affected by any of the treatments • Regarding the starch properties, water absorption index (WAI), water solubility index (WSI), pasting properties, and gel strength were clearly more affected by the ozone treatment as compared with the ultrasound treatment • The paste clarity was significantly higher when the combined treatments were applied, especially when US was used before O_3	Castanha, Lima, Junior, Campanella, & Augusto, 2019
Cassava starch	• Ozone generated by coronal-discharge method from industrial oxygen (95% purity) • Flow rate – 1 L/min • Ozone concentration – 45 mg/L	• Even though the dry heat treatment (DHT) alone did not change the surface of starch granules, this process possibly weakened the surface structure, making it more susceptible to cracking with the next treatment (ozone) • However, this behavior was not observed in the inverse sequence of treatments (ozone and later DHT)	Lima, Maniglia, et al., 2020
Wheat starch	• Ozone gas (1,500 mg/kg at 2.5 L/min) • Time – 30, 45 min	• Chemical analysis of starch isolates indicated depolymerization of high molecular weight amylopectins, with a subsequent increase in low-molecular-weight starch polymers as a result of starch hydrolysis • Ozone treatment resulted in elevated levels of carboxylic groups and decreased total carbohydrate content in amylopectin fractions	Sandhu, Manthey, & Simsek, 2012

(Continued)

TABLE 6.3 (CONTINUED)   STARCH MODIFICATION BY OZONATION

Source	Treatment Conditions	Result of Ozone Treatment	References
Arracacha starch	• Ozone generated by coronal-discharge method from industrial oxygen (95% purity) • Flow rate – 1 L/min • Ozone concentration – 47 mg/L	• Native arracacha starch presents some drawbacks that limit its further application, such as a weak gel and a high consistent paste. Therefore, modification techniques can expand its utilization • The ozonation process promoted structural changes – reduction in the size of the molecules, formation of carbonyl and carboxyl groups, and an increased reduction of sugar content – that significantly affected the arracacha starch properties • Among the main results, it is worth highlighting the improved stability of paste clarity and hydrogels up to 250% stronger than the native one	Lima, Villar, et al., 2020
Cocoyam and yam starches	• Ozone exposure time – 5, 10, 15 min	• The correlations between the amount of reacted ozone and carbonyl and carboxyl contents of the starches were positive, as ozone generation time (OGT) increased • The DSC data showed lower transition temperatures and enthalpies for retrograded gels compared to the gelatinized gels of the same starch types	Oladebeye, Oshodi, Amoo, & Abd Karim, 2013
Corn, sago, and tapioca starch	• Ozone generated by coronal-discharge method	• Carboxyl and carbonyl contents increased markedly in all starches with increasing ozone generation times • Oxidation significantly decreased the swelling power of oxidized sago and tapioca starches but increased that of oxidized corn starch • The solubility of tapioca starch decreased, and sago starch increased after oxidation. However, there was an insignificant change in the solubility of oxidized corn starch • Intrinsic viscosity of all oxidized starches decreased significantly, except for tapioca starch oxidized at 5 min ozone generation times • Pasting properties of the oxidized starches followed different trends as ozone generation times increased	Chan, Bhat, & Karim, 2009

(Continued)

TABLE 6.3 (CONTINUED)   STARCH MODIFICATION BY OZONATION

Source	Treatment Conditions	Result of Ozone Treatment	References
Waxy rice flour and starch	• Ozone generated by coronal-discharge method	• Compared with untreated waxy rice flour, the peak viscosities of waxy rice flour for 0.5, 1, and 2 h of ozone treatments were increased by 27.4%, 32.8%, and 45.5%, respectively • The alpha-amylase in waxy rice flour was inactivated during the treatment. The gelatinization temperature and enthalpy of waxy rice flour were kept unchanged after the treatment • For waxy rice starch, pasting viscosity, swelling power, and molecular weight were increased after 0.5 h of treatment but decreased as treatment time extended • The ozone treatment decreased gelatinization temperature and enthalpy of waxy rice starch	Ding, Wang, Zhang, Shi, & Wang, 2015
Rice starch	• Ozone generated by coronal-discharge method	• Ozonated starch exhibited similar pasting properties to those from oxidized starches treated with low concentrations of chemical oxidizing agents • The combination of lysine with ozonation resulted in pasting properties similar to starches treated with high levels of chemical oxidizing agents • The ozonated starch could be used as a thickening agent, whereas ozonated starch with lysine might be an alternative for a highly chemically oxidized starch	An & King, 2009

concentrations, it is highly effective against a broad spectrum of microorganisms. Ozone treatment of food products depends on the number and type of microorganisms, the nature of the product, the method of ozone generation and its application system, temperature, pH, and other factors. This method is preferable to other chemical agents because it leaves no hazardous residues on food or food-contact surfaces. Another advantage is the considerable reduction in the transportation cost and storage of sanitizers since ozone only needs to be produced on-site. With further research and development as well as innovation in ozone generation and application systems, the technique will be applied in food processing more effectively in the future.

Planning successful studies to decontaminate food by ozone requires (a) a good understanding of the physical and chemical properties of this unique compound, (b) availability of adequate ozone generation, containment, and destruction equipment, (c) expertise in microbiology so that pathogen monitoring in food is performed correctly, and (d) well-trained analysts who can work safely with this hazardous chemical.

6.10 ACKNOWLEDGMENTS

The authors would like to thank the partial funding received from the USDA NIFA through the South Dakota Agricultural Experiment Station at South Dakota State University.

REFERENCES

Allen, B., Wu, J., & Doan, H. (2003). Inactivation of fungi associated with barley grain by gaseous ozone. *Journal of Environmental Science and Health, Part B, 38*(5), 617–630.

Amoah, B. A., & Mahroof, R. M. (2018). Susceptibility of the life stages of cigarette beetle, *Lasioderma serricorne* (f.) (Coleoptera: Anobiidae) to ozone. *Journal of Stored Products Research, 78*, 11–17.

An, H., & King, J. (2009). Using ozonation and amino acids to change pasting properties of rice starch. *Journal of Food Science, 74*(3), C278–C283.

Asokapandian, S., Periasamy, S., & Swamy, G. J. (2018). Ozone for fruit juice preservation. In: *Fruit Juices* (pp. 511–527). Elsevier. https://www.elsevier.com/books/fruit-juices/rajauria/978-0-12-802230-6

Bonjour, E., Opit, G., Hardin, J., Jones, C., Payton, M., & Beeby, R. (2011). Efficacy of ozone fumigation against the major grain pests in stored wheat. *Journal of Economic Entomology, 104*(1), 308–316.

Castanha, N., da Matta Junior, M. D., & Augusto, P. E. D. (2017). Potato starch modification using the ozone technology. *Journal of Food Hydrocolloids, 66*, 343–356.

Castanha, N., Lima, D. C., Junior, M. D. M., Campanella, O. H., & Augusto, P. E. D. (2019). Combining ozone and ultrasound technologies to modify maize starch. *International Journal of Biological Macromolecules, 139*, 63–74.

Castanha, N., Santos, D. N., Cunha, R. L., & Augusto, P. E. D. (2019). Properties and possible applications of ozone-modified potato starch. *Journal of Food Research International, 116*, 1192–1201.

Chan, H. T., Bhat, R., & Karim, A. A. (2009). Physicochemical and functional properties of ozone-oxidized starch. *Journal of Agricultural and Food Chemistry, 57*(13), 5965–5970.

Chaudhry, M. (1997). Review A review of the mechanisms involved in the action of phosphine as an insecticide and phosphine resistance in stored-product insects. *Journal of Pesticide Science, 49*(3), 213–228.

Choi, L., & Nielsen, S. S. (2005). The effects of thermal and nonthermal processing methods on apple cider quality and consumer acceptability. *Journal of Food Quality, 28*(1), 13–29.

Cullen, P. J., Tiwari, B. K., O'Donnell, C. P., & Muthukumarappan, K. (2009). Modelling approaches to ozone processing of liquid foods. *Trends in Food Science and Technology, 20*(3–4), 125–136.

Dalmázio, I., Almeida, M. O., Augusti, R., & Alves, T. M. (2007). Monitoring the degradation of tetracycline by ozone in aqueous medium via atmospheric pressure ionization mass spectrometry. *Journal of the American Society for Mass Spectrometry, 18*(4), 679–687.

Diels, A. M., & Michiels, C. W. (2006). High-pressure homogenization as a non-thermal technique for the inactivation of microorganisms. *Critical Reviews in Microbiology, 32*(4), 201–216.

Ding, W., Wang, Y., Zhang, W., Shi, Y., & Wang, D. (2015). Effect of ozone treatment on physicochemical properties of waxy rice flour and waxy rice starch. *International Journal of Food Science Technology, 50*(3), 744–749.

Dubois, M., Canadas, D., Despres-Pernot, A.-G., Coste, C., & Pfohl-Leszkowicz, A. (2008). Oxygreen process applied on nongerminated and germinated wheat: role of hydroxamic acids. *Journal of Agricultural and Food Chemistry, 56*(3), 1116–1121.

Dubois, M., Coste, C., Despres, A.-G., Efstathiou, T., Nio, C., Dumont, E., & Parent-Massin, D. (2006). Safety of Oxygreen®, an ozone treatment on wheat grains. Part 2. Is there a substantial equivalence between Oxygreen-treated wheat grains and untreated wheat grains? *Journal of Food Additives and Contaminants, 23*(1), 1–15.

Heim, C., & Glas, K. (2011). Ozone I: Characteristics/generation/possible applications. *Brewing Science, 64*, 8–12.

Hunt, N. K., & Mariñas, B. J. (1997). Kinetics of *Escherichia coli* inactivation with ozone. *Water Research, 31*(6), 1355–1362.

Isikber, A. A., & Athanassiou, C. G. (2015). The use of ozone gas for the control of insects and micro-organisms in stored products. *Journal of Stored Products Research, 64*, 139–145.

Işikber, A. A., & Öztekin, S. (2009). Comparison of susceptibility of two stored-product insects, Ephestia kuehniella Zeller and Tribolium confusum du Val to gaseous ozone. *Journal of Stored Products Research, 45*(3), 159–164.

Kells, S. A., Mason, L. J., Maier, D. E., & Woloshuk, C. P. (2001). Efficacy and fumigation characteristics of ozone in stored maize. *Journal of Stored Products Research, 37*(4), 371–382.

Khadre, M., Yousef, A., & Kim, J. (2001). Microbiological aspects of ozone applications in food: A review. *Journal of Food Science -Chicago, 66*(9), 1242–1253.

Kim, J.-G., Yousef, A. E., & Dave, S. (1999). Application of ozone for enhancing the microbiological safety and quality of foods: a review. *Journal of Food Protection, 62*(9), 1071–1087.

Klein, B., Vanier, N. L., Moomand, K., Pinto, V. Z., Colussi, R., da Rosa Zavareze, E., & Dias, A. R. G. (2014). Ozone oxidation of cassava starch in aqueous solution at different pH. *Journal of Food Chemistry, 155*, 167–173.

Leesch, J. (2003). The mortality of stored product insects following exposure to gaseous ozone at high concentrations. Paper presented at the Advances in stored product protection. *Proceedings of the 8th International Working Conference on Stored Product Protection*, York, UK, 22–26 July 2002.

Lemke, S. L., Ottinger, S. E., Ake, C. L., Mayura, K., & Phillips, T. D. (2001). Deamination of fumonisin B1 and biological assessment of reaction product toxicity. *Chemical Research in Toxicology, 14*(1), 11–15.

Lima, D. C., Maniglia, B. C., Junior, M. D. M., Le-Bail, P., Le-Bail, A., & Augusto, P. E. D. (2020). Dual-process of starch modification: Combining ozone and dry heating treatments to modify cassava starch structure and functionality. *Journal of Biological Macromolecules, 167*, 894–904.

Lima, D. C., Villar, J., Castanha, N., Maniglia, B. C., Junior, M. D. M., & Augusto, P. E. D. (2020). Ozone modification of arracacha starch: Effect on structure and functional properties. *Journal of Food Hydrocolloids, 108*, 106066.

Lu, B., Ren, Y., Du, Y.-z., Fu, Y., & Gu, J. (2009). Effect of ozone on respiration of adult *Sitophilus oryzae* (L.), *Tribolium castaneum* (Herbst) and *Rhyzopertha dominica* (F.). *Journal of Insect Physiology, 55*(10), 885–889.

Mahapatra, A. K., Muthukumarappan, K., & Julson, J. L. (2005). Applications of ozone, bacteriocins and irradiation in food processing: A review. *Critical Reviews in Food Science and Nutrition, 45*(6), 447–461.

Mahroof, R. M., Amoah, B. A., & Wrighton, J. (2018). Efficacy of ozone against the life stages of Oryzaephilus mercator (Coleoptera: Silvanidae). *Journal of Economic Entomology, 111*(1), 470–481.

Mason, L. J., Woloshuk, C., & Maier, D. (1997). Efficacy of ozone to control insects, molds, and mycotoxins. Paper presented at the *International Conference on Controlled Atmosphere and Fumigation in Stored Products*, ed by E. J. Donahaye, S. Navarro, and A. Varnava, Printco, Ltd, Nicosia, Cyprus.

McDonough, M. X., Mason, L. J., & Woloshuk, C. P. (2011). Susceptibility of stored product insects to high concentrations of ozone at different exposure intervals. *Journal of Stored Products Research, 47*(4), 306–310.

Mendez, F., Maier, D., Mason, L., & Woloshuk, C. (2003). Penetration of ozone into columns of stored grains and effects on chemical composition and processing performance. *Journal of Stored Products Research, 39*(1), 33–44.

Mishra, G., Palle, A. A., Srivastava, S., & Mishra, H. N. (2019). Disinfestation of stored wheat grain infested with *Rhyzopertha dominica* by ozone treatment: Process optimization and impact on grain properties. *Journal of the Science of Food Agriculture, 99*(11), 5008–5018.

Muthukumarappan, K., Halaweish, F., & Naidu, A. S. (2000). Ozone. In: A. Naidu (Ed.), *Natural Food Antimicrobial Systems* (pp. 783–800). Boca Raton, FL: CRC LLC Press.

Muthukumarappan, K., O'Donnell, C. P., & Cullen, P. (2008). Ozone utilization. In: *Encyclopedia of Agricultural, Food, and Biological Engineering* (pp. 1–4). https://www.scopus.com/record/display.uri?eid=2-s2.0-68949172471&origin=inward&txGid=142f33bc9063f50bcb06e3c2162078cd

Muthukumarappan, K., O'Donnell, C., & Cullen, P. (2009). *Ozone Treatment of Food Materials.* In: (pp. 263–280): Boca Raton, FL: CRC Press. https://www.routledge.com/Food-Processing-Operations-Modeling-Design-and-Analysis-Second-Edition/Jun-Irudayaraj/p/book/9781420055535

O'Donnell, C., Tiwari, B. K., Cullen, P., & Rice, R. G. (2012). *Ozone in Food Processing.* John Wiley & Sons. https://onlinelibrary.wiley.com/doi/book/10.1002/9781118307472

Oladebeye, A. O., Oshodi, A. A., Amoo, I. A., & Abd Karim, A. (2013). Functional, thermal and molecular behaviours of ozone-oxidised cocoyam and yam starches. *Journal of Food Chemistry, 141*(2), 1416–1423.

Pandiselvam, R., Manikantan, M., Divya, V., Ashokkumar, C., Kaavya, R., Kothakota, A., & Ramesh, S. V. (2019). Ozone: An advanced oxidation technology for starch modification. *Journal of the International Ozone Association, 41*(6), 491–507.

Patil, S., & Bourke, P. (2012). Ozone processing of fluid foods. In: P. J. Cullen, B. K. Tiwari, & V. P. Valdramidis (Eds.), *Novel Thermal and Non-Thermal Technologies for Fluid Foods* (pp. 225–261). Academic Press. https://www.elsevier.com/books/novel-thermal-and-non-thermal-technologies-for-fluid-foods/cullen/978-0-12-381470-8

Proctor, A., Ahmedna, M., Kumar, J., & Goktepe, I. (2004). Degradation of aflatoxins in peanut kernels/flour by gaseous ozonation and mild heat treatment. *Journal of Food Additives and Contaminants, 21*(8), 786–793.

Raila, A., Lugauskas, A., Steponavicius, D., Railiene, M., Steponaviciene, A., & Zvicevicius, E. (2006). Application of ozone for reduction of mycological infection in wheat grain. *Journal of Annals of Agricultural and Environmental Medicine, 13*(2), 287–294.

Sandhu, H. P., Manthey, F. A., & Simsek, S. (2012). Ozone gas affects physical and chemical properties of wheat (*Triticum aestivum* L.) starch. *Journal of Carbohydrate Polymers, 87*(2), 1261–1268.

Shunmugam, G., Jayas, D., White, N., & Muir, W. (2005). Diffusion of carbon dioxide through grain bulks. *Journal of Stored Products Research, 41*(2), 131–144.

Sousa, A. D., Faroni, L. D. A., Guedes, R., Tótola, M., & Urruchi, W. (2008). Ozone as a management alternative against phosphine-resistant insect pests of stored products. *Journal of Stored Products Research, 44*(4), 379–385.

Suslow, T. (2004). *Ozone Applications for Postharvest Disinfection of Edible Horticultural Crops.* UCANR Publications. https://anrcatalog.ucanr.edu/pdf/8133.pdf

Tiwari, B., Brennan, C. S., Curran, T., Gallagher, E., Cullen, P., & O'Donnell, C. (2010a). Application of ozone in grain processing. *Journal of Cereal Science, 51*(3), 248–255.

Tiwari, B., Brennan, C. S., Curran, T., Gallagher, E., Cullen, P., & O'Donnell, C. J. (2010b). Application of ozone in grain processing. *Journal of Cereal Science, 51*(3), 248–255.

United States Food and Drug Administration (FDA). (2001). Hazard analysis and critical control point (HACCP): Procedures for the safe and sanitary processing and importing of juice. Retrieved from Federal Register. https://www.federalregister.gov/documents/2001/01/19/01-1291/hazard-analysis-and-critical-control-point-haacp-procedures-for-the-safe-and-sanitary-processing-and

Young, J. C., Zhu, H., & Zhou, T. (2006). Degradation of trichothecene mycotoxins by aqueous ozone. *Food Chemical Toxicology, 44*(3), 417–424.

Yvin, J.-C., Bailli, A., Joubert, J.-M., & Bertaud, O. (2005). Method and installation for making flour from ozone-treated grains. In: Google Patents.

Zhu, F. (2018). Effect of ozone treatment on the quality of grain products. *Food Chemistry, 264*, 358–366.

Chapter 7

Effects of an Oscillating Magnetic Field on the Stability of Stored Grain Produce

M. Selvamuthukumaran

CONTENTS

7.1 INTRODUCTION

Magnetic fields can be successfully used to enhance food quality by preventing the growth of microorganisms and slowing down enzymic activities (Grigelmo-Miguel et al., 2011; Maki and Hirota, 2014; Gupta et al., 2015; Esitken and Turan, 2004). It has been reported that a magnetic field can function via a window effect (Sakdatorn et al., 2018) and also via a field gradient effect to achieve particular biological goals (Otero et al., 2016). The magnetic field can react with the food, which consists of a thermodynamic-cum-quantum effect. The occurrence of magnetic field interaction is not very clear, therefore our understanding of the magnetic field action model is quite complicated and it is a challenging task for food processors. Thus it is necessary to focus on its application aspects in the food processing industry with an emphasis on factors like magnetic flux frequency, time of exposure, and frequency effect.

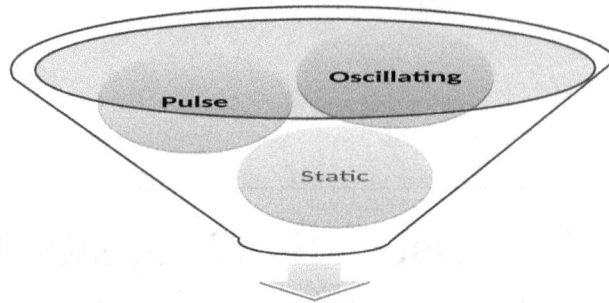

Figure 7.1   Classification of a magnetic field.

7.2 THE PRINCIPLES BEHIND MAGNETIC FIELDS

Magnetic fields can be generated by either AC or DC current in conductors or with the help of some permanent magnets. They can be classified into a pulsed, oscillating, or static magnetic field according to their behavior over time (Figure 7.1). When the field gradient is zero over space, the magnetic field spatial distribution will be homogeneous for the exposed samples, whereas it is heterogeneous in other cases (Zhao et al., 2017).

Based on magnetic flux density, the static magnetic field can be further classified as super-weak (100 nT to 0.5 mT), weak (5 T), moderate (1 mT to 1 T), strong (1 to 5 T), and ultra-strong (> 5 T) (Ali Hayder et al., 2015). With respect to frequency function, they can be classified as high frequency (> 10 GHz), intermediate frequency (300 Hz to 1 MHz), extremely low frequency (0 to 300 Hz), radio frequency (1 MHz to 500 MHz), and microwave frequency (500 MHz to 10 GHz) (Grigelmo-Miguel et al., 2011).

7.3 EFFECTS OF MAGNETIC FIELD INTERACTION ON PHYSICOCHEMICAL FOOD QUALITY

Food products comprise complex compounds containing water, carbohydrates, fats, proteins, minerals, and fibers. These compounds were created from molecules and atoms. The application of a magnetic field depends upon the magnetic susceptibility, i.e. X and isotropy. If the magnetic susceptibility is less than 0, it can be a diamagnetic field; elsewhere it is paramagnetic, thereby exhibiting different behavior. The carbon atoms of food substances show isotropic susceptibility, while organic molecules of food substances show anisotropic susceptibility (Grigelmo-Miguel et al., 2011). Therefore, it is essential to determine the molecular properties as well as compounds present in food, which makes it easier to study the interactive effect of magnetic field radiation on food's physicochemical characteristics. It was observed that by means of gradient magnetic field application the compounds like sucrose and sorbitol were separated in compressed oxygen gas and there was an occurrence of around 8.2% magnetic susceptibility between them (Maki and Hirota, 2014).

Gupta et al. (2015) explained that in the case of cherry tomatoes, the TSS, lycopene content, and average weight were all enhanced by exposure to a magnetic field and they were found to be higher within a pulsed magnetic field rather than a static magnetic field (Esitken and Turan, 2004). In the case of strawberries, the plants' metabolic growth and leaves' ions were significantly enhanced as a result of stimulation with a magnetic field (Esitken and Turan, 2004). Jia et al. (2015) also demonstrated that the shelf stability of cantaloupes was enhanced by the application of an oscillating magnetic field; they found that respiration was significantly slowed down. Goldschmidt Lins et al. (2016) explained that the color parameters of ground beef were retained for longer as a result of the application of a pulsed magnetic field, which successfully reduced the oxidation rate of atoms, i.e. iron atoms. The oxidation of Fe2+ to Fe3+ was decreased, which can significantly help retain the ground beef's color.

The application of a magnetic field to honey reduced the pH and enhanced the soluble solids content by disintegrating pollen, which leads to sugar and ascorbic acid release (Sakdatorn et al., 2018). Otero et al. (2016) reported that the magnetic field's frequency and strength should be enhanced only when it is used for biological product freezing; it is to be borne in mind that the application of a stronger magnetic field can destroy properties of cells and it should never be applied to fresh food or biological samples (Zhao et al., 2017). For food preservation and freezing applications, the intensity of the magnetic field applied can be greater than 1 mT.

7.4 FOOD ENZYME EFFECTS ON MAGNETIC FIELD APPLICATIONS

Enzymes are proteins that may have a particular activity in the food matrix and may affect the retention of quality aspects of food. They can be either beneficial or harmful, therefore one has to decide whether a magnetic field should be applied to enhance or to inactivate them. Its application can modify their structure, thereby significantly interrupting the biochemical processes. The modification of α-amylase's structure was confirmed by Jia et al. (2009) as a result of the application of a static magnetic field; similarly, enzymatic modification changes, i.e. peroxidase changes, were also confirmed by Ma et al. (2011) as a result of the application of a pulsed magnetic field.

Buchachenko (2016) suggests that biochemical processes can be controlled by energy and angular movement; the angular movement is important because selectivity of spin state should be identical in reactants and products, and particularly in an atomic forbidden state, it can be easily overcome by a synchronized spin with a magnetic field. It was noted that the involvement of more than a single unpaired electron in a biochemical process can be easily affected by a magnetic field (Celik et al., 2009). In order to enhance or stimulate the enzymatic activity in magnetic field applications, it is necessary to synchronize motion between substrate molecules and enzymes (Mizuki et al., 2013). The enzymes can be immobilized with the help of magnetic particles, thereby the enzyme movement can be controlled with the help of the frequency used in the magnetic field. It has been observed that by using a 5Hz frequency, the α-amylase activity can be significantly enhanced (Mizuki et al., 2010).

7.5 MICROORGANISMS' EFFECTS ON MAGNETIC FIELD APPLICATIONS

The major aim of applying a magnetic field to a food matrix is the destruction of harmful microbial species, thereby enhancing the growth of beneficial microbes that assist in a smooth fermentation process. For microbial inactivation in sealed foods, Barbosa-Cánovas et al. (2000) explained the use of pulsed magnetic field conditions, starting with pulses ranging from 1–100, 5–50, and 5–500 Hz with an exposure time and temperature ranging from 25–100 μs and 0–50°C. The microbial growth in CFUs (colony-forming units) is significantly reduced by 2 log.

Liu et al. (2017) showed that microbial cell growth and multiplication will be prohibited as a result of the interaction of a magnetic field with cell passivation. Even a low frequency level of magnetic field interaction may have a good impact on the cells and tissues of microbes, and a high level of pulsed magnetic field can prohibit the growth of microbes' cell membranes and organelles significantly. Ji et al. (2009) observed that after the successful application of a static magnetic field, the proliferation of *E. coli* cells was significantly diminished; similar observations were confirmed by the research findings of Strašák et al. (2002) and they came to the conclusion that the application of a static magnetic field had a good bactericidal effect by forming free radicals that can inhibit cell membrane growth.

It was demonstrated that the *Listeria* species can be destroyed by the application of a pulsed magnetic field; this can be effectively achieved by intracellular protein stability and gene expression changes (Wu et al., 2017). The microbial enzyme activity can be altered in *Saccharomyces cerevisiae* as a result of oscillating magnetic field applications, and thereby, the growth of this species is controlled with an enhanced death rate (Bayraktar, 2013). It was noted that after the successful application of a static magnetic field to *Saccharomyces cerevisiae*, the glucose–ethanol conversion process was significantly changed (Da-Motta et al., 2008).

For the fermentation process, it is essential to consider the oxygen paramagnetic behavior and carbon dioxide diamagnetic behavior of molecules after magnetic field interaction, which can even reduce the mass transfer (Iwasaka et al., 2004). In the case of *Lactococcus lactis*, it was observed that magnetic field application resulted in a metabolic pathway deviation for inhibiting the production of nisin (Alvarez et al., 2006). The various research studies show that particular parameters need to be triggered for specific microorganisms as they perform under a window effect when they are exposed to a magnetic field (Haile et al., 2008; Harte et al., 2001; He et al., 2014). Ahmed et al. (2013) prescribed the inhibition of *Staphylococcus aureus* by applying a pulsed magnetic field at 300 Hz and 1.5 mT.

7.6 MECHANISMS OF ACTION INVOLVED DURING MAGNETIC FIELD INTERACTION WITH FOOD MATRICES

Food is a heterogeneous mixture that comprises several atoms and molecules; therefore, magnetic field interaction is based on these two levels (Brizhik, 2014). The various mechanisms of action that are involved during a magnetic field's interaction with a food matrix are given in Figure 7.2 and explained below.

7.6.1 Radical Pair

Magnetic field interaction deals with radicals; for instance, in one study the reactivity was changed by the magnetic field interaction with spin singlet–triplet transitions (Brizhik, 2014). Therefore, a series of reactions occur in foods that are being changed by magnetic field exposure.

The radical pair effects will be non-deterministic if the diamagnetic anisotropy and radical pair recombination act on magnetic dipole elements (Albuquerque et al., 2016). This radical pair can influence four different types of interactions: inter-radical exchange interaction, hyperfine interaction, resonance effect, and electronic Zeeman interaction (Steiner and Ulrich, 1989). There are some restrictions during every interaction, which are solely based on radical lifetime and participating elements. Albuquerque et al. (2016) imposed some restrictions on radical pairing, correlated with cell growth as a result of exposure to a static magnetic field interaction effect.

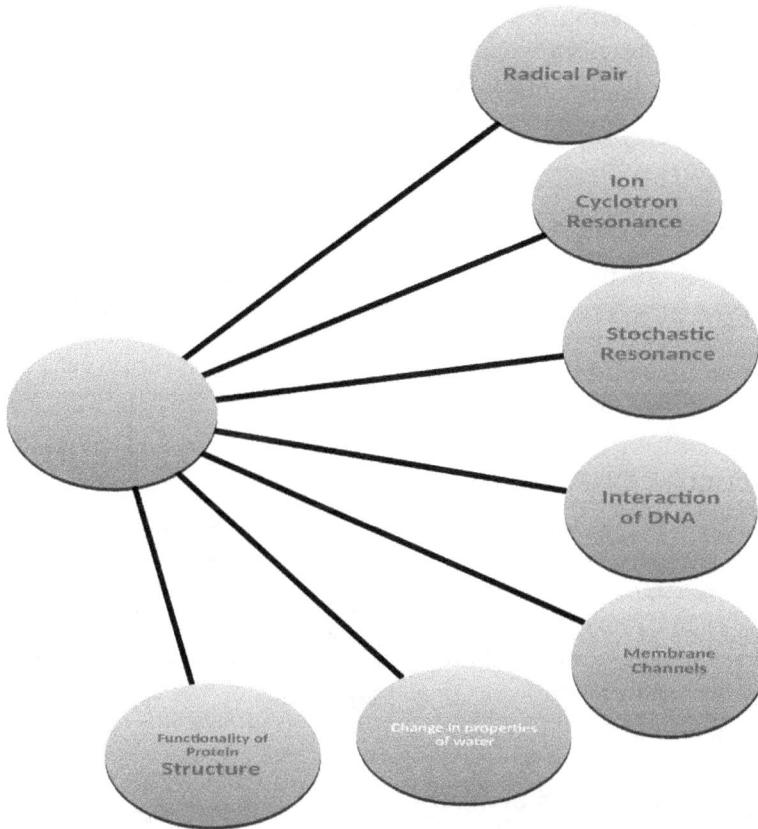

Figure 7.2   Mechanisms of action involved during a magnetic field's interaction with a food matrix.

It was identified that this model leads to the production of free radicals with a static magnetic field applied to cells, which can kindle oxidative stress; it also can damage ion channels, which leads to cell morphology changes as well as changes to various proteins and gene expressions (Ghodbane et al., 2013).

7.6.2 Ion Cyclotron Resonance

In an ion cyclotron resonance, the circulation of ions can occur on a plane that is perpendicular to either a static or oscillating magnetic field with Larmor frequencies and particular harmonics (Pothakamury et al., 1993). The different ions can circulate on a plane perpendicular to $Mg2+$, $Ca2+$, and $Na+$, which are involved in the biochemical process. The biological response is solely based on the oscillating magnetic field frequency's ratio to the static magnetic field flux density with either at weak values (Pazur et al., 2007). The application of an oscillating magnetic field can lead to rotation cessation that can effectively release bound ions, which can produce a biological response (Pazur et al., 2007); calcium-binding proteins will bind and regulate a large number of protein targets, which can significantly affect various cellular functions (Pothakamury et al., 1993).

7.6.2.1 Stochastic Resonance

Bingi and Savin (2003) described that a stochastic resonance system can be incorporated with noise tuning in order to enhance the noise signal ratio, thereby effectively considering underlying effects in bistable and mono systems. There was no such application of action models in the food industry, but it can be applied in the medical field in biological system enzymes.

7.6.3 Membrane Channels

Membrane channel properties can be influenced by a static magnetic field of medium intensity that alters the deforming of calcium ion flux or ion channel embedding and thereby alters activation kinetics (Rosen, 2010). For sodium channels, the relevant effect was reported to a lesser extent. The ion pore geometry will govern the ion channel sensitivity of a magnetic field, or a voltage sensor will be there in the case of voltage-gated ion channels (Tolosa et al., 2011).

7.6.4 Interaction of DNA

A magnetic field can have direct interaction with DNA, thereby breaking its structure. It was reported that when a pulsed electric field of 3T and around 30 pulses were applied to microbes, i.e. *Bacillus subtilis* (Qian et al., 2016a), the free radicals indirectly reacted with nucleic acid and could cause DNA damage. It also reduced energy metabolism and carbohydrate metabolism through the creation of various kinds of down-regulated and express up-regulated proteins, which can be easily identified by proteomics and can further lead to cell death (Qian et al., 2016b).

7.6.5 Change in Properties of Water

A food matrix contains more moisture in its fresh form, therefore it can alter its structure as well as its properties when interacting with a magnetic field. It was described that such interaction can enhance protein hydration and chemical activities and also the structure of the cell wall (Torgomyan and Trchounian, 2013). Ratter (1997) reported that based on ion size, if they are Na+ and hydrated fully in dense phase, they become reactive and weakly bound with water; if they are k+ and are hydrated in water individually, they may become viscous and less dense. The following six types of cell behavior can be a result of magnetic field interaction: (1) partitioning of cytoplasmic ion inducement; (2) alteration of cytoplasmic organization; (3) cellular protein changes; (4) enzymatic activity alteration; (5) active transport behavior alteration; and (6) biopolymer formation alteration.

7.6.6 Functionality of Protein Structure

Brizhik (2014) explained that solitons, a part of the redox system, are influenced by magnetic fields, and that this kind of electrons were self-trapped under helical polypeptides of alpha-type because of strong electron-lattice interaction, which yields non-heat resonant effects that may affect organism metabolism. Bingi and Savin (2003) showed that in the dissociation of an ion protein, the ion may get into the cavity of a protein-containing ligand. The magnetic field can disturb ion equilibrium redistribution, which can cause various biological responses, and the alteration of structure can change biological functionalities (Ma et al., 2011; Buchachenko, 2016).

7.7 OSCILLATING MAGNETIC FIELD'S EFFECT ON STABILITY ENHANCEMENT OF OAT SPROUT

Oats contain health-related functional components and therefore should be included as part of a regular diet (Kowalczyk and Ratajczak, 1995). Oats are found to possess protein of around 15–20% and a high amount of non-starch polysaccharides, i.e. β-glucans, that can significantly reduce the level of cholesterol in the blood (Gąsiorowski, 1995). It was observed that some undesirable changes may occur during the sprouting process that can result in inferior quality of the food. For an efficient oat germination process to occur, there should be an optimum temperature and humidity, otherwise oats are prone to microbial attack. Various bacteria and molds can alter the quality and reduce the wholesomeness of the oat sprouts, which cannot be fortified or processed further to exert the functional effects; therefore it is necessary to adopt a sprout treatment procedure that can enhance the stability of sprouts, thereby retaining antioxidants and minimizing the microbial degradation during storage. There are various non-thermal processing methods like ultrasound, high pressure, pulsed-field, and microwave radiation (Senorans et al., 2003) that are available to enhance the stability of grain-based food products; among them, the oscillating magnetic field plays a major role in extending the stability of processed grain products.

Lipiec et al. (2005) conducted a study to observe the effect of oscillating magnetic fields on the stability enhancement of oat sprouts and the retention of antioxidative nutrients (Table 7.1). They started growing oat sprouts in Petri dishes under controlled lab conditions, subjecting them to darkness and maintaining a humidity of around 65% and a temperature of around 21°C. After sprouting, i.e. after day five, the sprouts were treated with an oscillating magnetic pulsed field. The magnetic field effects were produced by a specially designed generator. The induction of an oscillating pulsed magnetic field can discharge a high-voltage condenser via an apt solenoid. Through high voltage supply, the battery of a high-voltage condenser with a capacity of 1.5 mF can be charged up to sufficient voltage. The maximum working current of 100 mA and working voltage of 5 kV are used. The magnetic field produces solenoids, which are single-layer coils whose surface is wound with a copper wire of 20 to 40 mm² and a rectangular section. In this study, the solenoid of approximately 17 mm internal diameter was selected so that standard test tubes with analyzed material could be placed inside. The mechanical resistance was provided by

TABLE 7.1   EFFECT OF USING AN OSCILLATING MAGNETIC FIELD ON MICROBIAL GROWTH AND ANTIOXIDANTS OF GERMINATED OATS

Magnetic Field Pulse Treatment	Microbial Growth Results	Polyphenol Content mg/g of Oat Sprouts	Antioxidant Activity (%) for Treated Oat Sprouts
1 × 8T	Insignificant reduction in bacteria and fungi count is observed	7.1	37.8
3 × 8T	-	7.25	28.5
5 × 5T	Significant microbial reduction is achieved. Bacteria and fungi reduce by nearly half	-	-
10 × 3T	-	5.05	47.5

Source: Modified from Lipiec et al., 2005.

coils that were kept inside a special container. The induction probe, which was kept inside the solenoid, was used to measure the various parameters of the oscillating magnetic field and was also connected by means of the digital oscilloscope Voltcraft DSO 2100.

The following treatment combinations were used to ascertain the oscillating magnetic field pulse's effect on microorganism survivability: a single pulse of 8T amplitude and a pulse of 5T amplitude repeated five times. To ascertain the effect on retaining antioxidants, different treatments were used, namely a single pulse of 8T amplitude, three pulses of 8T amplitude, and ten pulses of 3T amplitude. In all the treatments, 250 μs was the oscillating period and the total duration time of each single pulse was approximately three periods.

It was found that with magnetic field treatment, oat sprout samples showed a significant reduction up to 50% under a pulse of 5 × 5T, while an insignificant reduction was noted for samples treated with a magnetic field of 1 × 8T amplitude. Threefold reduction of vegetative bacteria was achieved in oat sprouts when treated with magnetic pulses of amplitudes 5 × 5T and 1 × 8T. In oat sprouts treated with a magnetic pulse of amplitude 5 × 5T, the fungi were drastically reduced by a factor of nearly ten.

The polyphenol content had significantly enhanced 1.5-fold for samples treated with magnetic field pulses of 10 × 3T amplitude when compared to other exposed samples. A 30% antioxidative activity was achieved for oat sprout samples treated with magnetic field pulses of 10 × 3T amplitude. The study reported that treating oat sprouts with oscillating magnetic field pulses can modify the macromolecular structure of sprouts, which can help to facilitate the extraction of antioxidant fractions to a greater extent.

7.8 CONCLUSION

The oscillating magnetic field can be adopted to enhance the stability of grain-based products. It is an apt non-thermal processing method for successfully reducing microbial count and also enhancing antioxidant properties when it is applied in sprouted grains like oats. The enzyme activities can be modified and physicochemical properties can be enhanced as a result of the use of magnetic fields in several food applications.

REFERENCES

Ahmed, I.; Istivan, T.; Cosic, I.; Pirogova, E. Evaluation of the effects of extremely low frequency (ELF) pulsed electromagnetic fields (PEMF) on survival of the bacterium *Staphylococcus aureus*. *EPJ Nonlinear Biomed. Phys.* 2013, 1(1), 5.

Albuquerque, W.W.C.; Costa, R.M.P.B.; de Salazar e Fernandes, T.; Porto, A.L.F. Evidences of the static magnetic field influence on cellular systems. *Prog. Biophys. Mol. Biol.* 2016, 121(1), 16–28.

Ali Hayder, I.; Al-HilphyAsaad, R.S.; Al-Darwash, A.K. The effect of magnetic field treatment on the characteristics and yield of Iraqi local white cheese. *IOSR J. Agric. Vet. Sci.* 2015, 8, 63–69.

Alvarez, D.C.; Pérez, V.H.; Justo, O.R.; Alegre, R.M. Effect of the extremely low frequency magnetic field on nisin production by *Lactococcus lactis* subsp. lactis using cheese whey permeate. *Process Biochem.* 2006, 41(9), 1967–1973.

Barbosa-Canovas, G.V.; Schaffner, D.W.; Pierson, M.D.; Zhang, Q.H. Oscillating magnetic fields. *J. Food Sci.* 2000, 65, 86–89.

Bayraktar, V.N. Magnetic field effect on yeast Saccharomyces cerevisiae activity at grape must fermentation. *Biotechnol. Acta* 2013, 6(1), 125–137.

Bingi, V.N.; Savin, A.V. Effects of weak magnetic fields on biological systems: Physical aspects. *Phys. Usp.* 2003, 46(3), 259–291.

Brizhik, L. Biological effects of pulsating magnetic fields: Role of solitons. *arXiv* 2014, arXiv:1411.6576.

Buchachenko, A. Why magnetic and electromagnetic effects in biology are irreproducible and contradictory? *Bioelectromagnetics* 2016, 37(1), 1–13.

Çelik, Ö.; Büyükuslu, N.; Atak, Ç.; Rzakoulieva, A. Effects of magnetic field on activity of superoxide dismutase and catalase in *Glycine max* (L.) Merr. roots. *Pol. J. Environ. Stud.* 2009, 18, 175–182.

da Motta, M.A.; Muniz, J.B.F.; Schuler, A.; da Motta, M. Static magnetic fields enhancement of Saccharomyces cerevisae ethanolic fermentation. *Biotechnol. Prog.* 2008, 20(1), 393–396.

Esitken, A.; Turan, M. Alternating magnetic field effects on yield and plant nutrient element composition of strawberry (*Fragaria x ananassa* cv. *camarosa*). *Acta Agric. Scand. B* 2004, 54(3), 135–139.

Gąsiorowski, H. *Oats. Chemistry and technology* (in Polish). PWRiL, Poznań, 180 pp, 1995.

Ghodbane, S.; Lahbib, A.; Sakly, M.; Abdelmelek, H. Bioeffects of static magnetic fields: Oxidative stress, genotoxic effects, and cancer studies. *BioMed Res. Int.* 2013, 2013, 1–12.

Grigelmo-Miguel, N.; Soliva-Fortuny, R.; Barbosa-Cánovas, G.; Martín-Belloso, O. Use of oscillating magnetic fields in food preservation. In: H.Q. Zhang, Barbosa-Cánovas, G.V., Balasubramaniam, V.M., Dunne, C.P., Farkas, D.F., Yuan, J.T.C. (eds.), *Nonthermal Processing Technologies for Food*, 1st ed., 675. Wiley-Blackwell, Ames, IA, 2011.

Goldschmidt Lins, P.; Aparecida Silva, A.; Marina Piccoli Pugine, S.; Ivan Cespedes Arce, A.; José Xavier Costa, E.; De Pires Melo, M. Effect of exposure to pulsed magnetic field on microbiological quality, color and oxidative stability of fresh ground beef. *J. Food Process Eng.* 2016, 40, 1–9.

Gupta, M.K.; Anand, A.; Paul, V.; Dahuja, A.; Singh, A.K. Reactive oxygen species mediated improvement in vigour of static and pulsed magneto-primed cherry tomato seeds. *Indian J. Plant Physiol.* 2015, 20(3), 197–204.

Haile, M.; Pan, Z.; Gao, M.; Luo, L. Efficacy in microbial sterilization of pulsed magnetic field treatment. *Int. J. Food Eng.* 2008, 4(4). doi:10.2202/1556-3758.1177.

Harte, F.; Martin, M.F.S.; Lacerda, A.H.; Lelieveld, H.L.M.; Swanson, B.G.; Barbosa-Cánovas, G.V. Potential use of 18 tesla static and pulsed magnetic fields on *Escherichia coli* and *Saccharomyces cerevisiae*. *J. Food Process. Preserv.* 2001, 25(3), 223–235.

He, R.; Ma, H.; Wang, H. Inactivation of *E. coli* by high-intensity pulsed electromagnetic field with a change in the intracellular Ca_{2+} concentration. *J. Electromagn. Waves Appl.* 2014, 28(4), 459–469.

Iwasaka, M.; Ikehata, M.; Miyakoshi, J.; Ueno, S. Strong static magnetic field effects on yeast proliferation and distribution. *Bioelectrochemistry* 2004, 65(1), 59–68.

Ji, W.; Huang, H.; Deng, A.; Pan, C. Effects of static magnetic fields on *Escherichia coli*. *Micron* 2009, 40(8), 894–898.

Jia, J.; Wang, X.; Lv, J.; Gao, S.; Wang, G. Alternating magnetic field prior to cutting reduces wound responses and maintains fruit quality of cut *Cucumis melo* L. cv Hetao. *Open Biotechnol. J.* 2015, 9, 230–235.

Jia, S.; Liu, Y.; Wu, S.; Wang, Z. Effect of static magnetic field on α-amylase activity and enzymatic reaction. *Trans. Tianjin Univ.* 2009, 15(4), 272–275.

Lipiec, J.; Janas, P.; Barabasz, W.; Pysz, M.; Pisulewski, P. Effects of oscillating magnetic field pulses on selected oat sprouts used for food purposes. *Acta Agrophysica* 2005, 5, 357–365.

Liu, Z.; Gao, X.; Zhao, J.; Xiang, Y. The sterilization effect of solenoid magnetic field direction on heterotrophic bacteria in circulating cooling water. *Procedia Eng.* 2017, 174, 1296–1302.

Kowalczyk, C.; Ratajczak, P. Oats production in Poland and in the world (in Polish). In: H. Gąsiorowski (ed.), *Oats. Chemistry and Technology* (in Polish), 9–19. PWRiL, Poznań, 1995.

Ma, H.; Huang, L.; Zhu, C. The effect of pulsed magnetic field on horseradish peroxidase. *J. Food Process Eng.* 2011, 34(5), 1609–1622.

Maki, S.; Hirota, N. Magnetic separation technique on binary mixtures of sorbitol and sucrose. *J. Food Eng.* 2014, 120, 31–36.

Mizuki, T.; Sawai, M.; Nagaoka, Y.; Morimoto, H.; Maekawa, T. Activity of lipase and chitinase immobilized on superparamagnetic particles in a rotational magnetic field. *PLOS ONE* 2013, 8(6), e66528.

Mizuki, T.; Watanabe, N.; Nagaoka, Y.; Fukushima, T.; Morimoto, H.; Usami, R.; Maekawa, T. Activity of an enzyme immobilized on superparamagnetic particles in a rotational magnetic field. *Biochem. Biophys. Res. Commun.* 2010, 393(4), 779–782.

Otero, L.; Rodríguez, A.C.; Pérez-Mateos, M.; Sanz, P.D. Effects of magnetic fields on freezing: Application to biological products. *Compr. Rev. Food Sci. Food Saf.* 2016, 15(3), 646–667.

Pazur, A.; Schimek, C.; Galland, P. Magnetoreception in microorganisms and fungi. *Open Life Sci.* 2007, 2(4), 597–659.

Pothakamury, U.R.; Barbosa-Canovas, G.V.; Swanson, B.G. Magnetic-field inactivation of microorganisms and generation of biological changes. *Food Technol.* 1993, 47, 85–93.

Qian, J.; Zhou, C.; Ma, H.; Li, S.; Yagoub, A.E.A.; Abdualrahman, M.A.Y. Biological effect and inactivation mechanism of *Bacillus subtilis* exposed to pulsed magnetic field: Morphology, membrane permeability and intracellular contents. *Food Biophys.* 2016a, 11(4), 429–435.

Qian, J.; Zhou, C.; Ma, H.; Li, S.; Yagoub, A.E.A.; Abdualrahman, M.A.Y. Proteomics analyses and morphological structure of Bacillus subtilis inactivated by pulsed magnetic field. *Food Biophys.* 2016b, 11(4), 436–445.

Ratter, J. The Brazilian Cerrado vegetation and threats to its biodiversity. *Ann. Bot.* 1997, 80(3), 223–230.

Rosen, A.D. Studies on the effect of static magnetic fields on biological systems. *Piers Online* 2010, 6(2), 133–136.

Sakdatorn, V.; Thavarungkul, N.; Srisukhumbowornchai, N.; Intipunya, P. Improvement of rheological and physicochemical properties of longan honey by non-thermal magnetic technique. *Int. J. Food Sci. Technol.* 2018, 53(7), 1717–1725.

Senorans, F.J.; Ibanez, E.; Cifuentes, A. New trends in food processing. *Crit. Rev. Food Sci. Nutr.* 2003, 43(5), 507–526.

Steiner, U.E.; Ulrich, T. Magnetic field effects in chemical kinetics and related phenomena. *Chem. Rev.* 1989, 89(1), 51–147.

Strašák, L.; Vetterl, V.; Šmarda, J. Effects of low-frequency magnetic fields on bacteria Escherichia coli. *Bioelectrochemistry* 2002, 55(1–2), 161–164.

Tolosa, M.F.; Bouzat, C.; Cravero, W.R. Effects of static magnetic fields on nicotinic cholinergic receptor function. *Bioelectromagnetics* 2011, 32(6), 434–442.

Torgomyan, H.; Trchounian, A. Bactericidal effects of low-intensity extremely high frequency electromagnetic field: An overview with phenomenon, mechanisms, targets and consequences. *Crit. Rev. Microbiol.* 2013, 39(1), 102–111.

Wu, P.; Qu, W.; Abdualrahman, M.A.Y.; Guo, Y.; Xu, K.; Ma, H. Study on inactivation mechanisms of *Listeria grayi* affected by pulse magnetic field via morphological structure, Ca_{2+} transmembrane transport and proteomic analysis. *Int. J. Food Sci.Technol.* 2017, 52(9), 2049–2057.

Zhao, H.; Zhang, F.; Hu, H.; Liu, S.; Han, J. Experimental study on freezing of liquids under static magnetic field. *Chin. J. Chem. Eng.* 2017, 25(9), 1288–1293.

Chapter 8

Grain-Based Functional Food Production

Sultan Arslan-Tontul

CONTENTS

8.1 INTRODUCTION

Grains are an important energy source for human nutrition. It is known that more than 60% of the energy intake in a daily diet comes from grains because of their high carbohydrate content. Recently, attention has come to grain-based functional foods due to increasing health consciousness among consumers. Grains are useful substrates for producing probiotic functional foods since their sugar content is easily fermented by various probiotic *Lactobacillacea*. Additionally, their bran layers are accepted as dietary fibre and they show prebiotic properties, increasing the growth of probiotic colon microflora selectively. Moreover, grains also provide other functional compounds such as phytosterols, fatty acids, tocopherols, phenolic acids and antioxidant compounds.

Grain-based functional foods are processed after a serial heat application. During the thermal processing of grains, the gelatinisation of starch takes place, which provides sensory and textural development of the product. On the other hand, the product is sterilised by microbial inactivation. However, high temperatures which are applied in the process cause the degradation of functional food components of grains. Additionally, heat-sensitive water-soluble vitamins, essential amino acids and polyphenols can be lost during production, and caramelisation and burnt taste can occur during cooking. Therefore, recent advances in food processing activities include innovations that prevent the adverse effects of heating on organoleptic and nutritional properties of food (Jongyingcharoen & Ahmad, 2014).

In recent years, the adverse effects of heat-induced processes have been a driving factor in the application of non-thermal processes in grain-based functional food production. These techniques are summarised as ultrasonication, pulsed electrical field, high hydrostatic pressure, microfluidisation and non-thermal plasma. Each process has its own mechanism and creates different effects on the nutritional properties of grain products.

8.2 ULTRASOUND

Ultrasound is known as a high-efficiency, non-toxic and environmentally friendly non-thermal process in the food industry. It consists of sound waves at a frequency above the limit of human hearing, about 20 kHz. It is classified into three different groups: power ultrasound (16–100 kHz), high-frequency ultrasound (0.1–1 MHz) and diagnostic ultrasound (1–10 MHz) (Zhang et al., 2019). The power of ultrasound comes from cavitation phenomena, which enlarge and implode gas bubbles in the liquid medium. In this process, continuous compression and decompression effects are created in molecules. It was reported that the temperature and pressure in the cavitation bubbles could reach up to 5000 K and 1000 atm, respectively. Recently, ultrasound has been used in many food preparation processes, such as emulsification, extraction, homogenisation, filtration, crystallisation, hydrolysis and inactivation of enzymes and microorganisms (Jongyingcharoen & Ahmad, 2014). The applications of ultrasound in grain-based materials are summarised in Table 8.1.

Ultrasound treatment can be applied in order to decrease the cooking time of brown rice, which is known as a healthy food with abundant essential nutrients such as proteins, lipids, dietary fibres, minerals, vitamins and γ-aminobutyric acid (GABA). These functional properties of brown rice come from the bran that remains after the hull has been removed. Zhang et al. (2015) ultrasonically treated soaked brown rice grains at 40°C for 30 min. After ultrasonic treatment, a cellulase enzyme (1.6 IU/mL) was applied to brown rice grains. According to the results, ultrasound and enzyme treatment effectively decreased the optimal cooking time of the brown rice to 22.3 min compared with the control sample (28.7 min). The authors attributed this result to the change in the surface area conditions of grains by ultrasonic cavitation. Thus, the cooking water is easily diffused to the inner layers of rice and reduces the cooking time. Additionally, the hardness value of rice decreased from 1,192.3 to 946 g by combined application of ultrasound and the enzyme. On the other hand, the sensory acceptability of treated brown rice samples was close to that of white rice samples.

Bran is the most nutritious part of grains, and it is commonly used in formulations for gaining functional properties. Additionally, it contains highly bioactive components such as phenolic acids, carotenoids and antioxidants. However, these compounds are present in the cereal grains, mostly in complex bound form. Ultrasonic pretreatment can increase the bioavailability of bioactive food compounds in bran by promoting their release. Mustac et al. (2019) determined that dietary fibre solubility increased after an ultrasonic treatment at 100% amplitude for 20 min. In the same study, the enhancement of the phenolic content and antioxidant activity of proso millet bran fraction was pronounced. Pasqualone et al. (2015) researched the supplementation of pasta with wheat bran aqueous extract obtained by ultrasonication. According to the results of the study, the supplemented pasta samples showed significantly higher antioxidant activity and phenolic content. Hassan et al. (2020) ultrasonicated the sorghum grains at a different level of amplitude (40–60%) and time (5–10 min). The authors revealed that sprouts of ultrasonically treated grains were rich in radical scavenging activity and phenolic profile and had higher in vitro protein digestibility.

TABLE 8.1   ULTRASONIC TREATMENTS OF GRAINS

Matrix	Parameters	Major Findings	References
White rice	665W, 15 and 30 min, 40, 70 and 100% amplitude	The glycemic index of rice decreased by combination of ultrasound and chilling via rearrangement of starch molecules	Kunyanee & Luangsakul, 2020
Brown rice	400W, 28 kHz, 17.83 W cm^{-2} 100% amplitude	Treatment increased antioxidants, proline and GABA in germinated grains	Xia et al., 2020
Sorghum	750 W, 20 kHz, 5 and 10 min, 40 and 60% amplitude	The treatment increased phytochemical content of germinated grains	Hassan et al., 2020
Rice	200 W, 24 kHz, 10–70°C, 30–70% amplitude	Extraction of phenolic compounds from grain	Setyaningsih et al., 2019
Barley	10, 15, 20 and 25°C with and without the application of 0.75 W/mL and 1.5 W/mL of nominal power density at 20 kHz	Treatment accelerated malt production	De Carvalho et al., 2018
Corn	25 kHz and volumetric power of 41 W/L	Ultrasound reduced the time of hydration of grains without creating deteriorated effect on starch granules	Miano et al., 2017
Corn	600W, 40kHz, 25, 60 and 80°C	Ultrasound pretreatment improved ethanol production from cornmeal	Nikolic et al., 2010
Wheat	250W, 40kHz, 50°C for 20 min	Extraction of phenolic compounds from bran	Wang et al., 2008

Recently, ultrasound has been commonly used in the extraction of functional food compounds from grains. β-glucan is composed of β-D-glucose units which are linked through the glycosidic bonds. It has been reported to have a reducing effect on plasma cholesterol and postprandial serum glucose levels in humans and animals. β-glucans are important dietary fibres and are mostly obtained from barley bran. Benito-Roman et al. (2013) extracted β-glucan from barley grains by ultrasound. The authors declared the ultrasound-assisted extraction as an efficient process (extraction yield > 40.5%) for the extraction of high-molecular-weight β-glucan (> 260 kDa) from barley. By ultrasound, the extraction process of β-glucan could be finished in a very short time due to cavitation created during sonication. On the other hand, in ultrasound-assisted extraction, several parameters affect the yield of functional compounds. Therefore these parameters should be optimised. Lin et al. (2011) determined optimised ultrasound parameters to be 144 W at 70°C for 50 min for maximum yield (16.6%) of the pigment extraction from red rice. Hou et al. (2016) studied the optimisation of ultrasound-assisted extraction of phenolics from sorghum shells since sorghum is a rich source of phenolic compounds. The authors reported that the optimum extraction conditions were a solid to 80% ethanol solvent ratio of 1:15 at 50°C in a 0.32 W cm^2 ultrasonic intensity bath for 10 min. Under these conditions, the highest total phenolic content yield obtained was 52.23 mg GAE/g.

Another research area where ultrasound has been used recently is its effects on starch. Ultrasonic cavitation creates changes in the polymer structure of starch. Thus starch gains resistance against digestive enzymes. Candal ve Erbaş (2019) applied ultrasonication to increase the resistant starch ratio of wheat flour. Moreover, the authors used treated wheat flours in biscuit production and tested the decrease in the glycemic index of biscuits. According to the result of this study, it was indicated that ultrasonication increased the resistant starch content of flour from 3.15% to 3.45%, and increasing rates of treated flour in biscuit formulation led to the decrease of the glycemic index from 67.31 to 61.82. Kunyanee and Luangsakul (2020) reported that ultrasound treatment followed by chilling could lower the glycemic index of white rice grains depending on the time and the amplitude level of the ultrasound and the amylose content of the grain. Chang et al. (2021) attribute the resistant starch formation ability of ultrasound to the fact that ultrasound could trigger high-pressure gradients and high local speeds to cause shear force that leads to damage to the starch granules and further cuts the long starch chains into suitable lengths. Thus, double helix formation and recrystallisation increase in amylose and amylopectin polymers, and starch gains stability against amylases.

Ultrasound is commonly used for increasing the enzymatic activity of grains during germination. Several papers have reported that the number of bioactive functional food components is higher in ultrasonically treated germinated grains. Ding et al. (2018) reported that GABA, riboflavin, o-phosphoethanolamine and glucose-6-phosphate were found to be significantly increased after the germination of ultrasonically treated red rice grains.

8.3 HIGH HYDROSTATIC PRESSURE

Over the last few decades, pressure technology has been widely used as a processing instrument in the food industry, primarily because it provides many advantages compared to conventional food processing technologies. These involve the protection of sensory properties, inactivation of both enzymes and microorganisms, improvements in product functionality and even polymer alteration (Hurtado-Romero et al., 2020). In the high hydrostatic pressure process, various pressure limits are used, ranging from 300 to 700 MPa (Oms-Oliu et al., 2012). High hydrostatic pressure provides microbial inactivation and does not cause the decomposition of macromolecules in food matrices since it does not break down low-energy covalent bonds. However, there have been studies considering that micronutrients such as vitamins and essential amino acids can be affected negatively by high hydrostatic pressure (Jongyingcharoen & Ahmad, 2014). Some of the studies carried on high hydrostatic pressure are listed in Table 8.2.

Due to reduced extraction time and improved solvent permeability, high hydrostatic pressure increases the extraction performance of bioactive compounds (Park et al., 2016). A flavonoid that is found in the outer layers of different cereal crops is tricin. The health benefit effects of tricin such as antioxidant activity, immune-modulating activity and cancer prevention activity have been documented in previous studies. Park et al. (2016) aimed at enzymatic hydrolysis with high hydrostatic pressure treatment to extract tricin from rice hulls. It was revealed that the efficacy of rice hull extract obtained by high hydrostatic pressure was significantly greater than that of the extract obtained by traditional solvent extraction.

The use of high-pressure processing can also serve as a cost-effective way to increase the number of bioactive components in grain and grain-based functional foods. Kim et al. (2015) evaluated

TABLE 8.2   HIGH HYDROSTATIC PRESSURE TREATMENTS OF GRAINS

Matrix	Parameters	Major findings	References
Rice	450–600 MPa, 50–70°C, 15–30 min	Thiamine content of polished rice was increased by high hydrostatic pressure treatment	Balakrishna & Farid, 2020
Rice	100–600 MPa for 30 min	Quality characteristics of parboiled rice were influenced by treatment	Xu et al., 2019
Millet	200, 400 and 600 MPa; 20, 40, 60 and 80°C; 30, 60, 90 and 120 min	The amount of bioactive food component was increased while anti-nutrients were decreased after germination	Sharma et al., 2018
Rice	200–400 MPa for 15 min	Treatment increased water sorption capacity of grains	Meng et al., 2018
Brown rice	50–350 MPa for 20 min	High pressure improved both functional and quality characteristics of grains	Xia et al., 2018
OaOt and rice Oat and rice	5, 10 and 15 min at 300, 400, 500 and 600 MPa	Decrease in total bacteria count of grains	Gao et al., 2013
Oat Oat	10 min at 200, 300, 350, 400 or 500 MPa	Treatment at high HP significantly improved batter viscosity and elasticity	Hüttner et al., 2009

the enhancement of functional components of germinated rough rice. In this study, rice was germinated at 37°C for six days and subjected to a high hydrostatic pressure treatment at 30 MPa for 24 h and 48 h. According to the results of the study, the highest GABA, total arabinoxylan and tricin contents were 121.21 mg/100 g, 10.6% and 85.82 µg/g respectively for the 48 h high-pressure treated rice sample. Additionally, γ-oryzanol contents increased from 23.19–36.20 mg/100 g to 31.80–40.32 mg/100 g after ultra high-pressure treatment. Moreover, high pressure increased vitamin B and E contents. The authors concluded that the combination of high-pressure treatment applied with germination efficiently enhances the functional characteristics of rough rice.

Baipong et al. (2020) used high hydrostatic pressure as an alternative to the heat process in the production of probiotic-containing purple rice beverages. In this study, non-germinated or germinated purple rice beverages were subjected to the high pressure of 500 MPa at 25°C for 20 min. After treatment, encapsulated *Lactobacillus casei* 01 was added into the beverage, and samples were stored at 4°C for four weeks. During storage, the changes in the probiotic count and phytochemical content of beverages were followed. According to the results, the anthocyanin, phenolic, GABA and γ-oryzanol content and the antioxidant activity of pressure-treated beverages were higher than those of heat-treated samples and the count of *L. casei* 01 remained stable at the level of 8 log CFU/ml in all treatments. Nguyen et al. (2007) fermented high-pressure homogenised rice/soybean slurries with *L. Plantarum* A6 to produce high-energy-density complementary food. In this study, it was shown that a mild pre-heating treatment combined with high-pressure homogenisation could substitute for gelatinisation before fermentation by the *L. plantarum* A6 to prepare the fermented slurry.

8.4 MICROFLUIDISATION

Microfluidisation is a kind of high-pressure homogenisation process. There has been a growing interest in microfluidisation due to its technological superiority when compared to traditional milling and grinding methods. Microfluidisation leads to creating uniform particle or emulsion droplets by the effects of intense shear rates, high-velocity impact forces, ultra-high pressure, instantaneous pressure drop and hydro-dynamic cavitations (Demirkesen-Mert, 2019). Therefore, its usage area is increasing day by day. From the technological point of view, the microfluidisation process is generally performed on cereal brans for decreasing particle size, increasing water/oil holding capacity and improving the rheological behaviour of dough. On the other hand, by microfluidisation, the bioavailability of bioactive compounds present in the bran can be enhanced.

He et al. (2016) declared that the physicochemical and antioxidant properties of corn bran were improved by the microfluidisation method and the degree of such improvements depended on processing pressure and the number of passes. Wang et al. (2018) researched the influence of microfluidisation on rice bran in regard to the sorption characteristics of Pb(II), cholesterol and sodium cholate. The microfluidisation process was performed at five different pressures (30, 60, 90, 120 and 150 MPa). The results of the study showed that the increasing microfluidisation pressure up to 150 MPa had a higher ability to absorb cholesterol and sodium cholate.

In the literature, the various microfluidised cereal brans have been applied to decrease the calorie value of products. In one of these studies, Mert et al. (2014) used microfluidised wheat bran to increase the dietary fibre content of the cake. Storage experiments with cake samples showed that the addition of microfluidised wheat bran reduced the loss of moisture and improved firmness. The results of this study suggest that microfluidisation can be used to produce fibrous materials with enhanced physical properties as a novel milling technology. Erinc et al. (2018) used microfluidised wheat bran to produce low-fat biscuits. In this study, it was concluded that the amount of shortening used in the biscuit formula was lowered by the utilisation of microfluidised wheat bran fibre. However, the increase in fibre content and/or decrease in fibre size resulted in harder biscuits with a lower spread ratio.

8.5 PULSED ELECTRIC FIELD

This technology involves the application of high voltage electrical energy at 20–80 kV/cm. The food material is placed between two electrodes, and electrical energy is applied for a very short time. In a pulsed electric field, the temperature of the sample does not exceed 40°C, so it is deemed a non-thermal process (Oms-Oliu et al., 2012; Zhang et al., 2019). A pulsed electric field is able to inactivate microorganisms and enzymes and simultaneously preserve health-related compounds. The inactivation mechanisms of a pulsed electric field come from the electroporation of cells (Jongyingcharoen & Ahmad, 2014).

There are very few studies in which electric fields are used in the production of grain-based functional food. These studies have mostly focused on reducing the microbial load of grain flours. A similar study was conducted by Ketih et al. (1998) with dark rye flour. Dark rye flour is rich in dietary fibres, vitamins and minerals. During storage of flour, microbial contamination can cause loss of flour. To prevent this problem, Ketih et al. (1998) applied a pulsed electric field to inactivate microorganisms in dark rye flour, and they achieved a 0.6 log reduction in aerobic bacteria count.

8.6 IRRADIATION

Radiation is generally used as a non-thermal sterilisation technique with the application doses of 2–7 kGy (Oms-Oliu et al., 2012). In this technique, to inactivate microorganisms or insects, food products can be exposed to either ionising or non-ionising radiation. Ionising radiation is created by electron beams, X-rays or gamma rays obtained by Cobalt-60 or Caesium-137. In contrast, non-ionising radiation is carried out by ultraviolet (UV), visible, microwave or infrared light. The main inactivation mechanism of radiation is to damage the DNA. Thus the growth or proliferation of microorganisms can be limited. There have been three different application doses of radiation: up to 1 kGy is applied for disinfestations or delay in ripening, 1–10 kGy is applied for pasteurisation and 10–50 kGy is applied for industrial sterilisation and elimination of viruses. Radiation has been largely used to preserve processed or non-processed food products, according to the outstanding function of radiation processing as cold pasteurisation (Jongyingcharoen & Ahmad, 2014).

Saengshik powder is an uncooked and ready-to-eat grain-based functional food. It is mostly known and consumed in Korea. It has been found to have beneficial effects on weight loss with decreased total- and LDL-cholesterol levels. The total bacteria count of Saengshik is high because of using several different ingredients. On the other hand, during production, cooking is not applied. To decrease bacterial population, irradiation can be used as a non-thermal preservation method. In the study by Kim et al. (2020), Saengshik powder was exposed to different irradiation treatments involving electron-beam and gamma-ray dosages of 0, 1, 3, 5 and 10 kGy. The total sugar reduction, total antioxidant activity and total phenolic, carotenoid, chlorophyll and amino nitrogen contents of samples were analysed. According to the results, it was determined that irradiation made an insignificant effect on reducing sugar and amino nitrogen contents. Additionally, irradiation increased the total phenolic content of samples while it decreased carotenoids and total chlorophyll content. Moreover, radiation doses up to 10 kGy resulted in a highly significant improvement in the antioxidant activity of samples (see Figure 8.1).

8.7 PLASMA

Plasma is defined as the fourth state of matter which is different from solid, liquid and gas. The state of matter can be changed by pressure and heat (Zhang et al., 2019). Plasma technology is one of the latest green technologies currently used in different industries, especially in the food industry. This technology has shown promising applications for preserving nutritional, functional and sensory. In addition to managing the microbial load, the technology helps in the beneficial structural modification of food and packaging materials (Mir et al., 2016).

Chen et al. (2012) applied the low-pressure plasma method to increase the cooking and eating qualities of brown rice. The microstructure of the brown rice surface and the cooking, textural and iodine-staining properties of plasma-treated brown rice were researched. The cooking time of the brown rice was shortened after plasma treatment, and the cooked brown rice had a smooth texture and was easier to chew. In another study by Chen et al. (2015), brown rice was exposed to low-pressure plasma ranging from 1 to 3 kV for 10 min and led to germination at 25°C for 12, 18 and 24 h. It was determined that plasma treatment increased GABA levels from 19 to 28 mg/100 g. Additionally, a significant increase was reported after plasma treatment in germinated brown rice grains. The authors also concluded that plasma treatment was an effective way to increase the nutritional value of grains.

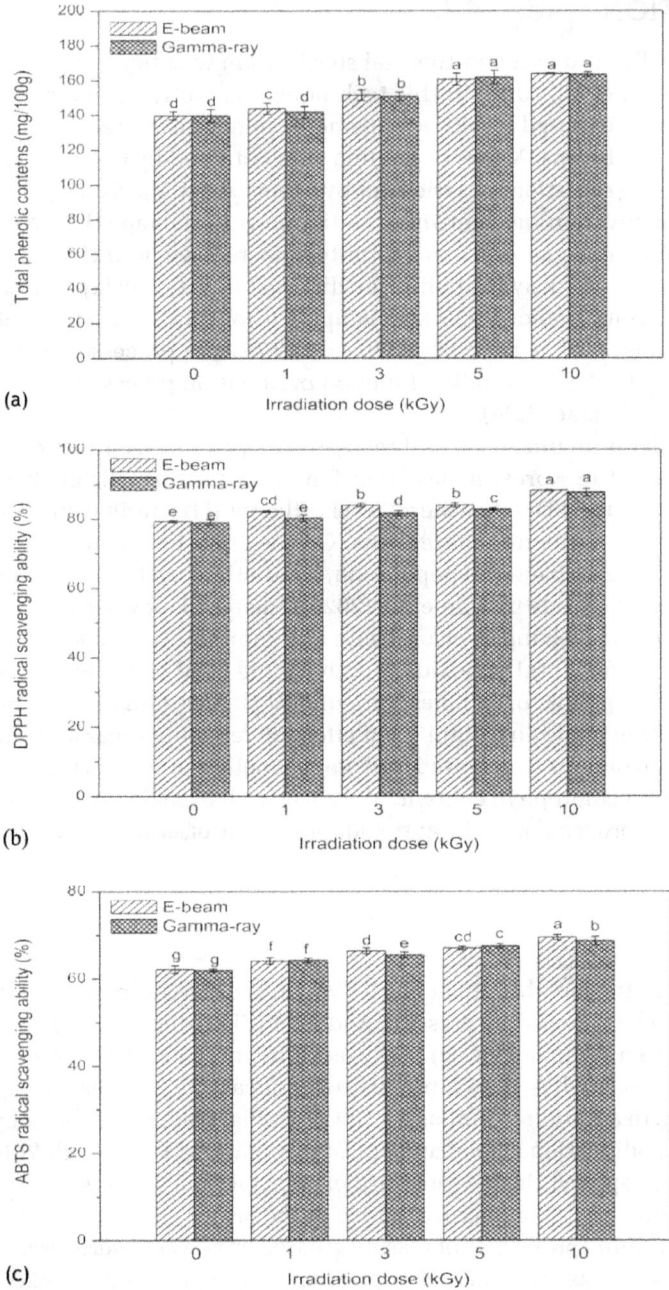

Figure 8.1 The change in (a) total phenolic content, (b) DPPH antioxidant activity and (c) ABTS antioxidant activity of Saengshik by irradiation (Kim et al., 2020).

8.8 CONCLUSION

Thermal processes have still been frequently used in the production of grain-based functional foods. The most important reason for this is that grains contain considerable amounts of starch. For the production of grain-based products, the starch must be partly or fully gelatinised. The heat-induced process accelerates the starch gelatinisation and makes production easier. However, high temperatures applied in the various stages of production have adverse effects on functional food components and limit bioavailability. From the nutritional point of view, non-thermal processes are gaining importance in the functional food industry. Among them, ultrasound and high-pressure homogenisation stand out in the production of grain-based functional foods.

REFERENCES

Baipong, S., Apichartsrangkoon, A., Worametrachanon, S., Tiampakdee, A., Sriwattana, S., Phimolsiripol, Y., Sintuya, P., Sintuya, P. 2020. Effects of germinated and nongerminated rice grains on storage stability of pressurised purple rice beverages with *Lactobacillus casei* 01 supplement. *Journal of Food Processing and Preservation*, 44: e14442.

Balakrishna, A.K., Farid, M. 2020. Enrichment of rice with natural thiamine using high-pressure processing (HPP). *Journal of Food Engineering*, 283: http://www.ncbi.nlm.nih.gov/pubmed/110040

Benito-Roman, O., Alonso, E., Cocero, M.J. 2013. Ultrasound-assisted extraction of beta-glucans from barley. *LWT-Food Science and Technology*, 50(1): 57–63.

Candal, C., Erbas, M. 2019. The effects of different processes on enzyme resistant starch content and glycemic index value of wheat flour and using this flour in biscuit production. *Journal of Food Science and Technology*, 56(9): 4110–4120.

Chang, R., Lu, H., Bian, X., Tian, Y., Jin, Z. 2021. Ultrasound assisted annealing production of resistant starches type 3 from fractionated debranched starch: Structural characterisation and in-vitro digestibility. *Food Hydrocolloids*, 110: http://www.ncbi.nlm.nih.gov/pubmed/106141

Chen, H.H., Chang, H.C., Chen, Y.K., Hung, C.L., Lin, S.Y., Chen, Y.S. 2015. An improved process for high nutrition of germinated brown rice production: Low-pressure plasma. *Food Chemistry*, 191: 120–127.

Chen, H.H., Chen, Y.K., Chang, H.C. 2012. Evaluation of physicochemical properties of plasma treated brown rice. *Food Chemistry*, 135(1): 74–79.

Čukelj Mustač, N., Voučko, B., Novotni, D., Drakula, S., Gudelj, A., Dujmić, F., Ćurić, D. 2019. Optimisation of high intensity ultrasound treatment of proso millet bran to improve physical and nutritional quality[$]. *Food Technology and Biotechnology*, 57(2): 183–190.

de Carvalhoa, G.R., Polachinia, T.C., Darros-Barbosaa, R., Bon,J., Telis-Romero, J. 2018. Effect of intermittent high-intensity sonication and temperature on barley steeping for malt production. *Journal of Cereal Science*, 82: 138–145.

Demirkesen-Mert, I. 2019. The applications of microfluidization in cereals and cereal-based products: An overview. *Critical Reviews in Food Science and Nutrition*, 60(6): 1007–1024.

Ding, J.Z., Ulanov, A.V., Dong, M.Y., Xiong, S., Dubat, A., Feng, H. 2018. Enhancement of gama-aminobutyric acid (GABA) and other health-related metabolites in germinated red rice (*Oryza sativa* L.) by ultrasonication. *Ultrasonocal Sonochemistry*, 40: 791–797.

Erinc, H., Mert, B., Tekin, A. 2018. Different sized wheat bran fibers as fat mimetic in biscuits: Its effects on dough rheology and biscuit quality. *Journal of Food Science and Technology*, 55(10): 3960–3970.

Gao, J., Yang, H., Rong, A., Bao, X., Zhang, M. 2013. Effects of HHP on microorganisms, enzyme Inactivation and physicochemical properties of instant oats and rice. *Journal of Food Process Engineering*, 37(2): 191–198.

Hassan, S., Imran, M., Ahmad, M.H., Khan, M.I., Xu, C.M., Khan, M.K., Muhammad, N. 2020. Phytochemical characterisation of ultrasound-processed sorghum sprouts for the use in functional foods. *International Journal of Food Properties*, 23(1): 853–863.

He, F., Wang, T., Zhu, S., Chen, G. 2016. Modeling the effects of microfluidization conditions on properties of corn bran. *Journal of Cereal Science*, 71: 86–92.

Hou, F., Su, D., Xu, J., Gong, Y., Zhang, R., Wei, Z., Chi, J., Mingwei, Z. 2016. Enhanced extraction of phenolics and antioxidant capacity from sorghum (*Sorghum bicolor* L. moench) shell using ultrasonic-assisted ethanol–water binary solvent. *Journal of Food Processing and Preservation*, 40(6): 1171–1179.

Hurtado-Romero, A., Toro-Barbosa, M.D., Garcia-Amezquita, L.E., García-Cayuela, T. 2020. Innovative technologies for the production of food ingredients with prebiotic potential: Modifications, applications, and validation methods. *Trends in Food Science and Technology*, 104: 117–131.

Hüttner, E.K., Bello, F.D., Poutanen, K., Arendt, E.K. 2009. Fundamental evaluation of the impact of high hydrostatic pressure on oat batters. *Journal of Cereal Science*, 49(3): 363–370.

Jongyingcharoen, J.S., Ahmad, I. 2014. Thermal and non-thermal processing of functional foods. In: A. Noomhorm, I. Ahmad & A.K. Anal (Eds.), *Functional Foods and Dietary Supplements* (First ed., pp. 295–324): 4. West Sussex, UK: John Wiley & Sons, Ltd.

Keith, W.D., Harris, L.J., Griffiths, M.W. 1998. Reduction of bacterial levels in flour by pulsed electric fields. *Journal of Food Process Engineering*, 21(3): 263–269.

Kim, M.Y., Lee, S.H., Janga, G.Y., Parka, H.J., Lia, M., Kim, S., Lee, Y.R., Noh, Y.H., Lee, J., Jeong, H.S. 2015. Effects of high hydrostatic pressure treatment on the enhancement of functional components of germinated rough rice (*Oryza sativa* L.). *Food Chemistry*, 166: 86–92.

Kim, G.R., Ramakrishnan, S.R., Ameer, K., Chung, N., Kim, Y.R., Kwon, J.H. 2020. Irradiation effects on chemical and functional qualities of ready-to-eat Saengshik, a cereal health food. *Radiation Physics and Chemistry*, 171: http://www.ncbi.nlm.nih.gov/pubmed/108692

Kunyanee, K., Luangsakul, N. 2020. The effects of ultrasound – Assisted recrystallisation followed by chilling to produce the lower glycemic index of rice with different amylose content. *Food Chemistry*, 323: http://www.ncbi.nlm.nih.gov/pubmed/126843

Lin, Z., Zhang, Z., Chen, X., Ming, Q. 2011. Ultrasonic-assisted extraction of red rice pigment. *Modern Food Science and Technology*, 27: 296–298.

Meng, L., Zhang, W., Wu, Z., Hui, A., Gao, H., Chen, P., He, Y. 2018. Effect of pressure-soaking treatments on texture and retrogradation properties of black rice. *LWT-Food Science and Technology*, 93: 485–490.

Mert, B., Tekin, A., Demirkesen, İ., Kocak, G. 2014. Production of microfluidized wheat bran fibers and evaluation as an ingredient in reduced flour bakery product. *Food and Bioprocess Technology*, 7(10): 2889–2901.

Miano, A.C., Ibarz, A., Augusto, P.E.D. 2017. Ultrasound technology enhances the hydration of corn kernels without affecting their starch properties. *Journal of Food Engineering*, 197: 34–43.

Mir, S.A., Shah, M.A., Mir, M.M. 2016. Understanding the role of plasma technology in food industry. *Food and Bioprocess Technology*, 9: 734–750.

Nguyen, T.T.T., Guyot, J.P., Icard-Vernière, C., Rochette, I., Loiseau, G. 2007. Effect of high pressure homogenisation on the capacity of Lactobacillus plantarum A6 to ferment rice/soybean slurries to prepare high energy density complementary food. *Food Chemistry*, 102(4): 1288–1295.

Nikolic, S., Mojovic, L., Rakin, M., Pejin, D., Pejin, J. 2010. Ultrasound-assisted production of bioethanol by simultaneous saccharification and fermentation of corn meal. *Food Chemistry*, 122(1): 216–222.

Oms-Oliu, G., Odriozola-Serrano, I., Soliva-Fortuny, R., Elez-Martinez, P., Martin-Belloso, O. 2012. Stability of health-related compounds in plant foods through the application of non thermal processes. *Trends in Food Science and Technology*, 23(2): 111–123. doi: 10.1016/j.tifs.2011.10.004

Park, C.Y., Kim, S., Lee, D., Park, D.J., Imm, J.Y. 2016. Enzyme and high pressure assisted extraction of tricin from rice hull and biological activities of rice hull extract. *Food Science and Biotechnology*, 25(1): 159–164.

Pasqualone, A., Delvecchio, L.N., Gambacorta, G., Laddomada, B., Urso, V., Mazzaglia, A., Di Miceli, G., Di Miceli, G. 2015. Effect of supplementation with wheat bran aqueous extracts obtained by ultrasound-assisted technologies on the sensory properties and the antioxidant activity of dry pasta. *Natural Product Communications*, 10(10): 1739–1742.

Setyaningsih, W., Saputro, I.E., Carrera, C.A., Palma, M. 2019. Optimisation of an ultrasound-assisted extraction method for the simultaneous determination of phenolics in rice grains. *Food Chemistry*, 288: 221–227.

Sharma, N., Goyala, S.K., Alam, T., Fatma, S., Chaoruangrit, A., Niranjan, K. 2018. Effect of high pressure soaking on water absorption, gelatinization, and biochemical properties of germinated and non-germinated foxtail millet grains. *Journal of Cereal Science*, 83: 162–170.

Wang, J., Sun, B., Cao, Y., Tian, Y., Li, X. 2008. Optimisation of ultrasound-assisted extraction of phenolic compounds from wheat bran. *Food Chemistry*, 106(2): 804–810.

Wang, L., Wu, J., Luo, X., Li, Y., Wang, R., Li, Y., Li, J., Chen, Z. 2018. Dynamic high-pressure microfluidization treatment of rice bran: Effect on Pb(II) ions adsorption in vitro. *Journal of Food Science*, 83(7): 1–11.

Xi, Q., Wang, L., Li, Y. 2018. Exploring high hydrostatic pressure-mediated germination to enhance functionality and quality attributes of wholegrain brown rice. *Food Chemistry*, 249: 104–110.

Xia, Q., Tao, H., Li, Y., Pan, D., Cao, J., Liu, L., Zhou, X., Barba, F.J. 2020. Characterizing physicochemical, nutritional and quality attributes of wholegrain Oryza sativa L. subjected to high intensity ultrasound-stimulated pre-germination. *Food Control*, 108: http://www.ncbi.nlm.nih.gov/pubmed/106827

Xu, X., Yan, W., Yang, Z., Wang, X., Xiao, Y., Du, X. 2019. Effect of ultra-high pressure on quality characteristics of parboiled rice. *Journal of Cereal Science*, 87: 117–123.

Zhang, X.X., Wang, L., Cheng, M.Y., Wang, R., Luo, X.H., Li, Y.N., Chen, Z.X. 2015. Influence of ultrasonic enzyme treatment on the cooking and eating quality of brown rice. *Journal of Cereal Science*, 63: 140–146. doi: 10.1016/j.jcs.2015.03.002

Zhang, Z.H., Wang, L.H., Zeng, X.A., Han, Z., Brennan, C.S. 2019. Non-thermal technologies and its current and future application in the food industry: A review. *International Journal of Food Science and Technology*, 54(1): 1–13. doi: 10.1111/ijfs.13903

Effects of Dense Phase CO$_2$ Application on Microbial Stability in Grain-Based Beverages and Food Products

M. Selvamuthukumaran

CONTENTS

9.1 INTRODUCTION

Pressurized carbon dioxide is commonly employed for preserving several food products due to its purity, low cost, and safety. Especially for liquid food products that can be preserved, their stability can be enhanced by using dense phase carbon dioxide. It is one of the most efficient non-thermal processing and preservation methods for enhancing the stability of food products. It successfully inactivates microorganisms and retains the majority of nutrients when compared to the conventional heat method of pasteurizing products. Food quality is not at all affected as a result of the application of dense phase CO$_2$. In the case of dense phase CO$_2$ applications in liquid foods, the food matrix will come into contact with either supercritical CO$_2$ or pressurized sub-CO$_2$, which can be operated in either a batch, semi-batch, or continuous process. The various treatment conditions used for dense phase CO$_2$ application to food products include CO$_2$ pressure ranging from 7 to 40 MPa and a processing time ranging from 2 to 2.5 hours, especially for a batch or a semi-continuous application process.

9.2 MECHANISMS INVOLVED IN MICROBIAL INACTIVATION BY THE DENSE PHASE CO_2 APPLICATION PROCESS

The mechanisms involved in the inactivation of microorganisms by dense phase CO_2 application are a critical process. Many researchers have explained that there are several mechanisms involved in microbial inactivation of food products as a result of dense phase CO_2 applications (Daniel et al., 1985; Gunes et al., 2005; Jones and Greenfield, 1982; Meyssami et al., 1992; Shimoda et al., 2001; Weder et al., 1992). Carbonic acid will be formed as a result of dissolving CO_2 in liquid foods, which can reduce the pH and release H+, carbonate, and bicarbonate ions, which can lower the pH even further.

Parton et al. (2007) state that dense phase CO_2 application leads to a better correlation between liquid food products and water by pH prediction and thereby in comparison with the resultant pH. The microbial cells' internal pH highly influences the reduction of microorganisms. The CO_2 can diffuse through the microbial cell membrane, which contains phospholipid layers, thereby lowering the inside pH and simultaneously enhancing the buffering capacity of cells. Cells can establish a pH gradient maintenance through the membrane and regulate the inside as well as outside environments, including pumping H+ ions out of cells. Microbial enzymes and their metabolic processes are prohibited by CO_2 permeation in cells, which can further minimize internal pH, and thereby the microorganisms are significantly inactivated (Castor and Hong, 1992).

9.2.1 Bicarbonate Ion and Molecular CO_2 Inhibitory Effects

CO_2 will prohibit the growth of several bacterial enzymes. The mechanism involved in its growth stoppage is the interaction of CO_2 with arginine to produce a bicarbonate complex, which can deactivate the enzymes (Yagiz et al., 2005; Yuk et al., 2009). The metabolic chain or process will be broken as a result of the treatment of more CO_2, which can prohibit decarboxylases (Kincal, 2000). The use of a micro-bubble system with a temperature of 35°C and a pressure of 15 MPa can deactivate enzymes like lipase and alkaline protease. There is a mechanism involved that includes intracellular ion precipitations, that is, Mg +2, Ca +2 from the bicarbonate (Park et al., 2002). The biological system will be subjected to lethal change when carbonate can precipitate Mg and Ca sensitive proteins.

9.2.2 Cell Membrane Modifications and Cellular Component Extraction

Other researchers have shown that CO_2 solvent characteristics, i.e. both hydrophobicity and lipophilic properties, can alter enzyme activity. Kincal et al. (2005) reported that phospholipid extraction can lead to inactivate microbes. There is also another mechanism explained, which is that CO_2 can penetrate into a cell membrane and accumulate because the internal layer is hydrophilic, and as a result, the CO_2 enhances membrane fluidity due to lipid chain order loss, which can be called the anesthesia effect; this can enhance permeability (Ishikawa et al. 1997). The microbes can be inactivated as a result of lipid and other prime component extraction in cells or a membrane. Ishikawa et al. (1995) explained that, when cells are subjected to dense phase CO_2 treatment at temperature and pressure conditions of 30°C, 7 MPa for a time period of 10 minutes, they can exhibit cell damages, which further lead to the release of intracellular ions, slat tolerance, loss, UV absorbing substance leakages, as well as proton permeability impairment.

9.3 INSTALLATION OF THE DENSE PHASE CO$_2$ SYSTEM

The dense phase CO$_2$ system is installed as shown in Figure 9.1 (Folkes, 2004). The erected system will blend pressurized and cooled CO$_2$ along with the liquid feed, which can be pressurized by its own pump. The mixture is then allowed to be passed through a holding tube 79 meters long for a specific time duration, which can be easily revised by altering the mixture flow rate. The temperature can be put under control during holding by means of insulation and electrical heating tapes, the system can be monitored and easily controlled by maintaining operating pressure, and the mixture can be collected by way of the holding tube; thus the mixture can be sent through the back-pressure valve after depressurization.

The machine can be operated by subjecting the feed mixture to 6.89 MPa with the help of a reciprocating pump with 30 mm of stroke length and a back-pressure valve, which is then brought to the required operating pressure with the help of another reciprocating pump and a second back-pressure valve. This reciprocating pump is used to incorporate CO$_2$ at even higher than 6 MPa. The liquid feed is already in pressurized form and since the CO$_2$ is also in pressurized form, both can be easily mixed in the right proportions, which can remain in liquid form. The feed mixture with CO$_2$ can be sent through the second or another reciprocating pump, which can maintain the working pressure of 6.89 MPa or even more. The pressure needs to be continuously maintained when the sample is in the holding period. The sample can be depressurized when the mixture comes out from the holding tube and drawn in sterilized bottles under aseptic conditions by using ethanol as a sterilant at the holding tube end so that the aseptic condition can be maintained, which can enhance the stability of the treated mixture.

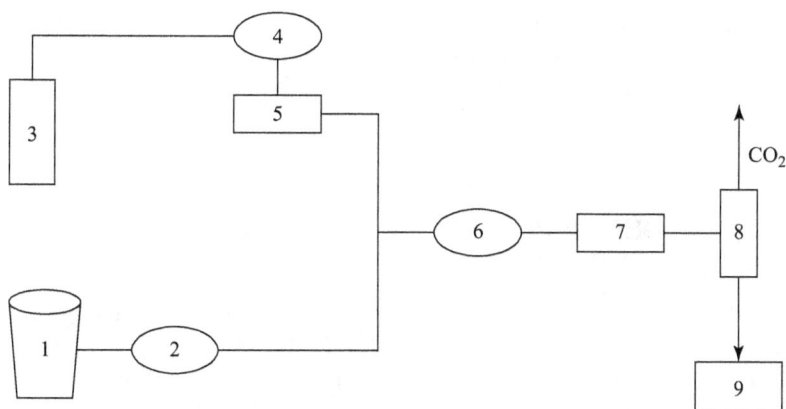

1. Juice	6. Main Pump
2. Pump	7. Holding Tube
3. CO$_2$ Tank	8. Vacuum
4. Pump	9. Treated Juice
5. Chiller	

Figure 9.1   Dense phase CO$_2$ system (Modified from Folkes, 2004).

9.4 EFFECTS OF DENSE PHASE CO$_2$ APPLICATION ON HAZE DEVELOPMENT

Folkes (2004) processed beer by applying dense phase CO$_2$ techniques in fermented alcoholic beverages. The quality effects, i.e. haze development as a result of dense phase CO$_2$ application, were studied. His result showed that the application of dense phase CO$_2$ had successfully reduced the development of haze or turbidity compared to a heat-processed beverage sample as well as a fresh sample. The lowest turbidity was recorded for the sample processed with dense phase CO$_2$ at 27.6 MPa, with the recorded turbidity of 95.3 NTU compared to heated samples with a turbidity of 120.7 NTU, which clearly indicates that the dense phase CO$_2$ process helps in significantly reducing haze development to a greater extent. The dense phase CO$_2$–processed beer sample exhibited less haze development ascribed to change in pH, i.e. reduction. As a result of this process, the pH was reduced to 3, which can impact polyphenol and protein conformation and thus interfere with the formation of protein-polyphenol complexes, which can further aid in haze formation in alcoholic beverages.

9.5 EFFECTS OF DENSE PHASE CO$_2$ APPLICATION ON FOAM STABILITY AND FOAM CAPACITY

The dense phase CO$_2$–processed sample exhibited a lesser foam capacity of 324% compared to the heat-processed sample, which recorded a greater foam capacity of 333%, while the heat-processed beverage sample recorded foam stability of 60%, which is found to be higher when compared to dense phase CO$_2$–processed sample, which contains foam stability of around 52%. The dense phase CO$_2$–processed sample showed a pronounced change in foam characteristics due to extraction of the cell membrane or part of the cell wall that may have altered the content of hydrophobic components found in beer, thereby significantly reducing the foaming capacity of the resultant product.

9.6 EFFECTS OF DENSE PHASE CO$_2$ APPLICATION ON MICROBIAL REDUCTION

The application of dense phase CO$_2$ further helps to significantly reduce microorganisms in various beverages (Table 9.1). The enhanced load reduction of yeast population was achieved in beer as a result of this technique at optimum conditions like a temperature of 21°C, a pressure of 26.5 MPa, a residence time of 4.77 min, and a CO$_2$ concentration of 9.6%. Kumagai et al. (1997) reported that reduction in the microbial count was directly proportional to CO$_2$ absorption. The microbial log reduction of seven-fold is achieved as a result of dense phase CO$_2$ applications, proving that it can be an effective process to significantly reduce the microorganisms when compared to heat pasteurization of beer. The stability of the beer can be extended further with CO$_2$ not only by reducing the yeast count but also by preventing the flavor damage to the beer caused by heat pasteurization. In the beer industry, dense phase CO$_2$ can be successfully employed to preserve the flavor, thereby skipping the prefiltration step during the processing of beer.

TABLE 9.1   EFFECTS OF DENSE PHASE CO_2 APPLICATIONS ON THE MICROBIAL REDUCTION OF BEVERAGES

Name of the Product	Pressure Used (MPa)	Log Reduction Achieved	Microorganisms	References
Apple juice	10	Full	APC	Gasperi et al. (2009)
Orange juice	< 6	4.0	TPC	Erkmen (2001)
Carrot juice	< 5	4.0	APC	Park et al. (2002)
Watermelon juice	34	6.5	APC	Lecky (2005)
Mandarin juice	41	3.5	APC	Yagiz et al. (2005)
Beer	> 26	7.3	Yeast	Folkes (2004)
Kava beverages	34	3.0	APC	Hsieh et al. (2007)
Lychee juices	8	5.0	APC	Guo et al. (2011)

Abbreviations: TPC: total plate count; APC: aerobic plate count.

9.7 EFFECTS OF DENSE PHASE CO_2 APPLICATION ON ORGANOLEPTIC CHARACTERISTICS OF BEER

The application of dense phase CO_2 doesn't show any organoleptic changes like a significant change in either aroma or flavor. Only a slight variation in aroma was noted among heat-processed and CO_2-processed beer samples, which suggests that application of or treating the beverage with CO_2 doesn't alter the sensory characteristics (Folkes, 2004). The stored samples also showed insignificant changes in sensory attributes (typically aroma and flavor) when compared to an untreated or fresh beer sample. Therefore, it can be concluded that dense phase CO_2 can be an efficient method to retain the organoleptic properties of beer samples, thereby significantly enhancing the overall acceptability of the stored beverage product.

9.8 CONCLUSIONS

The stability of a grain-based type of alcoholic beverage product like beer can be enhanced successfully by retaining the nutrients and reducing microorganisms by several-fold. Aspects that negatively affect product quality, like turbidity or haze formation, can be reduced without altering organoleptic attributes, thereby retaining consumer acceptability to a greater extent.

REFERENCES

Castor TP, Hong GT (1992) Supercritical fluid disruption of and extraction from microbial cells. *US Patent* 5: 380–826.
Daniels JA, Krishnamurti R, Rizvi SSH (1985) A review of effects of CO_2 on microbial growth and food quality. *Journal of Food Protection* 48(6): 532–537.

Erkmen O (2001) Effects of high-pressure carbon dioxide on *Escherichia coli* in nutrient broth and milk. *International Journal of Food Microbiology* 65(1–2): 131–135.

Folkes G (2004) *Pasteurization of Beer by a Continuous Dense Phase CO$_2$ System*. PhD thesis, University of Florida, United States.

Gasperi F, et al (2009) Effects of supercritical CO$_2$ and N$_2$O pasteurisation on the quality of fresh apple juice. *Food Chemistry* 115(1): 129–136.

Gunes G, Blum LK, Hotchkiss JH (2005) Inactivation of yeasts in grape juice using a continuous dense phase carbon dioxide processing system. *Journal of the Science of Food and Agriculture* 85(14): 2362–2368.

Guo M, Wu J, Xu Y, Xiao G, Zhang M, Chen Y (2011) Effects on microbial inactivation and quality attributes in frozen lychee juice treated by supercritical carbon dioxide. *European Food Research and Technology* 232(5): 803–811.

Hsieh M, et al. (2007) The quality improvement of Kava beverage by non-thermal dense phase CO$_2$ pasteurization. In: *Institute of Food Technologists Annual Meeting*, Chicago, IL, 189–158.

Ishikawa H, Shimoda M, Kawano T, Osajima Y (1995) Inactivation of enzymes in an aqueous solution by micro-bubbles of supercritical CO$_2$. *Bioscience, Biotechnology, and Biochemistry* 59(4): 628–631.

Ishikawa H, Shimoda M, Tamaya K, Yonekura A, Kawano T, Osajima Y (1997) Inactivation of *Bacillus* spores by the supercritical carbon dioxide microbubble method. *Bioscience, Biotechnology, and Biochemistry* 61(6): 1022–1023.

Jones RP, Greenfield PF (1982) Effect of carbon dioxide on yeast growth and fermentation. *Enzyme Microbiology Technology* 4(4): 210–223.

Kincal D (2000) *A Continuous High-Pressure Carbon Dioxide System for Cloud Retention, Microbial Reduction and Quality Change in Orange Juice*. MSc thesis, University of Florida, Gainesville, United States.

Kincal D, Hill S, Balaban MO, Marshall MR, Wei C (2005) A continuous high-pressure CO$_2$ system for microbial reduction in orange juice. *Journal of Food Science* 70(5): 249254.

Kumagai H, Hata C, Nakamura K (1997) CO$_2$ sorption by microbial cells and sterilization by high-pressure CO2. *Bioscience, Biotechnology, and Biochemistry* 61(6): 931–935.

Lecky M (2005) Shelf life evaluation of watermelon juice after processing with a continuous high-pressure CO$_2$ system. In: *IFT Annual Meeting Book of Abstracts*, Institute of Food Technologists, New Orleans.

Meyssami B, Balaban MO, Teixeira AA (1992) Prediction of pH in model systems pressurized with carbon dioxide. *Biotechnology Progress* 8(2): 149–154.

Park SJ, Lee JI, Park J (2002) Effects of a combined process of high pressure carbon dioxide and high hydrostatic pressure on the quality of carrot juice. *Journal of Food Science* 67(5): 1827–1834.

Parton T, Bertucco A, Elvassore N, Grimolizzi L (2007) A continuous plant for food preservation by high pressure CO$_2$. *Journal of Food Engineering* 79(4): 1410–1417.

Shimoda M, Cocunubo Castellanos J, Kago H, Miyake M, Osajima Y, Hayakawa I (2001) The influence of dissolved CO$_2$ concentration on the death kinetics of Saccharomyces cerevisiae. *Journal of Applied Microbiology* 91(2): 306–311.

Weder JKP, Bokor MV, Hegarty MP (1992) Effect of supercritical CO2 on arginine. *Food Chemistry* 44(4): 287–290.

Yagiz Y, Lim SL, Balaban MO (2005) *Continuous High-Pressure CO$_2$ Processing of Mandarin Juice: IFT Annual Meeting Book of Abstracts*. Institute of Food Technologists, New Orleans, United States.

Yuk GH, Geveke D, Zhang Q (2009) Efficacy of supercritical CO$_2$ for nonthermal inactivation of *Escherichia coli*- K12 in apple cider. *International Journal of Food Microbiology* 138(1–2): 9199.

Chapter 10

Packaging Requirements for Non-Thermal Processed Grain-Based Foods

Nese Basaran-Akgul

CONTENTS

10.1 INTRODUCTION TO GRAINS AND GRAIN-BASED FOODS

Grains, commonly referred to as 'cereals' or 'cereal grains', are the edible seeds of the grass family *Poaceae* (also known as *Gramineae*) which includes wheat, rice, barley, corn (maize), oats, sorghum, rye, and millet as major grain crops. Globally, cereals are very important crops, with maize, rice, and wheat being the predominant crops in terms of cultivated area and according to the FAO. Cereal foods contribute to food security, especially when fresh food is not available, and they are used as raw materials in a wide range of food varieties, such as fermented cereals. Wheat is utilized mainly for making flour and semolina, which are used in the preparation of a wide range of food products, including fermented foods, breakfast cereals, etc. Only wheat and rye are suitable for use in the preparation of leavened bread.

Pseudo-cereal grains are not 'true' grains; however, their overall nutritional composition is alike to 'true' grains and they can be used in similar ways in the manufacture of beverages or other foods. These minor or ancient grains include wild rice, red and black rice, buckwheat, sorghum, teff, millet, quinoa, amaranth, kamut, spelt, einkorn, emmer, and black corn. These ancient or minor cereals have drawn great interest, particularly in Western countries, because of their higher content in beneficial minor components (dietary fiber, resistant starch, minerals, vitamins, phenolic compounds). They have been named 'the grains of the twenty-first century' due to their nutritional qualities and bioactive compounds (FAO, 2011). They do not contain any prolamins that are toxic for people who suffer from celiac disease. Pseudocereal grains are known to contain high starch content but little or no gluten-like proteins, and thus these species are appropriate to consume for those intolerant to gluten (Alvarez-Jubete et al., 2010). Beverages from these grains can be integrated into gluten-free diets and could be a valuable contribution as an alternative nutrition source (Peñas et al., 2014). Recently, several gluten-free beers have been industrially produced from sorghum malt in the USA for the sake of celiac patients.

Cereals hold potential as a substitute for non-dairy probiotic foods. In addition to creating healthy food products and fortifying fruit and vegetable juices with probiotic bacteria, cereal beverages are valued for their nutrient content such as minerals, dietary fiber, antioxidants, and vitamins (Tuorila and Cardello, 2002; Heenan et al., 2004; Shori, 2016). Cereal beverages are becoming widely accepted among beverage consumers and are considered the most promising type of novel and functional foods because of the opportunity to improve nutrient content, convenient container options for refrigerated, shelf-stable storage, and relatively low cost potential of these products. Grain milk is a milk substitute prepared by using fermented grain or flour of oats, spelt, rice, rye, einkorn, wheat, or quinoa. As functional food ingredients, prebiotics have been attracting more and more attention, as they support probiotics to change the gut microbial balance toward health benefits, and cereal-based beverages are inexpensive sources of prebiotics. Furthermore, they contain nondigestible carbohydrates that act as prebiotic sources to stimulate the growth of probiotics (*Lactobacillus* and *Bifidobacterium*) present in the colon. Cereal-based probiotic drinks have been developed by fermentation of oat, barley, and malt substrates; examples include a probiotic oat flake beverage (Luana et al., 2014) and a probiotic ragi malt (VidyaLaxme et al., 2014). Oat, due to its functional attributes, is assumed to be one of the best choices among cereals as a potential probiotic carrier (Enujiugha and Badejo, 2015; Salmeron et al., 2014; Salmeron et al., 2015). In addition, nanoparticles can be utilized in cereal-based fermented beverages to improve organoleptic characteristics, increase absorption and intentioned delivery of nutrients and bioactive compounds, and stabilize active ingredients such as nutraceuticals in food structure (Ranjan et al. 2014; Salmeron et al., 2015).

Looking to scale up the processes and accomplish the industrialization of traditional grain-based beverages and foods, both the industry and researchers are assessing novel technologies and techniques including the modification of emerging non-thermal technologies to alter the sensory attributes of the products (Shiferaw and Augustin, 2020). However, due to the composition of bioactive substances, their health benefits to consumers in each beverage are scattered. Personal needs, cultural beliefs, allergies, intolerances, and processing of the food can change the consumers' attitudes to any food.

10.2 NON-THERMAL PROCESSING OF GRAINS AND GRAIN-BASED PRODUCTS

Traditional methods apply thermal processing including cooking, blanching, pasteurization, and sterilization to inactivate enzymes, microorganisms, and chemical reactions caused by them in food materials. This is done to extend the shelf life and the durability of the product at the desired quality (Kumar, 2012), but many chemical and physical changes occur after thermal processes. Health-conscious and educated consumers' demands for more natural, minimally processed, convenient high-quality food products have grown, as has the continuous search at industrial, retail, and distribution levels to meet these demands. Non-thermal processes do not utilize increased temperature to inactivate decomposing microorganisms and enzymes. This is the biggest advantage of non-thermal processes because this low-temperature pasteurization does not overcook and/or degrade foods. Non-thermal processes do not utilize increased temperature to achieve these effects and offer an alternative option. Several non-thermal techniques have evolved in recent times, including high-pressure processing, light (ultraviolet, pulsed light), pulsed electric field, ionizing radiation (gamma irradiation, electron beam), ultrasound, and gases (ozone, chlorine dioxide, cold plasma). As non-thermal processing technologies have been investigated for use in food processing (Pivarnik and Worobo, 2014), their application for grain-based foods and beverages specifically has been encouraging.

10.2.1 High Hydrostatic Pressure (HHP) Treatment

HHP is the most common emerging food processing technology with minimum quality loss in the industry (Farkas, 2016). In this process, foods undergo high hydrostatic pressure, generally in the range of 100–800 MPa, for periods ranging from seconds to minutes at a low or moderate temperature for food preservation and modification purposes (Cheftel and Culioli, 1997). HHP can be applied to both liquid and solid foods. During typical high-pressure treatment, the food product is treated in batches, but advances in technology have made a semicontinuous or continuous method of processing the product possible. In batch operations, the food material is vacuum packaged in flexible or semi-rigid packaging material and processed in a high-pressure chamber that is filled with pressure-transmitting fluid. In semicontinuous or continuous systems, the product is usually pumpable and after the pressure application is aseptically packaged (Han, 2007).

HHP causes the modification of plant cells by inducing the cleavage of chemical bonds of polysaccharides and disruption of the cellular network (Mateos-Aparicio et al., 2010; Allan et al., 2013). This property is used as an advantage for the extraction of functional molecules from a variety of grains, such as γ-oryzanol extraction from germinated rough rice (Kim et al. 2015), tricin extraction from rice hulls (Park et al., 2016), and higher levels of xanthohumol content in beer wort with HHP treatment (Santos et al., 2013). HHP (200–400 MPa/15 min) soaking of black rice improved the cooking properties and retrogradation with better water absorption capacity (Meng, 2018). In starchy foods, irreversible granule swelling (pressure-induced gelatinization) occurs and starches keep their granular forms or present a gel-like appearance; this can be treated with HHP application without affecting the quality of the food (Błaszczaka et al., 2005; Buckow et al., 2007; Li et al., 2012; Kieffer et al., 2007). HHP is applied to improve the functionality of oat batters (Huttner et al., 2009) and wheat doughs.

The first commercial application of HHP on rice and cereal grain was established by a Japanese company. This high-pressure-treated, RTE, brown or white rice is packaged in single-serve trays, and these products only need three minutes' heating in a microwave oven before consumption. In addition, when compared to conventional cooked brown rice, high-pressure treatment reduces the starch retrogradation and increases the gelatinization in brown rice, also giving it a better sensory quality and digestion properties. The shelf life of the RTE products is extended up to one year using a heat sterilization step after HHP. Single portion RTE cereal mix is also available, which is composed of eight types of grains including brown rice, black bean, soybean, azuki bean, oats, barley, millet, and red rice. The HHP treatment fastens the water hydration of the grains, which improves the cooking process of heat sterilization, and as a result, a sweeter and more digestible product is produced (Yamazaki, 2000, 2005a). HHP (200–400 MPa/15 min) soaking of black rice improved the cooking properties and retrogradation with better water absorption capacity (Meng, 2018). HHP is applied to brown rice to create a functional product with a short cooking time and modification of the cell wall structure. After the treatment, the product is dried and packed in pouches that are stable at room temperature (Yamazaki, 2005b). HHP is also applied to brown ice to increase the permeability of the cell wall and, following salt-extraction of allergenic proteins, to produce hypoallergenic rice (Yamazaki et al., 1998).

10.2.2 Pulsed Electric Field (PEF) Treatment

Pulsed electric field (PEF) treatment, also referred to as electroporation or electropermeabilization, is a non-thermal process with very short electric pulses (ms or µs) at high electric field strengths (10–80 kV/cm) and moderate temperatures (Martin-Belloso and Elez-Martinez, 2005). Typically, about one to 20 flashes per second are applied for the processing of liquid and semi-liquid food matrices, as well as the treatment of different solid foods (Barba et al., 2015). PEF is also applicable in other food processing zones such as the extraction of bioactive compounds or as a pretreatment prior to freezing, canning, and drying (Taiwo et al., 2002).

It is based on the application of electric field pulses applied in specially designed containers that change the structures of foods: cells are disrupted and cell walls become permeable, and as a result, structure and chewability characteristics are improved. The effectiveness of PEF on cells depends on electric field strength, treatment time, specific energy, pulse shape, pulse width, frequency and temperature, treatment type (batch or continuous), shape of the treatment chamber (collinear, coaxial, or parallel) (Zhang et al., 1995; Barbosa-Canovas et al., 1998; Clark, 2006; Van den Bosch, 2007), physicochemical characteristics of the treated matrix (density, water activity, pH, and conductivity) (Zhang et al., 1995; Wouters et al., 2001; Min et al., 2003), characteristics of the treated cells (size, shape, membrane, and envelope structure) (Raso et al., 1998; Wouters et al., 2001; Evrendilek and Zhang, 2003), and state (suspension, solid, or semi-solid) (Vorobiev and Lebovka, 2009).

Conventionally, beverages are commonly stabilized by thermal processes, causing loss of bio-functional content. PEF as a non-thermal technology is an alternative to heat for inactivating native enzymes of cereals or pathogenic microorganisms in order to avoid spoilage and extend the shelf life of beverages. PEF has the advantage of having less or little effect on the nutritional and sensory properties of food material; for example, beverages treated with PEF have higher levels of polyphenols, carotenoids, and vitamins compared to those treated with heat pasteurization (Odriozola-Serrano et al., 2013; Saldaña et al., 2014; Tokusoglu et al., 2014). Some of the current applications of PEF include mold inactivation in baked products, non-thermal pasteurization of grain-based drinks, and termination of the fermentation (Kumar et al., 2012; Morris et al., 2007).

10.2.3 Pulse Light Technology

Pulsed light processing can be described as a sterilization or decontamination technique used mainly to inactivate surface microorganisms on foods, packaging material, and processing equipment (Oms-Oliu et al., 2010). Pulsed light has an advantage over chemical sterilization by H_2O_2 or peracetic acid for package sterilization since it does not leave any chemical residue. The effectiveness of the PL depends on the transmissivity of the product, the shape of the reactor, the duration and number of pulses, the intensity and wavelength of light, the distance (between the product and the light source), the thickness of the product, and the type of package.

10.2.4 Irradiation

Irradiation of foods comprises exposure of bulk or pre-packaged foods to ionizing radiations such as gamma rays emitted from the radioisotopes 60Co and 137Cs, X-rays, or electron beams (Farkas, 2004). The applied radiation inactivates microorganisms by damaging their genetic materials. The effect of irradiation on foods depends on the absorbed dose, expressed in Gray (kGy). One Gy equals 1 kJ/kg of product. Food products could either be pasteurized (1–5 kGy) or sterilized (10–74 kGy) depending on the dose of radiation (Morris et al., 2007). Food irradiation is a 'cold' process for preserving food that has been extensively used for over 50 years (Junqueira-Goncalves et al., 2011). Commercial-scale applications of food irradiation have increased significantly in the United States of America (USA) and the European Union (EU) (Ihsanullah and Rashid, 2017). Food irradiation is approved in more than 60 countries, and there has been notable growth in the production and trade of irradiated foods since 2010 (Eustice, 2017). It has purposes in a wide variety of foodstuffs, mostly as a postharvest phytosanitary measure (Breidbach and Ulberth, 2016). In the USA, all radiation processes, including their packaging materials, should be approved by the US Food and Drug Administration (FDA), since radiation sources are considered food additives. The FDA also requires a statement 'treated with radiation' or 'treated by irradiation' along with a 'Radura' symbol on the package of all irradiated foods (FDA, 21 CFR 179). Factors such as the dose, temperature, presence of oxygen, and food type have an impact on the amount of vitamin loss. Besides such inherent abilities, several factors such as composition, moisture content, temperature during irradiation, presence or absence of oxygen, and fresh or frozen state influence radiation resistance, particularly in the case of vegetative cells (Farkas, 2004). Treatment doses less than 3.0 kGy at low temperatures in the absence of oxygen and the storage of treated foods in airtight packages at low temperatures minimize the vitamin loss. However, the sensitivity of all vitamins is not the same (Fanaro et al., 2015; Roberts, 2016).

Radiation can be applied through packaging materials including those that cannot withstand heat (Farkas, 2004; Farkas and Mohacsi-Farkas, 2011). Low doses (0.2–1 kGy) are effective at preventing losses caused by insect pests in stored grains, pulses, cereals, flour, coffee beans, spices, dried fruits, dried nuts, dried fishery products, and other dried food products. The application of radiation for insect disinfestation is important. The use of a radiation dose in the 0.2–1 kGy range is found effective for preventing losses associated with insect pests in stored grains, pulses, cereals, flour, coffee beans, spices, dried fruits, dried nuts, dried fishery products, and other dried food products. The minimum dose of 300 Gy could prevent insects from establishing in non-infested areas while a minimum dose of 150 Gy could be effective for fresh fruit and vegetables. In most cases, irradiation either kills or inhibits further development of different life-cycle stages of insect pests (Sunil et al., 2018).

The effect of irradiation technology on the viability of fungi and on preventing the formation of mycotoxins was studied (Calado et al., 2018; Junqueira-Goncalves et al., 2011) and suggested that gamma irradiation can be generally considered to significantly improve the mycotoxicological safety of food and feed. Jeong and Kang (2017) reported that low-dose (< 3 kGy) irradiation treatment can be sufficient to control pathogens in cookie dough without affecting quality. Processors employing irradiation for handling the food are liable to follow the regulations properly. Currently, FDA regulations do not permit the use of irradiation to treat cookie dough to eliminate foodborne pathogens (FDA, 2014), but it is permitted for control of microbial pathogens on seeds for sprouting, not to exceed 8.0 kGy. Gamma irradiation was found to preserve nutritive content and prolong shelf life by preventing post-harvest insect and pest infestation of beans and grains (Armelim et al., 2006). As the level of radiation dose increases, the structure of plant foods appears to break down, and in some cases, this can improve the digestibility of the food.

10.2.5 Plasma

Plasma is described as the fourth state of matter other than solid, liquid, and gaseous states of matter (Niemira, 2012). In general, plasma can be classified according to its temperature into thermal and non-thermal plasmas. In comparison to thermal plasma, non-thermal ones are only partially ionized. Cold plasma (CP) is generated at or near room temperature and therefore considered a novel non-thermal technology for the improvement of food safety and quality and a new discipline in food processing (Fernandez et al., 2013). There is continued and growing interest in CP for microbial inactivation in the food sector, whereby the possible applications are manifold, such as the treatment of fresh and dry products, or the in-package treatment of food (Bourke et al., 2017; Niemira, 2012; Schlüter et al., 2013). Plasma processing parameters include power or voltage applied, frequency, electrode contact space, discharge gap, distance from the plasma source, treatment time, discharge gas type, gas pressure, gas flow rate, type of target compound, and type of material.

Plasma treatment is also investigated for its effect on seed germination, inactivation of enzymes, physicochemical modification of food properties including modification of starch and protein, and in-pack sterilization of products (Thirumdas et al., 2015). Although CP offers a promising technology in the different fields of food and feed processing, presently, the only commercial application of CP technology in food industries is limited to polymer processing used for food packaging applications (Bubler, 2016; Pankaj et al., 2014). CP is capable of modifying wet and dry surfaces of agricultural and food products (Grzegorzewski et al., 2010; Khanal et al., 2014; Hertwig et al., 2018). In the last decade, various technologies have been evaluated for seed germination such as ultrasound treatment (Yaldagar, 2009), magnetic field (Vashissth et al., 2010), electric field (Tkalec et al., 2009), and plasma (Hati, 2018). The plasma treatment has been found efficient for prompting an increase in surface permeability and a quicker seed germination process for wheat seeds with an increase in root length and germination rate (Dobrynin et al., 2009). Besides the improvement in germination and other growth parameters, plasma treatment is capable of decreasing cooking time and enhancing dough quality, texture, and nutritional values (Sarangapani et al., 2017). Lee et al. (2016) reported an increase in α-amylase activity (1.21-fold) and water absorption in brown rice due to plasma etching and depolymerization. In addition to these, plasma is also applicable for removing the chemical residues from foods (Sarangapani et al., 2017). Two major soy allergens, β-conglycinin and glycinin (Meinlschmidt et al., 2016), became less visible on protein gel after cold plasma treatment.

10.2.6 Ultrasound

Ultrasound, another emerging technology, was established for controlling the microstructure and modifying textural characteristics of fat products (sonocrystallization), emulsification, defoaming, modifying the functional properties of different food proteins, inactivation or acceleration of enzymatic activity to improve shelf life and quality of food products, microbial inactivation, freezing, thawing, freeze-drying and concentration, drying, and facilitating the extraction of various food and bioactive components (Awad et al., 2012).

The applications of ultrasound in food processing, analysis, and quality control based on the frequency can be separated into two groups (low and high energy). Low-energy (low-power, low-intensity) ultrasound with frequencies higher than 100 kHz at intensities below 1 W/cm^2 can be applied for the treatment of bread and cereal products (Awad et al., 2012), while high-energy (high-power, high-intensity) ultrasound uses intensities higher than 1 W/cm^2 (10–1000 W/cm^2) at lower frequencies (20–100 kHz) (McClements, 1995), which are disruptive and induce undesired effects on the physical, mechanical, or chemical/biochemical properties of foods. Higher-power ultrasound has been applied in order to modify the functional properties of proteins of different foods such as dairy, animal products, cereals, legumes, tubers, and fruits (O'Sullivan et al., 2017), generate emulsions, disrupt cells, promote chemical reactions, inhibit enzymes (McClements, 1995), and inactivate microorganisms by causing cavitation (implosion of gas bubbles) (Feng and Yang, 2005) and the mechanical and chemical degradation of polymers (Chemat et al., 2017). These effects are promising in food processing, preservation, and safety.

Important processing factors include probe design, geometry, and characteristics (e.g., frequency), suitable ultrasound processing system, operation conditions, physicochemical properties, and functional properties of the food product (Knorr et al., 2011; Awad et al., 2012). The practical uses of ultrasound in food processing include extraction and separation, sterilization and fresh-keeping, detection and analysis in food, freezing, thawing, crystallization, filtration, drying, and osmotic dehydration (OD) (Ma et al., 2016). Ultrasound is also used in processing lines to detect leaks in packages and control the microbiological quality of several foodstuffs. Ultrasonic size reduction, including cutting, slicing, slitting, and chopping, has found commercial applications in the food industry (Liu et al., 2015; Cardoni and Lucas, 2005), as have the acceleration of microbial fermentation, mixing, homogenization, emulsification of oil/fat in a liquid stream, spraying, degassing, crystallization of fats and sugars, foam breaking, effluent treatment, humidifying, and fogging. Ultrasonic cutting equipment for size reduction uses a horn-shaped titanium-made knife to cut sticky and brittle foods such as cereal bars, nuts, raisins, and peanuts with minimum waste (Liu et al., 2015). Ultrasound is also used for grain-based products for the purposes of accelerating microbial fermentation, mixing, homogenization, degassing, and foam breaking. An ultrasound defoamer was fabricated and commercially utilized to control the excess foam formed during the filling process of bottles and cans on high-speed canning lines and in fermenting vessels and other reactors of large sizes (Gallego-Juárez et al., 2010; Juárez et al., 2010; Rodríguez et al., 2010; AWAD 2012).

Ultrasound-assisted mixing (2.5 kW for 9 min) was used to enhance batter characteristics, which resulted in higher overrun and viscosity by lowering the batter density and flow behavior index, and the cake produced from that batter had much better quality (lower hardness but higher springiness, cohesiveness, and resilience) (Tan et al., 2011). Ultrasound-assisted extraction (UAE) was developed to remove the desired active biomolecules from specific food sources. The frequencies from 20 to 2,000 kHz are utilized to increase the permeability of cell walls and follow their breakage, favoring the removal of biologically active compounds. UAE is considered a green method of extraction of compounds of interest due to shorter extraction time, lower energy and solvent requirements, and decreased CO_2 emissions (Chemat et al., 2017). Red pigment from red rice is extracted with a 16.66% yield at the optimum settings (144 W, 70°C, 50 min) (Lin et al., 2011). The ultrasound-assisted soaking and gelatinization of rice (75°C with sonication) reduced the rice parboiling time (70%), and compared to conventional soaking (10 hours), moisture content was reached up to 48% after only 3 hours (Wambura et al., 2008).

10.2.7 Ozone

Ozone in gaseous and aqueous states has been applied for the surface disinfection of packages and equipment (Pascual et al., 2007). It is effective against a wide range of microorganisms including bacteria, viruses, fungi, spores, and their secondary metabolites (mycotoxins). The antimicrobial effect of gaseous ozone on dry food and ingredients depends on the humidity, treatment temperature and concentration, and product (water activity, surface property) (Kim et al., 2003). The use of gaseous ozone is reported as an effective antifungal fumigant to preserve wheat during storage (Wu et al., 2006; McKenzie et al. 1997; De Alencar et al., 2012). A higher concentration and longer treatment time are required for cereal flour than whole cereal treatment (Naitoh et al., 1988). The mycotoxins produced by *Aspergillus* and *Penicillum* species were reduced in barley, peanut, and wheat grains after 3 min treatment (Ciccarese et al., 2007). Savi et al. (2014) studied the effect of ozone treatment (40–60 μmol/mol exposed up to 3 h) on wheat grains artificially contaminated with *Fusarium graminearum* and observed total inhibition with no changes in lipid peroxidation or total protein profile. Ozone also can be used against insect infestation. However, the effectiveness of the ozone is dependent not only on the concentration but also on the species and life cycles of insects (McDonough et al., 2011; Isikber and Oztekin, 2009). Ozone applications at high concentrations may also be involved in affecting the nutritional and functional quality of the grains. The effect of the ozone treatment on quality was studied by Mendez et al. (2003) and no unfavorable effect was reported on the milling characteristics and fatty and amino acid compositions of wheat and maize, the baking characteristics of wheat, or the stickiness of rice.

10.2.8 Chlorine Treatment

Grain is produced in fields. Therefore, it may bear foreign vegetative material in addition to plant leaves, dust, and stalks and be contaminated with bird and rodent droppings. The milling process removes the grain from all foreign material and grinds it into flour. During milling, water is added to increase the grain's moisture content from below 12% to approximately 15% since dry wheat is too hard to grind into flour. Chlorine is added to the water to control bacteria and mold during this process.

Throughout the exploration of novel functional components, the enhancement of health-promoting effects of functional beverages, and the use of innovative food processing technologies, the sensory and physicochemical parameters of the local beverages should be preserved during storage with proper packaging.

10.2.9 Packaging Requirements for Non-Thermal Processing

Packaging has a major impact on the shelf life of food and beverages (Robertson, 2011). Traditionally, packaging is used to protect the product and guarantee its safe distribution without physical damages, chemical contamination, and biological (microorganisms, insects, and other animals) or environmental threats (i.e. exposure to gases [typically oxygen], moisture [gain or loss], or light) that can threaten the life of the consumer. Beyond this simple barrier function, research and development efforts have been focused on creating new roles for food packaging systems, including active packaging, modified atmosphere packaging (MAP), edible films and coatings, and packaging that addresses environmental issues (Ramos, 2015). The package, the product, and its environment interact to extend the food's shelf life and/or to improve its safety or sensory properties while maintaining food quality. Additionally, packaging material should serve as a moisture barrier to inhibit moisture gain in the product. An extensive variety of packaging materials and packaging designs are utilized to handle, store, and distribute fresh and processed food products from farm to consumer.

However, selecting the appropriate packaging material will be very crucial to preserve high quality during the shelf life of the food products. Consumer acceptance and regulatory requirements for these non-thermal processes will govern their commercialization in the future. Currently, regulations are in place only for packaging materials required for irradiation.

10.2.9.1 Packaging Materials

Depending on the non-thermal process, the product is mostly treated in batches; however, with improvements in technology, semicontinuous and continuous methods have also become available. The quality of the packaged food is directly related to the attributes of the food and packaging materials. Materials that have traditionally been used in food packaging include glass, metals (aluminum, foils and laminates, tinplate, and tin-free steel), paper and cardboard, and plastics. The selection of material depends on the nature of the food product because different packaging materials possess a range of performance characteristics that wield significant impacts on shelf life (Robertson, 2011).

Metal cans can be considered traditional beverage-packaging material; aluminum is commonly used in making cans for carbonated beverages (beer and soft drinks). Foil and laminated paper for seafood are also often packed on aluminum packaging material. Pouches with multilayer film are also on the market for non-carbonated beverages. Traditional packaging materials for grains, flour, and rice include film bags, PP woven bags, and paper bags. Plastic is used as a packaging material for many types of foods including fruits and vegetables, grains, seeds and nuts, cereals, baked goods, confectionery, dairy products, meat, poultry, and seafood (Yam and Lee, 2012).

10.2.9.2 Modified Atmosphere/Reduced Oxygen Packaging

One of the packaging systems used for the extension of the shelf life of bread is modified-atmosphere packaging (MAP). For this purpose, pure or combinations of the gases CO_2 and nitrogen are used in this system. CO_2 is effective against yeast and bacteria growth, thus, it is the most common gas utilized for bakery products. Gas mixtures of carbon dioxide and nitrogen 40:60% (Moore et al., 2004, 2005) and 80:20% (Gallagher et al., 2003) are examples used for gluten-free bread and wheat bread (Rasmussen and Hansen, 2001).

10.2.9.3 Active Packaging

Active packaging is defined as a package system designed to intentionally integrate components that would release or absorb substances into or from the packaged food or the environment surrounding the food, and it is intended to lengthen the shelf life or to sustain or improve the condition of the packaged food (Rooney, 2005). Active packaging is mostly used for beverage preservation (beer, malt-based drinks) and for improving their organoleptic quality in flavor, taste, and color (Ramos, 2015).

Grains (wheat, rice, maize, etc.) and flour derived from them are very susceptible to microbiological and insect infestation that causes substantial physicochemical and organoleptic changes that reduce the quality of the product and cause subsequent economic losses (Baik and Donelson, 2018). Generally, paper bags with no other protective agent added are used as a material for grain and flour packaging, but this will not be enough to block any possible infestation agents. There is a growing interest in the use of natural antioxidants in active food packaging. A wide range of natural extracts, essential oils from herbs and spices, natural antioxidants, and particularly tocopherols have been integrated into biodegradable materials in order to establish antioxidant properties (Valdes et al., 2014; Valdes et al., 2015). These act as a barrier to protect the product from exterior harm and in addition play an active role in food preservation and quality (Gómez-Estaca et al., 2014; Emamifar et al., 2010). Muratore et al. (2019) developed a paper-based active packaging by crafting eugenol onto cellulose using polycarboxylic acid as a linking agent for wheat flour. The developed material showed good preservation of the mechanical and water absorption properties of wheat flour and other grain-based food products, and no migration of odors and flavors to the packaged product.

Antioxidant active packaging is not only applied in dry grain and grain products packaging but also in beer and malt-based drinks packaging (Ramos, 2015), where it significantly increases the bioactive properties (insecticide/insectifuge) of the products.

10.2.9.4 Oxygen-Scavenging Packaging

Oxygen-scavenging packaging is based on the mechanism of the metal catalyst, ascorbic/metallic salts, and enzymes. This packaging is widely utilized for beer by the integration of scavenging agents inside of the closure, either in a sachet that is separated with a membrane from the beer or a polymer coating on the closure (crown) (Brody, 2001; Foster, 2003). Crowns that had a combination of ascorbate salts and copper and iron metals established an oxygen reduction after the bottling of the beer and maintained the level for 12 months (Foster, 2003). Other oxygen scavengers have been developed with the combination of two enzymes, glucose oxidase and catalase, that would react with some substrate to scavenge incoming oxygen; they would be part of the packaging structure or put in an independent sachet in beer and wine bottles (Broady, 2001). The activated yeast is also used to eliminate the oxygen in the headspace of hermetically sealed beer bottles while producing carbon dioxide and alcohol as a result (Edens, 1992). Another way of scavenging oxygen is using a combination of the enzymes glucose oxidase and catalase as a part of the structure of the package or placed in a sachet in beer and wine bottles. Similar packaging systems are also used for bread, cake, cooked rice, biscuits, cured meats and fish, dried foods, and beverages (Conte et al., 2013).

Self-cooling containers for beer and soft drinks are developed by using a gel layer coating that contains water in a separate container inside the can, in contact with the beverage. When the base is twisted, the system will be activated and water will come into contact with a desiccant that is held in a separate evacuated outer chamber. The cooling effect is achieved once the water is evaporated at room temperature by removing the heat from the system (Cork, 2020).

10.2.9.5 Intelligent Packaging Systems

Intelligent packaging systems (including sensors, indicators, and radio frequency identification [RFID] systems) are also commonly used for the packaging of beverages (Vanderroost et al., 2014). Color-changing indicators provide visual information about the package. Thermochromic ink is used in metal- and glass-packaging containers of beer as a temperature sensor (Coors Light, 2020), showing that the beer is at the desired temperature for consumption.

10.3 PACKAGING FOR NOVEL FOOD PROCESSING TECHNOLOGY

Traditional thermal processes can be problematic not only for the nature of the food products themselves but also with regard to the chemical interactions between thermally processed foods and common food packaging materials, which have been well identified.

Non-thermal processes are usually applied at (or below) room temperature and have minimal impact on the quality and freshness of food products (Barba et al., 2012, 2015), including preserving nutrients and other health-promoting components, and they have different mechanisms of preserving foods compared to thermal processes. Therefore, the material and design of packaging should be changed accordingly to assist the preservation of food products by these non-thermal processes. Packaging materials for non-thermal processes should have good resistance to the specific non-thermal processing mechanisms and maintain their properties during and after non-thermal processing. Therefore, understanding of packaging material properties (mechanical and barrier) and interaction between packaging material and food components for each of these non-thermal processes is needed. For example, packaging material must be transparent for pulsed light emission to penetrate the product, or chemically stable at the applied radiation dose with no polymer degradation or significant changes in the packaging materials (Han, 2014; Marsh and Bugusu, 2007).

Non-thermal processes such as pulsed electric field, high pressure, irradiation, pulsed light emission, and microfiltration can be operated with aseptic processing units for pumpable foods to maintain the sterility of the food throughout the process (Lelieveld, 2006). Innovations in the packaging industry have led to the development of novel sustainable materials as an alternative to classic packaging systems. Starch-based packaging materials have attracted much interest because of their biodegradability, which could help ease the environmental crisis and the petroleum shortage arising from the consumption of traditional polymers. Starch has become an interest as a component for bioplastics. It has many advantages, including being inexpensive, easily reachable in large amounts from a broad variety of plants, and entirely degradable. Even it is not appropriate for making films by itself, if it is combined with glycols (e.g., glycerol, sorbitol), polyethers, urea, or water it can be plasticized. Thermoplastic starch (TPS) can be produced by the application of heat and mechanical power to the starch in an extruder (Bastioli, 2001; Petersen et al., 1999). Examples include starch-based packages for dry products such as nuts and grains. The improvements in plastic packaging have opened a new area of production of natural polymer-based systems that have some benefits such as biodegradability, environmental friendliness, low cost, high competence as active supports, and similar processing conditions to synthetic polymers (Ramos et al., 2015).

Unfortunately, there is no general requirement for packaging materials for all non-thermal processes. Currently, regulations are in place only for the packaging materials required for irradiation.

10.3.1 Packaging Materials for HHP

While HHP is currently limited to the pasteurization of food and beverages that must be kept refrigerated, it is the most common emerging food-processing technology in the industry (Farkas, 2016). During typical high-pressure treatment, the food product is treated in batches, but semi-continuous and continuous methods of processing the product exist. In batch operations, the food material is vacuum packaged in flexible packaging material (Mensitierire et al., 2013) and processed in a high-pressure chamber that is filled with pressure-transmitting fluid. The packaging prevents the interaction of the food with the fluid in the chamber as well as the recontamination of the product after HHP. To prevent recontamination of the product after HHP, the food is usually prepackaged or aseptically packaged post-processing. Due to the high cost of aseptic packaging installation, it is not a common application of the HHP process.

When selecting appropriate packaging for HHP foods, besides the material type, one must also consider the package geometry and fabrication method, the nature of the product, the location and volume of the package, and the headspace gas, dissolved oxygen, and deformation characteristics of packaged foods (Balasubramaniam et al., 2004). The thermal, mechanical, and physical properties are also important in the selection of the desired packaging material (Hogan et al., 2005). However, the selection of the packaging material is crucial since the pressure of HHP will impact the package of the product and any damage to the container may allow recontamination and create microbiological risks.

The product is compressed by uniform pressure from every direction and then returns to its initial shape, maintaining sealing and barrier properties when the pressure is removed. Therefore the packaging material used for HHP must be able to handle a 15% reduction in size. The material of plastic pouches, containers, semirigid trays, jars, and bottles should be flexible enough to handle the pressure. Metal cans collapse permanently and glass bottles are likely to break, hence, packages made using metal, glass, or paper may not be good packaging candidates for use during pressure treatment (Caner et al., 2004). Minimum headspace in the package is also an important packaging requirement for HHP as air or gases are compressible, so the larger headspace can create bigger deformation strains on the packaging material during pressure treatment (Lambert et al., 2000). HHP is a non-thermal process. However, due to the adiabatic heating in pressure applications at higher levels, the temperature of the food increases around 2–3°C for every 100 MPa depending on the composition of the product (Caner et al., 2004).

Polypropylene (PP), polyethylene pouches (PE), polyester tubes, nylon PP pouches, PET, EVOH films, and combinations thereof are some of the commonly used packaging materials for high-pressure pasteurization (Juliano et al., 2010). Packaging material should also retain good mechanical and barrier properties after the HHP treatment. High-barrier packaging materials with oxygen- and light-impermeable properties may be desired for extended refrigerated product storage. Lambert et al. (2000) studied mechanical (tensile strength, seal strength, and delamination) properties of six multilayered plastics with different combinations of polyamide (PA), polyethylene (PE), polyethylene terephthalate (PET), and polyvinylidene chloride (PVDC). The laminates became more rigid and less flexible when the change in seal strength and delamination was within 25%. But the degree of delamination was correlated with the amount of air present in the package or product (Lambert et al., 2000). Galotto et al. (2009) investigated the mechanical properties of a biopolymer (polylactic acid [PLA] coated with silicon oxide) and a synthetic polymer (PET coated with aluminum oxide) and reported increased tensile strength of the PET films during HHP treatment, whereas elongation decreased due to the formation of pinholes and cracks in the film after 15 min (500MPa) of HHP treatment.

In addition, any migration of package components to the food product must be avoided during high-pressure treatment. Commercial polymeric films are widely used for food packaging applications, such as LDPE, LLDPE, PP, PLA, and PET. The processing of pre-packaged foods, normally in flexible films, is typically carried out in batch-type systems. Layered films such as co-extruded films with polymeric barrier layers, adhesive laminated films on a polymer base, or inorganic layers such as aluminum foil (a few micrometers thick) or vacuum-deposited coating (nanometers thick) are also used (Richter et al., 2010). Flexible, laminate films have been identified as potentially suitable packaging materials for HHP (Caner et al., 2004). To deliver premium, long-shelf-life products with excellent nutritional and organoleptic properties, the retention of barrier properties and freshness is essential (Bull, 2010). Ayvaz et al. (2012) reported that nylon/EVOH/EVA packaging material was among the best (along with nylon/EVA and MetPET/PE) for preserving the color and b-carotene content of the product after HHP treatment and during storage. Thus, the effect of HHP on the barrier properties of selected packaging materials should be considered when designing packaging materials for HHP.

10.3.2 Packaging Materials for PEF

This process works well with flexible packaging. For PEF treatment of food products, the most common packaging process design is aseptic packaging integrated with PEF treatment systems. Packaging materials can absorb flavor compounds and may degrade the flavor, color, and nutrients by diffusion of oxygen (Ayhan et al., 2001) or change the moisture content of food (Min and Zhang, 2007). Therefore, the choice of packaging materials for PEF treatments is essential in order to preserve the quality throughout storage. The most common aseptic packaging materials are plastic containers and paper-laminated material. The shelf life of products depends on the barrier properties (OP and WVP) of the aseptic packaging material. Packaging materials with low OP values should be preferred for PEF-treated foods that are susceptible to oxidation. For example, the changes in the flavor, color, and nutrient quality of a product packaged in different materials (glass, PET, high-density polyethylene [HDPE], and low-density polyethylene [LDPE]) were evaluated and the results indicated that the concentration of ascorbic acid, affecting flavor, in glass and PET bottles was significantly higher than that in HDPE and LDPE bottles during storage at 4°C (Ayhan et al. 2001). The color also changes during storage due to the high OP values of HDPE and LDPE bottles.

10.3.3 Packaging for Pulsed Light

For pulsed light treatment, transparent packaging materials should be used because the light cannot penetrate opaque surfaces (Han, 2007). Materials should have resistance to heat since pulsed light treatment may result in excessive heating at the surface of the food. Microbial inactivation by PL depends on the surface topography of the packaging. Smooth surfaces can lead to a clustering of microbial cells while rough surfaces with pores can be protective for the microorganisms, therefore they are not appropriate for PL (Oms-Oliu et al., 2010).

10.3.4 Packaging Materials for Irradiation

In many countries, regulatory approval is not needed for the irradiation of packaging. However, the application of irradiation on sensitive health-care materials (hydrogels, herbs, drugs, etc.) and food items in packages are exceptions due to the possible migration of any package components (or products of the degradation of those components) to food (Arvanitoyannis and Tserkezou, 2010). Irradiation can cause chemical changes like crosslinking, which would likely reduce migration, but it can also cause the breakdown of lower-molecular-weight bodies that have increased migration features. Therefore, all radiation processes, including their packaging materials, should be approved by the US Food and Drug Administration (FDA) because the FDA considers the radiation sources to be food additives – or by any related regulatory agencies in other countries since governmental regulations on the irradiation of food differ substantially from country to country. Most countries approve irradiation in individually food-based scenarios. The main advantage is that the technology can be used to treat packaged foods, which will remain safe and protected from microbial contamination after treatment. Packaging material can either affect the irradiation process with its barrier properties or can add radiolysis products into the product (Hammad et al., 2006). Ionizing radiation affects the polymeric material in two ways, either by increasing the molecular weight (cross-linking) or decreasing it (chain scission – degradation). Cross-linking is generally the leading reaction for most plastic materials (polyethylene [PE], polypropylene [PP], and polystyrene [PS]) used as a food packaging material. Marathe et al. (2002) reported that as little as 0.25 kGy was sufficient to increase the shelf life of whole wheat flour packaged in polyethylene pouches by up to six months without any effect on moisture, free fatty acids, starch, sugars, or gelatinization viscosity. Meanwhile, up to 1kGy irradiation showed no adverse effect on total proteins, fat, carbohydrates, vitamin B1 and B2 content, sedimentation value, dough properties, and total bacterial and mold count during storage up to six months.

10.3.5 Ozone Packaging

Ozone mainly reacts with the surfaces of polymers. The rate of oxidation and the changes in the properties of polymers as a result of exposure to ozone depend on the chemical structure of the polymer. Polarity and surface tension can be changed because of the formation of oxygen-containing functional groups and the degradation of the polymer chains. After ozone treatment, PE, PP, and polybutylene oxidation caused a reduction in the melting point, an increase in solubility, a decrease in intrinsic viscosity, and a change of color to yellow, while the films of some polymers also became brittle and opaque. Studies showed that the effect of ozone on the mechanical properties of plastic films depends on the polymer and the treatment conditions, such as ozone concentration and temperature on biaxially oriented nylon films. However, changes after the ozone treatment do not always have adverse results. These modifications could be turned into an advantage, such as the increased adhesivity of some plastics (Ozen, 2001).

10.3.6 Plasma

While cold plasma is currently in the research and developmental stages, it holds great potential for the surface treatment of many foods and even flexible packaging materials. Plasma treatment is also being investigated for its effect on seed germination, inactivation of enzymes, physicochemical modification of food properties including starch and protein, and in-pack sterilization of products (Thirumdas et al., 2015).

10.4 FUTURE TRENDS AND CONCLUSIONS

Adoption of suitable packaging technologies by the food industry can be useful for extending shelf life, improving quality and safety, and providing information about the product. Non-thermal processing will continue to develop and its applications will become more common with the increasing consumer demand for minimally processed, convenient, and high-quality grain-based food products. Introducing the application of a new non-thermal food processing technique that patrons are not aware of will always be challenging regardless of the positive results of scientific risk assessments; therefore, the commercialization of non-thermal processes is strongly recommended. For the extended commercialization of non-thermal processes, it is necessary to study consumers' acceptance of non-thermally processed packaged foods. New trends in the area of food packaging will be focusing on the development of new materials with enhanced properties to control food–package–environment interactions. Advances in packaging technology are determining the finest packaging materials for grain-based food products processed by non-thermal procedures such as high pressure, pulsed electric fields, ultraviolet (UV), irradiation, microfiltration, active packaging (oxygen-scavenging or antimicrobial packaging), etc. to preserve the benefits and the quality of the product. Currently, regulations are in place only for packaging materials required for irradiation.

Therefore, there is a need for collaboration between industrial, academic, and governmental groups to research and develop packaging strategies for new non-thermal processing methods for grain and grain-based products. Future research will also place emphasis on escalating the use of non-thermal processes by merging these processes with other thermal or non-thermal processes. Nonetheless, the selection of the appropriate packaging material will always be very crucial in order to maintain high quality until the end of the shelf life of the food products.

Further research is necessary to help understand and predict the behavior of compounds during non-thermal food processing and storage, taking into account both the effects of the food matrix and the packaging.

BIBLIOGRAPHY

Allan, B., Woolf, A. B., Wibisono, R., Farr, J., Hallett, I., Richter, L., Oey, I., Wohlers, M., Zhou, J., Fletcher, G. C., and Requejo-Jackman, C. (2013). Effect of high pressure processing on avocado slices. *Innovative Food Science and Emerging Technologies*, 18, 65–73.

Alvarez-Jubete, L., Arendt, E. K., and Gallagher, E. (2010). Nutritive value of pseudocereals and their increasing use as functional gluten-free ingredients. *Trends in Food Science & Technology*, 21(2), 106–113. doi:10.1016/j.tifs.2009.10.014.

Armelim, J. M., Canniatti, B., and Piedade, S. (2006). Quantitative descriptive analysis of common bean (*Phaseolusvulgarius L.*) under gamma radiation. *Journal of Food Science*, 71(1), 8–12.

Arvanitoyannis, I. S., and Tserkezou, P. (2010). Legislation on food irradiation European Union, United States, Canada, and Australia. In: Arvanitoyannis, I. S. (Ed.), *Irradiation of Food Commodities: Techniques, Applications, Detection, Legislation, Safety and Consumer Opinion* (pp. 3–22). Boston, MA: Academic Press.

Awad, T., Moharram, H., Shaltout, O., Asker, D., and Youssef, M. (2012). Applications of ultrasound in analysis, processing and quality control of food: A review. *Food Research International*, 48(2), 410–427.

Ayhan, Z., Yeom, H. W., Zhang, H., and Min, D. B. (2001). Flavor, color, vitamin C retention of pulsed electric field processed orange juice in different packaging materials. *Journal of Agricultural and Food Chemistry*, 49(2), 669–674.

Ayvaz, H., Schirmer, S., Parulekar, Y., Balasubramaniam, V. M., Somerville, J. A., and Daryaei, H. (2012). Influence of selected packaging materials on some quality aspects of pressure-assisted thermally processed carrots during storage. *LWT – Food Science and Technology*, 46(2), 437–447. doi:10.1016/j.lwt.2011.12.004.

Baik, B. K., and Donelson, T. (2018). Postharvest and postmilling changes in wheat grain and flour quality characteristics. *Cereal Chemistry*, 95, 141–148.

Balasubramaniam, V. M., Ting, E. Y., Stewart, C. M., and Robbins, J. A. (2004). Recommended laboratory practices for conducting high pressure microbial inactivation experiments. *Innovative Food Science and Emerging Technologies*, 5(3), 299–306.

Barba, F. J., Jäger, H., Meneses, N., Esteve, M. J., Frígola, A., and Knorr, D. (2012). Evaluation of quality changes of blueberry juice during refrigerated storage after high-pressure and pulsed electric field processing. *Innovative Food Science and Emerging Technologies*, 14, 18–24.

Barba, F. J., Parniakov, O. Pereira, S. A. Wiktor, A., Grimi, N., Boussetta, N., Saraiva, J. A., Raso, J., Martin-Belloso, O., Witrowa-Rajchert, D., Lebovka, N., and Eugène Vorobiev, E. (2015). Current applications and new opportunities for the use of pulsed electric fields in food science and industry. *Food Research International*, 77(4), 773–798.

Barbosa-Canovas, G. V., Pothakamury, U. R., Palou, E., and Swanson, B. (1998). *Nonthermal Preservation of Foods*. New York: Marcel Dekker.

Bastioli, C. (2001). Global status of the production of biobased packaging materials. *Starch*, 53, 351–355.

Błaszczaka, W., Fornala, J., Valverdeb, S., and Garridoc, L. (2005). Pressure induced changes in the structure of corn starches with different amylose content. *Carbohydrate Polymers*, 61(2), 132–140.

Bourke, P. Zuizina, D., Han, L., Cullen, P. J., and Gilmore, B. (2017). Microbiological interactions with cold plasma. *Journal of Applied Microbiology*, 123(2), 308–324.

Breidbach, A., and Ulberth, F. (2016). Comparative evaluation of methods for the detection of 2-alkylcyclobutanones as indicators for irradiation treatment of cashew nuts and nutmeg. *Food Chemistry*, 201, 52–58.

Brody, A. L., Strupinsky, E. R., and Kline, L. R. (2001). Oxygen scavengers. *Active Packaging for Food Applications*. Boca Raton, FL: CRC Press. pp. 11–30.

Bußler, S., Rumpold, B. A., Fröhling, A., Jander, E., Rawel, H. M., and Schlüter, O. K. (2016). Cold atmospheric pressure plasma processing of insect flour from Tenebrio molitor: Impact on microbial load and quality attributes in comparison to dry heat treatment. *Innovative Food Science and Emerging Technologies*, 36, 277–286. doi:10.1016/j.ifset.2016.07.002.

Buckow, R., Heinz, V., and Knorr, D. (2007). High pressure phase transition kinetics of maize starch. *Journal of Food Engineering*, 81(2), 469–475.

Bull, M. K., Steele, R. J., Kelly, M., Olivier, S. A., and Chapman, B. (2010). Packaging under pressure: Effects of high pressure, high temperature processing on the barrier properties of commonly available packaging materials. *Innovative Food Science and Emerging Technologies*, 11(4), 533–537.

Calado, T., Fernández-Cruz, M. L., Cabo Verde, S., Venâncio, A., and Abrunhosa, L. (2018). Gamma irradiation effects onochratoxin A: Degradation, cytotoxicity and application in food. *Food Chemistry*, 240, 463–471.

Caner, C., Hernandez, R. J., and Harte, B. R. (2004). High-pressure processing effects on the mechanical, barrier and mass transfer properties of food packaging flexible structures: A critical review. *Packaging Technology and Science*, 17(1), 23–29.

Cardoni, A., and Lucas, M. (2005). Strategies for reducing stress in ultrasonic cutting systems. *Strain*, 41, 11–18.

Cheftel, J. C., and Culioli, J. (1997). Effects of high pressure on meat: A review. *Meat Science*, 46, 211–236.

Chemat, F., Rombaut, N., Meullemiestre, A., Turk, M., Perino, S., Fabiano-Tixier, A. S., and Abert-Vian, M. (2017). Review of green food processing techniques. Preservation, transformation, and extraction. *Innovative Food Science and Emerging Technologies*, 41, 357–377.

Ciccarese, F., Sasanelli, N., Ciccarese, A., Ziadi, T., and Mancini, L. (2007). Seed disinfestation by ozone treatments. In: *Proceedings of the IOA Conference and Exhibition*. October 29–31, 2007, Valencia, Spain.

Clark, J. (2006). Pulsed electric field processing. *Journal of Food Technology*, 60(1), 66–67.

Conte, A., Angiolillo, L., Mastromatteo, M., and Del Nobile, A. (2013). Technological options of packaging to control food quality. *Food Industry*, 16, 354–379. doi:10.5772/53151.

Coors Light Company. Available online: http://www.coorslight.com/ (accessed on October 2, 2020).

Cork, C., and Seal Packaging Europe GmbH, Switzerland. Available online: http://www.crowncork.com/ (accessed on October 2, 2020).

De Alencar, E. R., Faroni, L. R. D. A., Soares, N. D. F. F., da Silva, W. A., and da Silva Carvalho, M. C. (2012). Efficacy of ozone as a fungicidal and detoxifying agent of aflatoxins in peanuts. *Journal of the Science of Food and Agriculture*, 92(4), 899–905.

Dobrynin, D., Fridman, G., Friedman, G., and Fridman, A. (2009). Physical and biological mechanisms of direct plasma interaction with living tissue. *New Journal of Physics*, 11, 2–26.

Edens, L., Farin, F., Ligtvoet, A. F., and van der Platt, J. B. (1992). Dry yeast immobilized in wax or paraffin for scavenging oxygen. U.S. Patent 5,106,633.

Emamifar, A., Kadivar, M., Shahedi, M., and Soleimaian-Zad, S. (2010). Evaluation of nanocomposite packaging containing Ag and ZnO on shelf life of fresh orange juice. Innov. *Food Science & Emerging*, 11(4), 742–748.

Enujiugha, V. N., and Badejo, A. A. (2015). Probiotic potentials of cereal-based beverages. *Critical Reviews in Food Science and Nutrition*, 57(4), 790–804.

Eustice, R. F. (2017). Global status and commercial applications of food irradiation. In: Ferreira, I. C. F. R., Antonio, A. L., and Verde, S. C. (Eds.), *Food Irradiation Technologies: Concepts, Applications and Outcomes* (pp. 397–424). Wales: Royal Society of Chemistry.

Evrendilek, G. A., and Zhang, Q. H. (2003). Effects of pH, temperature, and pre-pulsed electric field treatment on pulsed electric field and heat inactivation of Escherichia coli O157:H7. *Journal of Food Protection*, 66(5), 755–759.

FDA. (2014). 179.26 Ionizing radiation for the treatment of food. Edited by Services DoHaH. gpo.gov. Government Publishing Office. Title 21.

Fanaro, G., Hassimotto, N., Bustos, D., and Villavicencio, A. (2015). Effects of g-radiation on microbial load and antioxidant proprieties in green tea irradiated with different water activities. *Radiation Physics and Chemistry*, 107, 40–46.

FAO. (2011). Quinoa: An ancient crop to contribute to world food security. Technical report of the 37th FAO Conference. Rome, Italy. Food and Agriculture Organization Regional Office for Latin America, and the Caribbean PROINPA. http://www.fao.org/alc/file/media/pubs/2011/cultivo_quinua_en.pdf.

Farkas, J. (2004). Food irradiation. In: Mozumder, A., and Y. Hatano (Eds.), *Charged Particle and Photon Interactions with Matter* (pp. 785–812). New York: Dekker.

Farkas, J., and Mohácsi-Farkas, C. (2011). History and future of food irradiation. *Trends in Food Science and Technology*, 22, 121–126.

Farkas, D. F. (2016). A short history of research and development efforts leading to the commercialization of high-pressure processing of food. In: Balasubramaniam, V., Barbosa-Canovas, G., and Lelieveld, H. (Eds.), *High Pressure Processing of Food. Food Engineering Series*. New York: Springer.

Feng, H., and Yang, W. (2005). Power ultrasound. In: Hui, Y. H. (Ed.), *Handbook of Food Science, Technology and Engineering* (pp. 121–127). New York: CRC Press.

Fernandez, A., Noriega, E., and Thompson, A. (2013). Inactivation of *Salmonella enterica* serovar Typhimurium on fresh produce by cold atmospheric gas plasma technology. *Food Microbiology*, 33(1), 24–29.

Foster, T., and Vasavada, P. C. (2003). *Beverage Quality and Safety*. Boca Raton, FL: CRC Press.

Gallagher, E., Kunkel, A., Gormley, T. R., and Arendt, E. K. (2003). The effect of dairy and rice powder addition on loaf and crumb characteristics, and on shelf life (intermediate and long-term) of gluten-free breads stored in a modified atmosphere. *European Food Research and Technology*, 218(1), 44–48.

Gallego-Juárez, J., Rodriguez, G., Acosta, V., and Riera, E. (2010). Power ultrasonic transducers with extensive radiators for industrial processing. *Ultrasonics Sonochemistry*, 17(6), 953–964.

Galotto, M. J., Ulloa, P. A., Guarda, A., Gavara, R., and Miltz, J. (2009). Effect of high pressure food processing on the physical properties of synthetic and biopolymer films. *Journal of Food Science*, 74(6), E304–E311.

Gómez-Estaca, J., López-de-Dicastillo, C., Hernández-Muñoz, P., Catalá, R., and Gavara, R. (2014). Advances in antioxidant active food packaging. *Trends in Food Science and Technology*, 35(1), 42–51.

Grzegorzewski, F., Rohn, S., Kroh, L. W., Geyer, M., and Schleuter, O. (2010). Surface morphology and chemical composition of lamb's lettuce (*Valerianella locusta*) after exposure to a low-pressure oxygen plasma. *Food Chemistry*, 122(4), 1145–1152.

Hammad, A. A., Abo Elnour, S. A., and Salah, A. (2006). Use of irradiation to ensure hygienic quality of minimally processed vegetables and fruits. *IAEA-TECDOC*, 1530, 106–129.

Han, J. H. (2007). *Packaging for Nonthermal Processing of Food*. Oxford: Blackwell Publishing.

Han, J. H. (2014). A review of food packaging technologies and innovations. *Innovations in Food Packaging*, Second edition (pp. 3–12). San Diego, CA: Academic Press. doi:10.1016/B978-0-12-394601-0.00001-1.

Hati, S., Patel, M., and Yadav, D. (2018). Food bioprocessing by non-thermal plasma technology. *COFS*, 19, 85–91. doi:10.1016/j.cofs.2018.03.011.

Heenan, C. N., Adams, M. C., Hosken, R. W., and Fleet, G. H. (2004). Survival and sensory acceptability of probiotic microorganisms in fermented frozen vegetarian dessert. *LWT Food Science and Technology*, 37, 461–466.

Hertwig, C., Meneses, N., and Mathys, A. (2018). Cold atmospheric pressure plasma and low energy electron beam as alternative nonthermal decontamination technologies for dry food surfaces: A review. *Trends in Food Science and Technology*, 77, 131–142.

Hogan, E., Kelly, A. L., and Sun, D. W. (2005). High pressure processing of foods: An overview. In: *Emerging Technologies for Food Processing* (pp. 3–32). London: Elsevier Academic Press.

Hüttner, E. K., Dal Bello, F., Poutanen, K., and Arendt, E. K. (2009). Fundamental evaluation of the impact of high hydrostatic pressure on oat batters. *Journal of Cereal Science*, 49, 363–370.

Ihsanullah, I., and Rashid, A. (2017). Current activities in food irradiation as a sanitary and phytosanitary treatment in the Asia and the Pacific Region and a comparison with advanced countries. *Food Control*, 72, 345–359.

Isikber, A. A., and Oztekin, S. (2009). Comparison of susceptibility of two stored-product insects, *Ephestia kuehniella Zeller* and *Triboliumconfusum* du Val to gaseous ozone. *Journal of Stored Products Research*, 45(3), 159–164.

Jeong, S. G., and Kang, D. H. (2017). Inactivation of *Escherichia coli* O157:H7, *Salmonella Typhimurium*, and *Listeria monocytogenes* in ready-to-bake cookie dough by gamma and electron beam irradiation. *Food Microbiology*, 64, 172–178.

Juárez, J. A. G., Corral, G. R., Vitini, F. M., Aparicio, V. A., De Sarabia, E. R. F., and Blanco, A. B. (2010). Macrosonic generator for the air-based industrial defoaming of liquids. US Patent (US 7719924 B2), United States.

Juliano, P., Koutchma, T., Sui, Q., Barbosa-Cánovas, G. V., and Sadler, G. (2010). Polymeric-based food packaging for high-pressure processing. *Food Engineering Reviews*, 2(4), 274–297.

Junqueira-Gonçalves, M. P., Galotto, M. J., Valenzuela, X., Dinten, C. M., Aguirre, P., and Miltz, J. (2011). Perception and view of consumers on food irradiation and the Radura symbol. *Radiation Physics and Chemistry*, 80, 119–122.

Khanal, B. P., Knoche, M., Bußler, S., and Schlüter, O. (2014). Evidence for a radial strain gradient in apple fruit cuticles. *Planta*, 240(4), 891–897.

Kieffer, R., Schurer, F., Kohler, P., and Wieser, H. (2007). Effect of hydrostatic pressure and temperature on the chemical and functional properties of wheat gluten: Studies on gluten, gliadin and glutenin. *Journal of Cereal Science*, 45, 285–292.

Kim, J. G., Yousef, A. E., and Khadre, M. A. (2003). Ozone and its current and future application in the food industry. *Advances in Food and Nutrition Research*, 45, 168–218.

Kim, S., Lee, Y. R., Noh, Y. H., Lee, J., and Jeong, H. S. (2015). Effects of high hydrostatic pressure treatment on the enhancement of functional components of germinated rough rice (*Oryza sativa* L.). *Food Chemistry*, 166, 86–92.

Knorr, D., Froehling, A., Jaeger, H., Reineke, K., Schlueter, O., and Schoessler, K. (2011). Emerging technologies in food processing. *Annual Review of Food Science and Technology*, 2, 203–235.

Kumar, P., and Han, J. H. (2012). Packaging materials for non-thermal processing of food and beverages. In: K. L. Yam and D. S. Lee (Eds.), *Emerging Food Packaging Technologies* (pp. 323–334). Sawston, Cambridge: Woodhead Publishing Limited.

Lambert, Y., Demazeau, G., Largeteau, A., Bouvier, J. M., Laborde-Croubit, S., and Cabannes, M. (2000). Packaging for high-pressure treatments in the food industry. *Packaging Technology and Science*, 13(2), 63–71.

Lee, K. H., Kim, H.-J., Woo, K. S., Jo, C., Kim, J.-K., Kim, S. H., Park, H. Y., Oh, S.-K., and Kim, W. H. (2016). Evaluation of cold plasma treatments for improved microbial and physicochemical qualities of brown rice. *LWT – Food Science and Technology*, 73, 442–447.

Lelieveld, H. L. M. (2006). Pulsed electric field pasteurization of foods. *New Food*, 9(1), 31–33.

Li, W., Bai, Y., Mousaa, S.A, Zhang, Q., and Shen, Q. (2012). Effect of high hydrostatic pressure on physicochemical and structural properties of rice starch. *Food and Bioprocess Technology*, 5, 2233–2241.

Lin, Z., Zhang, Z., Chen, X., and Ming, Q. (2011). Ultrasonic-assisted extraction of red rice pigment. *Modern Food Science and Technology*, 27, 296–298.

Liu, L., Jia, W., Xu, D., and Li, R. (2015). Applications of ultrasonic cutting in food processing. *Journal of Food Processing and Preservation*, 39, 1762–1769.

Luana, N., Rossana, C., Curiel, J.A., Kaisa, P., Marco, G., and Rizzello, G. (2014). Manufacture and characterization of a yogurt-like beverage made with oat flakes fermented by selected lactic acid bacteria. *International Journal of Food Microbiology*, 185, 17–26.

Ma, K. J., Si, J., and Yan-liang, P. (2016). Research status and prospect of ultrasonic technology in food development. *Food Industry*, 9, 207–211.

Marathe, S., Machaiah, J., Rao, B., Pednekar, M., and Rao, V. (2002). Extension of shelf-life of whole-wheat flour by gamma radiation. *International Journal of Food Science and Technology*, 37, 163–168. doi: 10.1046/j.1365-2621.2002.00553.x.

Marsh, K., and Bugusu, B. (2007). Food packaging – Roles, materials, and environmental issues. *Journal of Food Science*, 72(3), R39–R55.

Martín-Belloso, O., and Elez-Martínez, P. X. (2005). Enzymatic inactivation by pulsed electric fields. In: Da-Wen Sun (Ed.), *Emerging Technologies for Food Processing* (pp. 156–181). London, UK: Elsevier.

Mateos-Aparicio, I., Mateos-Peinado, C., and Rupérez, P. (2010). Highhydrostatic pressure improves the functionality of dietary fibre inokara by-product from soybean. *Innovative Food Science and Emerging Technologies*, 11, 445–450.

McClements, D. J. (1995). Advances in the application of ultrasound in food analysis and processing. *Trends in Food Science and Technology*, 6, 293–299.

McDonough, M. X., Mason, L. J., and Woloshuk, C. P. (2011). Susceptibility of stored product insects to high concentrations of ozone at different exposure intervals. *Journal of Stored Products Research*, 47(4), 306–310.

McKenzie, K. S., Sarr, A. B., Mayura, K., Bailey, R. H., Miller, D. R., Rogers, T. D., and Phillips, T. D. (1997). Oxidative degradation and detoxification of mycotoxins using a novel source of ozone. *Food and Chemical Toxicology*, 35(8), 807–820.

Meinlschmidt, P., Schweiggert-Weisz, U., Brode, V., and Eisner, P. (2016). Enzyme assisted degradation of potential soy protein allergens with special emphasis on the technofunctionality and the avoidance of a bitter taste formation. *LWT – Food Science and Technology*, 68, 707–716.

Meinlschmidt, P., Ueberham, E., Lehmann, J., Reineke, K., Schlüter, O., Schweiggert-Weisz, U., and Eisner, P. (2016). The effects of pulsed ultraviolet light, cold atmospheric pressure plasma, and gamma-irradiation on the immunoreactivity of soy protein isolate. *Innovative Food Science and Emerging Technologies*, 38, 374–383.

Mendez, F., Maier, D. E., Mason, L. J., and Woloshuk, C. P. (2003). Penetration of ozone into columns of stored grains and effects on chemical composition and processing performance. *Journal of Stored Products Research*, 39(1), 33–44.

Meng, L., Zhang, W., Wu, Z., Hui, A., Gao, H., Chen, P., and He, Y. (2018). Effect of pressure-soaking treatments on texture and retrogradation properties of black rice. *LWT – Food Science and Technology*, 93, 485–490.

Mensitieri, G., Scherillo, G., and Iannace, S. (2013). Flexible packaging structures for high pressure treatments. *Innovative Food Science and Emerging Technologies*, 17, 12–21. doi:10.1016/j.ifset.2012.10.007.

Min, S., Jin, Z. T., and Zhang, Q. H. (2003). Commercial scale pulsed electric field processing of tomato juice. *Journal of Agricultural and Food Chemistry*, 51(11), 3338–3344.

Min, S., and Zhang, Q. H. (2007). Packaging for high pressure processing, irradiation and pulsed electric field processing. In: Han, H. J. (Ed.), *Packaging for Non Thermal Processing of Food* (pp. 70–71). Ames, IA: Blackwell Publishing.

Moore, M. M., Heinbockel, M., Dockery, P., Ulmer, H. M., and Arendt, E. K. (2005). Network for in gluten-free bread with the application of transglutaminase. *Cereal Chemistry*, 83, 28–36.

Moore, M. M., Schober, T. J., Dockery, P., and Arendt, E. K. (2004). Textural comparison of gluten-free and wheat based doughs, batters and breads. *Cereal Chemistry Journal*, 81(5), 567–575.

Morris, C., Brady, A. L., and Wicker, L. (2007). Non-thermal food processing/preservation technologies: A review with packaging implications. *Packaging Technology and Science*, 20(4), 275–286.

Muratore, F., Barbosa, S. E., and Martini, R. E. (2019). Development of bioactive paper packaging for grain-based food products. *Food Packag. Shelf Life*, 20, 100317. doi:10.1016/j.fpsl.2019.100317.

Naitoh, S., Okada, Y., and Sakai, T. (1988). Studies on utilization of ozone in food preservation. V. Changes in microflora of ozone-treated cereals, grain, peas, beans, and spices during storage. *Journal of the Japanese Society for Food Science and Technology*, 35, 69–77.

Niemira, B. A. (2012). Cold plasma decontamination of foods. *Annual Review of Food Science and Technology*, 3, 125–142.

Odriozola-Serrano, I., Aguiló-Aguayo, I., Soliva-Fortuny, R., and Martín-Belloso, O. (2013). Pulsed electric fields processing effects on quality and health-related constituents of plant-based foods. *Trends in Food Science and Technology*, 29(2), 98–107,

Oms-Oliu, G., Martın-Belloso, O., and Soliva-Fortuny, R. (2010). Pulsed light treatments for food preservation. A review. *Food and Bioprocess Technology*, 3(1), 13–23.

O'Sullivan, J. J., Park, M., Beevers, J., Greenwood, R. W., and Norton, I. T. (2017). Applications of ultrasound for the functional modification of proteins and nanoemulsion formation: A review. *Food Hydrocology*, 71, 299–310.

Ozen, B. F., and Floros, J. D. (2001). Effects of emerging food processing techniques on the packaging materials. *Trends in Food Science and Technology*, 12(2), 60–67.

Pankaj, S., Bueno-Ferrer, C., Misra, N., Oneill, L., Jiménez, A., Bourke, P., and Cullen, P. (2014). Characterization of polylactic acid films for food packaging as affected by dielectric barrier discharge atmospheric plasma. *Innovative Food Science and Emerging Technologies*, 21, 107–113.

Park, C. Y., Kim, S., Lee, D., Park, D. J., and Imm, J. Y. (2016). Enzyme and high pressure assisted extraction of tricin from rice hull and biological activities of rice hull extract. *Food Science and Biotechnology*, 25(1), 159–164.

Pascual, A., Llorca, I., and Canut, A. (2007). Use of ozone in food industries for reducing the environmental impact of cleaning and disinfection activities. *Trends in Food Science and Technology*, 18, S29–S35.

Peñas, E., Uberti, F., Lorenzo, C., Ballabio, C., Brandolini, A., and Restani, P. (2014). Biochemical and immu-nochemical evidences supporting the inclusion of quinoa (*Chenopodium quinoa* Willd.) as a gluten-free ingredient. *Plant Foods for Human Nutrition*, 69(4), 297–303. doi: 10.1007/s11130-014-0449-2.

Petersen, K., Væggemose, Nielsen P., Bertelsen, G., Lawther, M., Olsen, M. B., Nilsson, N. H., and Mortensen, G. (1999). Potential of biobased materials for food packaging. *Trends in Food Science and Technology*, 10, 52–68.

Pivarnik, L. F., and Worobo, R. (2014). Non-thermal or alternative food processing methods to enhance microbial safety and quality. *USDA-NIFA*, 1, 1–2.

Ramos, M., Valdés, A., Mellinas, A. C., and Garrigós, M. C. (2015). New trends in beverage packaging systems: A review. *Beverages*, 1(4), 248–272.

Ranjan, S., Dasgupta, N., Chakraborty, A. R., Samuel, S. M., Ramalingam, C., Shanker, R., and Kumar, A. (2014). Nanoscience and nanotechnologies in food industries: Opportunities and research trends. *Journal of Nanoparticle Research*, 16(6), 1–23.

Rasmussen, P. H., and Hansen, A. (2001). Staling of wheat bread stored in modified atmosphere. *Lebensm. Wiss. Technol.*, 34, 487–491.

Raso, J., Calderón, M. L., Barbosa-Cánovas, G. V., and Swanson, B. G. (1998). Inactivation of *Zygosaccharomyces bailii* in fruit juices by heat, high hydrostatic pressure and pulsed electric fields. *Innovative Food Science and Emerging Technologies*, 1(1), 21–29.

Richter, T., Sterr, J., Jost, V., and Langowski, H.-C. (2010). High pressure-induced structural effects in plastic packaging. *High Pressure Research*, 30(4), 555–566.

Roberts, P. (2016). Food irradiation: Standards, regulations and world-wide trade. *Radiation Physics and Chemistry*, 129, 30–34.

Robertson, G. L. (2011). Packaging and food and beverage shelf life. In: Kilcast, D., and Subramaniam, P. (Eds.), *Food and Beverage Stability and Shelf Life* (pp. 244–272). Oxford, UK: Woodhead Publishing.

Rooney, M. L. (2005). Introduction to active food packaging technologies. In: Han, J. H. (Ed.), *Innovations in Food Packaging* (pp. 63–79). Oxford, UK: Elsevier Academic Press.

Saldaña, G., Álvarez, I., Condón, S., and Raso, J. (2014). Microbiological aspects related to the feasibility of PEF technology for food pasteurization. *Critical Reviews in Food Science and Nutrition*, 54(11), 1415–1426.

Salmeron, I., Thomas, K., and Pandiella, S. S. (2014). Effect of substrate composition and inoculum on the fermentation kinetics and flavour compound profiles of potentially nondairy probiotic formulations. *LWT – Food Science and Technology*, 55, 240–247.

Salmeron, I., Thomas, K., and Pandiella, S. S. (2015). Effect of potentially probiotic lactic acid bacteria on the physicochemical composition and acceptance of fermented cereal beverages. *Journal of Functional Foods*, 15, 106–115.

Sarangapani, C., Devi, R. Y., Thirumdas, R., Trimukhe, A. M., Deshmukh, R. R., and Annapure, U. S. (2017). Physico-chemical properties of low-pressure plasma treated black gram. *LWT – Food Science and Technology*, 79, 102–110.

Santos, M. C., Salvador, A. C., Domingues, F. M., Cruz, J. M., and Saraiva, J. A. (2013). Use of high hydrostatic pressure to increase the content of xanthohumol in beer wort. *Innovative Food Science and Emerging Technologies*, 13, 13–22.

Savi, G. D., Piacentini, K. C., Bittencourt, K. O., and Scussel, V. M. (2014). Ozone treatment efficiency on *Fusarium graminearum* and deoxynivalenol degradation and its effects on whole wheat grains (*Triticum aestivum* L.) quality and germination. *Journal of Stored Products Research*, 59, 245–253.

Schlüter, O., Ehlbeck, J., Hertel, C., Habermeyer, M., Roth, A., Engel, K. H., Holzhauser, T., Knorr, D., and Eisenbrand, G. (2013). Opinion on the use of plasma processes for treatment of foods. *Molecular Nutrition and Food Research*, 57(5), 920–927.

Shiferaw Terefe, N., and Augustin, M. A. (2020). Fermentation for tailoring the technological and health related functionality of food products. *Critical Reviews in Food Science and Nutrition*, 60(17), 2887–2913. doi:10.1080/10408398.2019.1666250.

Shori, A. B. (2016). Influence of food matrix on the viability of probiotic bacteria: A review based on dairy and non-dairy beverages. *Food Bioscience*, 13(1), 1–8.

Sunil, Neelash Chauhan, Singh, Jaivir, Chandra, Suresh, Chaudhary, Vipul, and Kumar, Vikrant. (2018). "Non-thermal techniques: Application in food industries: A review. *Journal of Pharmacognosy and Phytochemistry*, 7(5), 1507–1518.

Taiwo, K. A., Angersbach, A., and Knorr, D. (2002). Rehydration studies on pretreated and osmotically dehydrated apple slices. *Journal of Food Science*, 67(2), 842–847.

Tan, M. C., Chin, N. L., and Yusof, Y. A. (2011). Power ultrasound aided batter mixing for sponge cake batter. *Journal of Food Engineering*, 104(3), 430–437.

Thirumdas, R., Sarangapani, C., and Annapure, U. S. (2015). Cold plasma: A novel non-thermal technology for food processing. *Food Biophysics*, 10, 1–11.

Tkalec, M., Malarić, K., Pavlica, M., Pevalek-Kozlina, B., and Vidaković-Cifrek, Ž. (2009). Effects of radio-frequency electromagnetic fields on seed germination and root meristematic cells of *Allium cepa* L. *Mutation Research/Genetic Toxicology and Environmental Mutagenesis*, 672(2), 76–81.

Tokusoglu, O., Odriozola-Serrano, I., and Martín-Belloso, O. (2014). Quality, safety, and shelf-life improvement in fruit juices by pulsed electric fields. In: Tokusoglu, O., and Swanson, B. G. (Eds.), *Improving Food Quality with Novel Food Processing Technologies* (pp. 385–412). Boca Raton, FL: CRC Press.

Tuorila, H., Cardello, A. V. (2002). Consumer responses to an off-flavor injuice in the presence of specific health claims. *Food Quality and Preference*, 13, 561–569.

Valdes, A., Mellinas, A. C., Ramos, M., Burgos, N., Jimenez, A., and Garrigos, M. C. (2015). Use of herbs, spices and their bioactive compounds in active food packaging. *RSC Advances*, 5(50), 40324–40335.

Valdés, A., Mellinas, A. C., Ramos, M., Garrigós, M. C., and Jiménez, A. (2014). Natural additives and agricultural wastes in biopolymeformulations for food packaging. *Frontiers in Chemistry*, 2, 1–10.

Van den Bosch, H. F. M. (2007). Chamber design and process conditions for pulsed electric field treatment of food. In: Lelieveld, H. L. M., Notermans, S., and de Haan, S. W. H. (Eds.), *Food Preservation by Pulsed Electric Fields* (pp. 70–93). Woodhead Publishing Series in Food Science, Technology and Nutrition, Cambridge, UK: Woodhead Publishing.

Vanderroost, M., Ragaert, P., Devlieghere, F., and De Meulenaer, B. (2014). Intelligent food packaging: The next generation. *Trends in Food Science and Technology*, 39(1), 47–62.

Vashisth, A., and Nagarajan, S. (2010). Effect on germination and early growth characteristics in sunflower (*Helianthus annuus*) seeds exposed to static magnetic field. *Journal of Plant Physiology*, 167(2), 149–156.

VidyaLaxme, B., Rovetto, A., Grau, R., and Agrawal, R. (2014). Synergistic effects of probiotic *Leuconostoc mesenteroides* and *Bacillus subtilis* in malted ragi (*Eleucine corocana*) food for antagonistic activity against *V. cholerae* and other beneficial properties. *Journal of Food Science and Technology*, 51(11), 3072–3082.

Vorobiev, E., and Lebovka, N. (2009). *Electrotechnologies for Extraction from Food Plants and Biomaterials*. Food Engineering Series. New York: Springer New York.

Wambura, P., Yang, W., and Wang, Y. (2008). Power ultrasound enhanced one-step soaking and gelatinization for rough rice parboiling. *International Journal of Food Engineering*, 4(4), 1–12.

Wouters, P., Álvarez, I., and Raso, J. (2001). Critical factors determining inactivation kinetics by pulsed electric field food processing. *Trends in Food Science and Technology*, 12, 112–121. doi:10.1016/S0924-2244(01)00067-X.

Wu, J., Doan, H., and Cuenca, M. A. (2006). Investigation of gaseous ozone as an anti-fungal fumigant for stored wheat. *Journal of Chemical Technology & Biotechnology: International Research in Process, Environmental & Clean Technology*, 81(7), 1288–1293.

Yaldagard, M., Mortazavi, S. A., Mousavi, S. M., and Tabatabaie, F. (2009). Investigation of the effects of ultrasound on extraction of alpha-amylase from the flour of malted barley. *Journal of the American Society of Brewing Chemists*, 67(3), 141–145.

Yam, K. L., and Lee, D. S. (2012). *Emerging Food Packaging Technologies: Principles and Practice*. Cambridge, UK: Woodhead Publishing Limited.

Yamazaki, A. (2000). Processed food and cooking the same. Japanese patent: JP2000217526(A). Echigo Seika.

Yamazaki, A. (2005a). Advanced in the use of high pressure for food processing and preservation: Application of high pressure and its effects on rice grains and rice starch. *Foods and Food Ingredients Japan*, 210, 29–36.

Yamazaki, A. (2005b). Brown rice processing method. Japanese patent: JP2005117982. Echigo Seika.

Yamazaki, A., Itou, M., and Sasagawa, A. (1998). Allergenic rice and production of processed food. Japanese patent: JP10150935. Echigo Seika.

Zhang, Q. H., Barbosa-Canovas, G. V., and Swanson, B. G. (1995). Engineering aspects of pulsed electric fields pasteurization. *Journal of Food Engineering*, 25, 261–281.

Chapter 11

Impact of Organoleptic and Consumer Acceptability for Non-Thermally Processed Grain-Based Food Products

Esra Dogu-Baykut and Celale Kirkin

CONTENTS

11.1 INTRODUCTION

Grains form a major part of the daily diets of many populations around the world. They are the seeds of plants, belonging mainly to the botanical cereal, pseudo-cereal, and legume groups (Carcea, 2020). Grain-based bakery products are produced in many different varieties, including breads (rolls, bagels, buns, and loaves), pastas and noodles, tortillas, pizza crusts, cakes, muffins, cookies, crackers, biscuits, chips, pastries, and breakfast cereals. These products are mostly made from wheat, oats, rice, maize, rye, barley, millet, quinoa, and sorghum. Among them, wheat, rice, barley, maize, and sorghum are also widely used in the manufacture of fermented foods and alcoholic and non-alcoholic beverages (Blandino, Al-Aseeri, Pandiella, Cantero, & Webb, 2003).

Non-thermal food processing techniques have attracted the attention of many researchers in recent years because of their potential benefits in ensuring food safety and quality. Non-thermal techniques are highly effective in inactivating pathogenic and spoilage microorganisms at low processing temperatures without causing off-flavors, nutrient loss, and changes in the food's physical and chemical properties. Moreover, in addition to improving the shelf life and safety of foods, these non-thermal techniques provide other advantages such as shorter treatment times and higher energy efficiency (Zhang, Wang, Zeng, Han, & Brennan, 2019). For these reasons, many studies are investigating their application potential for different foods. Because each non-thermal technology has limitations and specific applications according to the type of food processed, it needs to be investigated separately for each type of food (Barbosa-Cánovas, Góngora-Nieto, Rodriguez, & Swanson, 2005).

Non-thermal technologies used in food production should primarily ensure food safety, but at the same time, their effects on the nutritional composition, sensorial properties, and physical structure and properties of the food are also important criteria in terms of consumer acceptability. This chapter presents an overview of recent studies on the application of non-thermal technologies to cereal grain-based food products and their impact on organoleptic quality and consumer acceptability. Non-thermal technologies, such as high-pressure processing (HPP), ultrasound (US), ionizing irradiation (IR), ultraviolet (UV) irradiation, pulsed electric fields (PEF), cold plasma (CLP), and ozonation are discussed in the following sections.

11.2 HIGH-PRESSURE PROCESSING

High-pressure processing (HPP) is a non-thermal food preservation method in which microbial inactivation is achieved due to pressure. This technology is also called high hydrostatic pressure or ultra-high pressure. It is a uniform process and generally applied at 100–600 MPa levels (Welti-Chanes et al., 2004). Pressure, temperature, and exposure time are the main processing parameters (Heinz & Buckow, 2010).

Food can be treated at low temperatures by HPP. Thus, undesired alterations in the product caused by the thermal effect of conventional heat treatments are minimized. However, a compression-induced temperature increase of 3°C per 100 MPa for most food products and up to 9°C per 100 MPa for high-fat products can be observed during processing, but the temperature drops to the initial level after pressure release (Toepfl, Mathys, Heinz, & Knorr, 2006).

Vegetative bacteria are inactivated by the damage to the cell membrane and inactivation of metabolic enzymes caused by HPP (Heinz & Buckow, 2010; Smelt, Rijke, & Hayhurst, 1994). It has been claimed that microbial inactivation is also related to solute losses during pressure treatment, denaturation of proteins, inactivation of key enzymes, and damage to the recovery mechanism (Mañas & Pagán, 2005). In addition, the main virus inactivation mechanism of HPP has been suggested as the denaturation of the proteins of viruses, which play a role in binding to the host cell (Kahdre & Yousef, 2002).

HPP permits the in-package processing of food in a shorter time compared to traditional methods with minimum effects on color, flavor, and nutritional properties; however, it has a limited inactivation effect on enzymes and spores (Priyadarshini, Rajauria, O'Donnell, & Tiwari, 2019). The limited inactivation activity against bacterial spores can be overcome by combining HPP with thermal sterilization (Farkas & Hoover, 2000; Reineke, Mathys, Heinz, & Knorr, 2013).

HPP can be used for the microbial decontamination of liquid and solid foods. The HPP application has also been a promising process that can be used in the production of reduced-salt food products while improving color and texture (Crehan, Troy, & Buckley, 2000; Ros-Polski, Koutchma, Xue, Defelice, & Balamurugan, 2015; Yang et al., 2015). Moreover, high-pressure (HP) treatment can lead to modifications in functional properties, such as denaturation of proteins, inactivation of enzymes, and gelatinization of starches (Kim, Kim, & Baik, 2012; Rastogi, Raghavarao, Balasubramaniam, Niranjan, & Knorr, 2007).

11.2.1 Utilization of HPP during Processing of Grain-Based Foods

HPP can improve the digestibility, sensory properties, gelation, and gelatinization of cereal-based food products while decreasing cooking losses. It has also been reported that HP-assisted gelatinization of starch can increase the amounts of slowly digestible starch and resistant starch and reduce retrogradation and the amount of rapidly digestible starch (Liu, Fan, Cao, Blanchard, & Wang, 2016; Tian, Li, Zhao, Xu, & Jin, 2014).

It is possible to utilize HPP in the production of low-gluten or gluten-free grain-based foods (Collar & Angioloni, 2017), as it changes structures and functional properties of proteins and starches (Knorr, Heinz, & Buckow, 2006; McCann, Leder, Buckow, & Day, 2013; Vallons & Arendt, 2009; Vallons, Ryan, & Arendt, 2011). Thus, dough properties can be improved by HP modification of starches and proteins. It has been suggested that the gelatinization and viscoelastic properties of wheat dough can be enhanced when incorporated with HP-processed oat, millet, sorghum, and wheat hydrated flours depending on the amount of the added grain flour and the pressure level (Angioloni & Collar, 2012b).

HPP possibly affects the sensory properties of cooked rice by changing the composition of volatiles, as it was claimed that the treatment of 200 and 400 MPa increased the number of alcohols, ketones, esters, and olefins while decreasing heterocycles, alkanes, and arenes depending on the rice variety (Deng et al., 2013).

The pasteurization of beer to inactivate microorganisms is crucial for stability, but thermal treatments cause decreases in organoleptic acceptability and freshness of the product (Fischer, Russ, & Meyer-Pittroff, 2002). HPP can minimize the quality losses due to thermal processing of beer (Buzrul, 2012; Santos, Oliveira, Ferreira, & Rosenthal, 2017). It was suggested as a fast and energy-efficient alternative to thermal pasteurization for the inactivation of *Saccharomyces cerevisiae* ascospores (Milani, Ramsey, & Silva, 2016). It was also reported that the loss of volatiles which is observed in the thermal processing of beer can be prevented by HP treatment, but it can result in a lower amount of bitter compounds, as the isomerization of hop compounds due to high pressure was less than boiling (Fischer et al., 2002). Buzrul, Alpas, and Bozoglu (2005) also reported that the ethanol, density, real extract, apparent extract, original extract, fermentation degree, and pH values of lager beer were not affected by HP pasteurization. Changes in color were noted depending on the pressure level and treatment time. They observed increases in protein sensitivity and chill haze after HPP. The authors also reported that the bitterness of HP-treated samples was similar to that of the untreated.

11.2.2 Effects of HPP on the Organoleptic Quality and Consumer Acceptance of Grain-Based Foods

Several other studies have investigated the utilization of HP treatment and its effects on organoleptic properties of cereal-based products. It has been used in the processing of cereal-based products, such as bread (Bárcenas, Altamirano-Fortoul, & Rosell, 2010; Collar & Angioloni, 2017), rice (Boluda-Aguilar, Taboada-Rodríguez, López-Gómez, Marín-Iniesta, & Barbosa-Cánovas, 2013), beer (Milani & Silva, 2016), and rice wine (Yang et al., 2019) without negatively altering sensory properties. A summary of the studies – conducted in the last decade – evaluating the effects of HP on the sensory properties of grain-based foods is given in Table 11.1.

11.3 ULTRASOUND

Ultrasound is sonic waves with a high frequency above the human audible range. Ultrasound technology is divided into two categories as high-frequency (low-intensity) ultrasound and low-frequency (high-intensity) ultrasound. High-frequency ultrasound is applied at frequencies above 100 kHz at intensities below 1 W/cm^2, whereas low-frequency ultrasound applications include a frequency range between 18 and 100 kHz with intensities above 1 W/cm^2 (Knorr, Zenker, Heinz, & Lee, 2004).

Microbial inactivation during ultrasound processing occurs due to cavitation, disruption of the cell membrane, loss of cellular material, free radical generation, and heat (Bermudez-Aguirre, Mobbs, & Barbosa-Canovas, 2011; Gao, Lewis, Ashokkumar, & Hemar, 2014; Piyasena, Mohareb, & McKellar, 2003). The amplitude of ultrasound waves, treatment time, temperature, and volume and composition of the product also affect microbial inactivation (Piyasena et al., 2003).

Potential applications of ultrasound technology in food processing are plenty, including but not limited to microbial and enzymatic inactivation, cooking, crystallization, cutting, drying, emulsification, extraction, filtration, freezing, and thawing (Bermudez-Aguirre et al., 2011; Chemat, Zill-e-Huma, & Khan, 2011). Moreover, ultrasound technology has a wide array of applications in food analysis and quality control, as it enables the estimation of physicochemical, mechanical, and structural properties, contamination, and the composition of the product (Awad, Moharram, Shaltout, Asker, & Youssef, 2012).

11.3.1 Utilization of Ultrasound during Processing of Grain-Based Foods

Ultrasound has been successfully used in grain processing during germination (Ding & Feng, 2018), hydration (Miano, Ibarz, & Augusto, 2017), fortification (Bonto, Jearanaikoon, Sreenivasulu, & Camacho, 2020; Tiozon Jr., Camacho, Bonto, Oyong, & Sreenivasulu, 2021), hydrolysis (Zhu, Su, Guo, Peng, & Zhou, 2011), and nixtamalization (Janve et al., 2013). It can also be utilized in mixing of doughs and batters (Pa, Chin, Yusof, & Aziz, 2014; Tan, Chin, & Yusof, 2011), fermentation (Luo et al., 2018), cutting (Zahn, Schneider, Zucker, & Rohm, 2005), and pre-gelatinization (Jalali, Sheikholeslami, Elhamirad, Haddad Khodaparast, & Karimi, 2020). Ultrasound technology generally improves the efficiency and reduces the treatment time of these processes. Applications of ultrasound technology in the bakery and snacks foods sector are comprehensively reviewed in Chapter 5.

TABLE 11.1   STUDIES EVALUATING THE EFFECTS OF HIGH-PRESSURE TREATMENT ON THE SENSORY PROPERTIES OF GRAIN-BASED FOODS (CONDUCTED IN THE LAST TEN YEARS)

Product	Research Objective	HP Treatment Conditions	Key Findings	References
Bread	• Investigation of the effects of HP on some properties of dough • Evaluation of the effects of HP on the texture of wheat bread	50, 100, 150, 200, 250 MPa for 1, 2, 3, 4 min	• HP-treated wheat breads, except for 200 MPa–treated samples, demonstrated similar sensory acceptability to non-treated samples • Panelists liked the croissant-like structure of the HP-treated bread	(Bárcenas et al., 2010)
	• Evaluation of the influences of HP on the microstructural and rheological properties of batters prepared with sorghum flour (at different levels)	200, 300, 400, 500, 600 MPa for 10 min at 20°C	• Staling of 2% sorghum flour–added bread was delayed at 600 MPa	(Vallons, Ryan, Koehler, & Arendt, 2010)
	• Evaluation of the effects of HP on some quality properties of wheat and oat-, millet-, and sorghum-based breads	200, 350, 500 MPa for 10 min at 20°C	• The overall acceptability of bread samples prepared with HP-treated (at 350 MPa) wheat, millet, and sorghum and the firmness of samples with HP-treated wheat and oat were higher compared to those of conventionally prepared samples	(Angioloni & Collar, 2012a)
	• Evaluation of the effects of HP on some properties of high-legume wheat-based breads	350 MPa for 10 min at 20°C	• The addition of HP-treated hydrated chickpea, pea, and soybean flours to the bread dough mix did not negatively alter the sensory acceptance	(Collar & Angioloni, 2017)
	• Evaluation of the breadmaking performance of HP-treated oats	200, 350, 500 MPa for 10 min at 20°C	• Staling of the bread samples containing 200 MPa–treated oat batter was decreased, but quality degradations were observed in the samples containing oat batters treated at 350–500 MPa	(Hüttner, Bello, & Arendt, 2010)

(Continued)

TABLE 11.1 (CONTINUED)   STUDIES EVALUATING THE EFFECTS OF HIGH-PRESSURE TREATMENT ON THE SENSORY PROPERTIES OF GRAIN-BASED FOODS (CONDUCTED IN THE LAST TEN YEARS)

Product	Research Objective	HP Treatment Conditions	Key Findings	References
Rice	• Utilization of HP treatment in the production of quick-cooking rice	• Single HP treatment: 300, 400 MPa for 2, 4 min at 25°C • Double HP treatment: 300, 400 MPa for 2, 4 min followed by 570 MPa for 20 min at 25°C	• The grain shape, roughness, texture, adhesiveness, and cohesiveness of single HPP–treated samples were more acceptable compared to those treated with double HPP • The results of sensory analysis (adhesiveness and cohesiveness) were in agreement with the results of instrumental texture analysis • The HP-processed samples were preferred over the unpressurized samples	(Boluda-Aguilar et al., 2013)
Beer	• Evaluation of the effects of HPP on the inactivation of *Saccharomyces cerevisiae* ascospores and sensory properties of beer	• Effect of carbonation on ascospore inactivation: 200 MPa for 10 min at ≤ 26.5°C • Effect of alcohol content on ascospore inactivation: 400 MPa for 1, 10 min • Modeling: 200, 300, 400 MPa for up to 40 min • Sensory evaluation: 600 MPa for 30 s	• There was no significant sensory preference between HP-processed (at 600 MPa for 30 s) beer and untreated beer • No significant taste difference was noted between the samples according to the triangle test • HPP can be used in the pasteurization of beer depending on the level of pressure and alcohol content of the beer	(Milani & Silva, 2016)

(Continued)

TABLE 11.1 (CONTINUED)   STUDIES EVALUATING THE EFFECTS OF HIGH-PRESSURE TREATMENT ON THE SENSORY PROPERTIES OF GRAIN-BASED FOODS (CONDUCTED IN THE LAST TEN YEARS)

Product	Research Objective	HP Treatment Conditions	Key Findings	References
	• Effects of HPP on some characteristics of unfiltered lager beer	250, 550 MPa for 5 min at 25°C	• Reductions in fullness, bitterness, and hoppy flavor of unfiltered lager beer treated at 550 MPa compared to those of the samples treated at 250 MPa and untreated • These changes can be linked to the breakdown of polyphenols, hop bitter acids, and aromatic compounds at 550 MPa	(Štulíková et al., 2020)
Chinese rice wine	• Evaluation and comparison of the effects of boiling, ultra-high temperature, and HP treatment on some quality properties of Chinese rice wine	200, 400, 600 MPa for 10, 20 min at 24–27°C	• HP treatment at 400–600 MPa for 10 min preserved the sensory characteristics of untreated samples • Ultra-high-temperature treatment (125–135°C for 5 s) and traditional boiling (90°C for 15 min) caused flavor losses	(Y. Yang et al., 2019)
	• Evaluation of the effects of HPP on the quality of Hongqu rice wines during storage	250–550 MPa for 30 min	• The sensory profiles of HP-treated rice wines were similar to those of untreated except for acidity and mellowness at the beginning of the storage period; however, the sensory quality of HP-treated samples decreased during 18-month storage	(Yuting Tian et al., 2016)

(Continued)

TABLE 11.1 (CONTINUED)   STUDIES EVALUATING THE EFFECTS OF HIGH-PRESSURE TREATMENT ON THE SENSORY PROPERTIES OF GRAIN-BASED FOODS (CONDUCTED IN THE LAST TEN YEARS)

Product	Research Objective	HP Treatment Conditions	Key Findings	References
Wheat beer	• Evaluation of the effects of HP on the physicochemical and sensory properties of wheat beer	400 MPa for 15 min, 500 MPa for 10 min, 600 MPa for 5 min	• Significant inactivation of initial microbial load • Original extract, ethanol content, pH, bitterness, viscosity, and color were unchanged after HP treatment • Slight increases in the contents of 4-vinylguaiacol and 4-vinylphenol depending on the pressure level • HP treatment demonstrated higher foam stability and colloidal haze stability • Heat-treated samples demonstrated a less fruity/banana-like aroma compared to HP-treated and untreated samples • HP-treated samples had higher fullness values according to sensory evaluation	(Yin et al., 2016)

11.3.2 Effects of Ultrasound Processing on the Organoleptic Quality and Consumer Acceptance of Grain-Based Foods

Chang (2005) utilized ultrasound technology for accelerating the aging process of a rice-based alcoholic beverage that is fermented with *Saccharomyces sake*. The author reported that the taste of wine samples treated at 20 kHz was equal to that of the control group (standard aging for one year in fired clay containers). They also added that as the number of treatments (at 20 kHz) increases, the taste of the wine is improved; however, an opposite trend was observed when a 1.6 MHz ultrasonic treatment was applied.

Janve, Yang, and Sims (2015) evaluated the effects of ultrasound-assisted nixtamalization on the quality and sensory properties of tortilla chips. Although the authors noted changes in some properties of the samples by physicochemical and instrumental methods, these differences were not detected during the sensory evaluation. Overall appearance, overall liking, overall flavor, and overall texture of the samples were not affected by the treatment.

In another study, Zhang et al. (2015) used ultrasound-assisted enzymatic treatment for producing quick-cooking brown rice. They reported that the treatment increased water absorption and shortened cooking time. The viscosity of the brown rice and the texture of the cooked samples were improved. The odor and flavor of the ultrasound enzyme–treated cooked samples were also enhanced, and the acceptance of the treated samples was similar to that of the untreated.

Taghinia, Ataye-Salehi, and Sheikholeslami (2015) added ultrasound-treated rice bran to a wheat dough. They observed that the water absorption, development time, and softening degree of the dough were increased, while the dough stability was decreased upon the addition of the ultrasound-treated rice bran. The addition of the bran increased the moisture content of the barbari bread samples, but no changes in the specific volume, porosity, and firmness were noted due to the addition of bran up to 10%. Up to 10% addition of the ultrasound-treated rice bran also increased the total acceptability of the bread samples.

Moreover, Pasqualone et al. (2016) compared the use of bran aqueous extract (obtained using ultrasound), bran oleoresin (obtained using supercritical CO_2), and powdery lyophilized tomato matrix in functional pasta production. The addition of bran aqueous extract that was obtained using ultrasound did not change the sensory properties of the pasta, but the authors recommended the addition of bran oleoresin as it improved antioxidant activity and did not alter the quality properties.

Also, the sensory acceptability and technological and visual characteristics of gluten-free bread prepared with pre-gelatinized corn flour treated with ultrasound at 30% intensity were better compared to the untreated samples, the samples treated with ultrasound at 70% intensity, and the samples treated with microwave (Jalali et al., 2020).

Furthermore, it was reported that emulsification with the aid of ultrasound treatment enhanced the porosity and size of cakes and delayed staling (Esmaeilzadeh Kenari & Nemati, 2020). Ultrasound-treated cupcake samples received higher sensory scores. The authors also reported that the use of probe-type ultrasound was more effective compared to the ultrasound bath.

In addition, ultrasound technology has also been effectively utilized in the extraction of nutrients and antioxidants. The extracted nutrients can be used in food production for enrichment or fortification purposes. For instance, Polat et al. (2020) added germinated lentil extract, which was obtained by ultrasound extraction, into the formulation of crackers by replacing some portion of the flour. The authors reported that the contents of protein, ash, phenolics, and antioxidants were increased by the addition of the extract. Sensory properties of the samples with

5% germinated lentil extract were not negatively affected, while those of the samples with the extract added at a level of 10% and 15% were lower.

11.4 IONIZING IRRADIATION

Irradiated foods are exposed to ionizing radiation, and the ionizing radiation is provided by gamma rays, X-rays, or electron beams. The method provides a fast and effective way of reducing microbial load, sprouting, and insect infestation.

Gamma irradiation is generated using ^{60}Co or ^{137}Cs, whereas electron-beam irradiation and X-ray irradiation use electricity as the power source (Fan & Niemira, 2020). Gamma irradiation has a lower operating cost and good penetration compared to X-rays and electron-beam irradiation. Electron-beam irradiation has a limited penetration ability and is suitable for smaller products, and X-ray irradiation has a higher operating cost (Fan & Niemira, 2020).

The unit of absorbed dose is a Gray (Gy) and is measured by a dosimeter. It has been claimed that gamma irradiation can be utilized in shortening the cooking time of cereals, and doses at or below 1 kGy can be used for disinfestation of insects whereas doses at or above 2 kGy can be used for bacteria (Andrews et al., 1998; Lacroix & Ouattara, 2000). The maximum dose that can be applied to foods is generally limited to 10 kGy in most countries; however, some food products such as spices and seasonings can be irradiated at a maximum dose of 30 kGy in the USA.

11.4.1 Utilization of Ionizing Irradiation during Processing of Grain-Based Foods

Aziz, Souzan, and Shahin Azza (2006) reported that gamma irradiation of grains up to 10 kGy was effective for the inactivation of microorganisms and did not cause adverse effects on nutritional quality. Agúndez-Arvizu et al. (2006) claimed that irradiation of wheat flour at 1 kGy sufficiently reduced microbial load without negatively affecting quality and nutritional properties.

Gamma irradiation up to 8 kGy did not affect the apparent amylose content of milled rice depending on the cultivar as reported by Bao et al. (2001). Changes in the gel consistency and viscosity of the milled rice and the texture of the cooked rice were observed due to the irradiation dose. The authors also noted a color change and off-odor development in cooked rice at doses over 2 kGy.

Moreover, Shao et al. (2013) reported that gamma irradiation of whole-grain rice at doses up to 10 kGy can improve phenolics content and antioxidant activity as well as eating quality by decreasing paste viscosity and gel hardness without causing visible color changes. Irradiation can also change the physicochemical, nutritional, functional, and antioxidant properties of starch (Bashir & Aggarwal, 2019; Zhu, 2016).

11.4.2 Effects of Ionizing Irradiation on the Organoleptic Quality and Consumer Acceptance of Grain-Based Foods

Although ionizing irradiation improves some quality properties and the microbial safety of foods, its effects on sensory properties and consumer acceptance are the most important factors that determine whether the technology is suitable for application to a particular product. Thus, there have been several studies investigating the influence of irradiation on the sensory properties of grain-based food products. For instance, Roy, Ghosh, and Chatterjee (1991) reported that cooked and uncooked rice were acceptable after gamma irradiation at up to 5 kGy, and irradiation at 3 kGy was better at preventing color and cooking losses. In another study, Rao, Srirangarajan, Kamat, Adhikari, and Nair (1994) reported that gamma irradiation prevented insect infestation while not affecting the moisture content and microbial counts, that and the overall acceptability of the 0.25 kGy–irradiated semolina samples was not affected by the treatment. In addition, the use of BOPP/LDP packages was recommended by the authors. Moreover, it was also reported that gamma irradiation (0.05–10 kGy) did not affect the sensory odor and instrumental color of wheat grains (Warchalewski, Gralik, Zawirska-Wojtasiak, Zabielski, & Kusnierz, 1998).

Furthermore, Marathe, Machaiah, Rao, Pednekar, and Rao (2002) reported that the quality of chapati made from 0.25 kGy–irradiated whole wheat flour was maintained during six-month storage. The authors also stated that the sensory evaluation scores (color, appearance, hand-feel, pliability, mouth-feel, taste, aroma, overall acceptability) of chapati samples made from 0.25 kGy–irradiated wheat flour were higher compared to those of the control samples (not irradiated) after the six-month storage period.

Effects of irradiation on the maturation of rice wine were investigated by Chang (2003). The author reported that the application of gamma irradiation (200–800 Gy) enhanced the taste while decreasing the defects.

Sung, Hong, and Chang (2008) evaluated the effects of gamma irradiation and storage on the pasting properties of japonica waxy starch and the quality of mochi (a product of waxy starch) and observed that the pasting viscosity of the waxy rice decreased due to irradiation and storage. They also added that sensory color, hardness, and overall likeness values were higher in 1 kGy–irradiated samples compared to the control samples, and the influence of storage on quality improvement was greater compared to that of irradiation.

Also, Sirisoontaralak and Noomhorm (2006) reported that gamma irradiation at 0.2–2 kGy significantly affected the sensory properties of cooked rice, and doses below 0.5 kGy were recommended for the insect disinfestation of aromatic milled rice. The sensory acceptance of appearance, color, odor, texture, and taste and the overall acceptability were decreased by irradiation.

In addition, Villavicencio, Araújo, Fanaro, Rela, and Mancini-Filho (2007) evaluated the effects of gamma irradiation doses applied for insect disinfestation on the sensory properties of cereal bars. They reported that the dose of 2.0 kGy was recommended for chocolate and banana cereal bars, and irradiation at 3.0 kGy was suggested for strawberry cereal bars to preserve the sensory quality. No adverse effects of irradiation on appearance, odor, and taste were noted by the researchers.

Azzeh and Amr (2009) reported that the sensory appearance, color, stickiness, firmness, and overall acceptability of lasagna samples made from semolina that was irradiated at 0.25 and 1 kGy did not change. However, the stickiness, firmness, and overall acceptability of the samples made from semolina irradiated at 2.5, 5, and 10 kGy were decreased.

Studies conducted in the last decade investigating the effects of ionizing irradiation on sensory properties of grain-based foods are summarized in Table 11.2.

11.5 UV IRRADIATION

Ultraviolet (UV) irradiation is a non-thermal, non-chemical, and safe technique to use for the decontamination of foods. The wavelength of UV irradiation lies between 100 and 400 nm and is further subdivided into UV-A (315–400 nm), UV-B (280–315 nm), and UV-C (200–280 nm). UV-C irradiation, also called short-wavelength UV or germicidal irradiation, effectively inactivates microorganisms on packaging materials, food contact surfaces, and some foodstuffs (Koutchma, 2019).

TABLE 11.2   STUDIES EVALUATING THE EFFECTS OF IONIZING IRRADIATION ON THE SENSORY PROPERTIES OF GRAIN-BASED FOODS (CONDUCTED IN THE LAST TEN YEARS)

Product	Research Objective	Irradiation Dose	Key Findings	References
Bread	• Effects of gamma irradiation of some properties of wheat grains and pan breads made with the flour of the irradiated wheat grains	0, 3, 6, 9 kGy	• Changes in technological and rheological properties of flour were observed after irradiation • The sensory odor, taste, appearance, and texture values were not affected by irradiation • Irradiation increased the loaf volume • Irradiation at 6 and 9 kGy caused a color change in bread according to the sensory evaluation	(Anwar, Asael, & El-Adly, 2015)
Rice	• Effects of electron-beam irradiation on some characteristics of white rice during storage	2, 4, 6, 8 kGy	• Electron-beam irradiation at 6–8 kGy decreased hardness, appearance, taste, and overall acceptability of white rice, while the scores of the samples irradiated at 2–4 kGy were higher	(Du et al., 2019)
	• Evaluation of different doses of irradiation on some quality properties of brown japonica rice during storage	0.2, 0.5, 1.0, 2.0 kGy	• Losses in sensory quality, especially in odor and overall acceptability, were observed during 18-month storage after 1–2 kGy gamma irradiation • Lower doses (0.2–0.5 kGy) had no effect on the sensory properties	(Chen et al., 2015)
	• Effects of irradiation on physicochemical and sensory properties of rice	1, 2, 5 kGy	• Irradiation up to 5 kGy did not affect the appearance, taste, and texture of rice • A decrease was noted in the aroma and color of the samples irradiated at 5 kGy • Irradiation of rice at a dose of 1 kGy was recommended, as no changes in other quality characteristics of the samples were observed	(Polesi et al., 2012)

(Continued)

TABLE 11.2 (CONTINUED)   STUDIES EVALUATING THE EFFECTS OF IONIZING IRRADIATION ON THE SENSORY PROPERTIES OF GRAIN-BASED FOODS (CONDUCTED IN THE LAST TEN YEARS)

Product	Research Objective	Irradiation Dose	Key Findings	References
Bibimbap (Korean rice-based food)	• Evaluation of the effects of irradiation in combination with frozen irradiation, vacuum packaging, and vitamin C addition on microbiological and sensory properties of *Bibimbap* as a space food	25 kGy	• The sensory scores (color, texture, taste, flavor, and overall acceptability) were decreased after irradiation • When the irradiation treatment was combined with frozen irradiation, vacuum packaging, and vitamin C addition (0.1%) the scores were increased	(Park et al., 2012)
Pulav (rice-based Indian food)	• Evaluation of the effects of irradiation on shelf life, nutritional and sensory properties, and antioxidant activity of *pulav*	5, 7.5, 5, 10, 15, 25 kGy	• Sensory scores (appearance, color, odor, texture, taste, aftertaste, and overall acceptability) of 25 kGy gamma-irradiated vegetable *pulav* were lower compared to non-irradiated, but the samples were still acceptable after 12 months • It was also noted by the authors that irradiation at 7.5–15 kGy did not significantly alter the sensory quality	(Marathe, Deshpande, Tripathy, & Jamdar, 2017)
Noodles	• Investigation of the inhibition *Listeria innocua* inoculated in fresh noodles by electron-beam irradiation • Effects of electron-beam irradiation on sterilization and quality of fresh noodles during refrigerated storage	0.5, 1.0, 1.5, 2.0, 2.5, 3.0, 4.0, 5.0 (inoculated samples) 1.0, 2.0, 3.0, 4.0, 5.0 kGy (non-inoculated samples)	• Electron-beam irradiation did not alter the sensory properties of fresh noodles • The sensory scores of irradiated noodles were higher compared to those of non-irradiated during refrigerated storage up to 28 days	(Shi et al., 2020)

(Continued)

TABLE 11.2 (CONTINUED)   STUDIES EVALUATING THE EFFECTS OF IONIZING IRRADIATION ON THE SENSORY PROPERTIES OF GRAIN-BASED FOODS (CONDUCTED IN THE LAST TEN YEARS)

Product	Research Objective	Irradiation Dose	Key Findings	References
	• Investigation of the effects of water activity and irradiation on the sensory properties and shelf life of fresh noodles	0, 2, 4, 6, 8 kGy	• Irradiation at 4 kGy was recommended, as it was the dose that provided the best sensory scores	(Li, Zhu, Guo, Peng, & Zhou, 2011)
Semolina	• Effects of irradiation of wheat grains and semolina on the quality of semolina and its products	0.25, 1.00, 2.50, 5.00, 10.00 kGy	• Losses were noted in stickiness, firmness, and overall acceptability of lasagna samples prepared with irradiated semolina and semolina from irradiated wheat grains at doses generally above 0.25 kGy • These changes were more pronounced after six-month storage of the semolina samples	(Azzeh, Amr, & Alazzeh, 2011)

UV-C can denature the DNA of microorganisms by altering the cross-linking between neighboring pyrimidine nucleotide bases (thymine and cytosine) in the same DNA strand (Gómez-López, Koutchma, & Linden, 2012). The UV-absorption characteristics of the medium, the target microorganism's resistance to UV exposure, and the applied UV dose are the main factors affecting the amount of cell damage (Gayán, Condón, & Álvarez, 2014; Guerrero-Beltrán & Barbosa-Cánovas, 2004).

The efficiency of UV irradiation especially on liquid products such as fruit juices, beverages, milk, and liquid egg products is studied by many researchers (Ochoa-Velasco et al., 2020). Moreover, UV irradiation is an approved technology by the Food and Drug Administration (FDA) for juice pasteurization and surface microorganism control (FDA, 2000). Promising results have also been obtained in studies seeking to reduce the microbial load of fruits and vegetables, meat and meat products, powdered foods, and some bakery products by UV irradiation (Canto et al., 2019; Dogu-Baykut & Gunes, 2019; Fan, Huang, & Chen, 2017; Liu, Gong, Shao, & Qu, 2014).

11.5.1 Utilization of UV Irradiation during Processing of Grain-Based Foods

UV irradiation was found effective for microbial inactivation of flours. Condón-Abanto, Condón, Raso, Lyng, and Álvarez (2016) designed a 2-m vertical tunnel with twelve 480-W UV-C lamps to treat flour by forming a continuous cloud of dust (0.05–2.4 kg/h). They reported that UV treatment inactivated 4-log10 cycles of both *Salmonella typhimurium* and *Lactobacillus plantarum* in refined wheat flour. The authors also noted that the effectiveness of UV irradiation in flour was depended on the exposure of the flour grains to UV light. In another study, the potential of UV irradiation to modify wheat protein and thus to improve the baking characteristics of wheat flour is evaluated (Kumar, Nayak, Purohit, & Rao, 2021). Protein–protein and protein–starch interactions in wheat flour determine the suitability of the flour for baking and can be modified by various physicochemical methods. The authors indicated that UV irradiation is capable of altering the solubility of proteins by leading to conformational changes. According to their study, the viscosity of soft wheat flour was decreased after mild UV treatment and increased after severe UV treatment, but in hard wheat flour, it was increased steadily with UV irradiation time.

UV irradiation has also been applied to grain-based products such as bread, cake, chips, and beer to prolong shelf life and improve quality in some studies.

11.5.2 Effects of UV Irradiation on the Organoleptic Quality and Consumer Acceptance of Grain-Based Foods

Liu, Gong, Shao, and Qu (2014) evaluated the influence of UV irradiation on the shelf life and quality of steamed bread. They reported that a UV treatment of 7 s with 3 UV lamps from a 5-cm distance increased the shelf life of steamed bread from 20 h to 40 h without any obvious change in sensorial quality.

UV irradiation (with an intensity of 3.636 mJ/m²s) of cake batter for up to 2 h before baking resulted in increased volume and improved sensory properties (appearance, texture, taste, and odor), whereas further increasing the duration resulted in lower values (Konak, Certel, & Karakaş, 2017). UV treatment (30 W, 30 min) of naan (a kind of crusty pancake) before deoxygenation packaging increased the shelf life of naan from 12 days (control) to 23 days at 25°C. Moreover, the sensory evaluation showed that UV-treated naan had better sensory characteristics (color, smell, taste, and texture) compared to control naan. The authors claimed that UV can be effectively employed on naan for improving the overall qualities and extending shelf life (Zhao et al., 2019). Consumer acceptance of angel cakes prepared with UV-C-treated eggs was also evaluated by de Souza and Fernández (2012). Consumers were asked to score the degree of liking for angel cake prepared with UV-treated (4.2 J/cm^2), heat-treated (56.6°C, 3.5 min), or untreated eggs in acceptance tests. Color, flavor, aroma, sponginess, humidity, appearance, and overall acceptability scores of all cakes containing UV-treated eggs were not significantly different, and no off-flavors due to UV-C treatments were reported.

The combination of UV (30 W intensity at a distance of 17 cm, 24 h exposure) and infrared (IR) (30 mW/cm^2 intensity for 1.5 h at a distance of 15 cm) treatments has been studied to reduce antinutritional factors of soybean snack chips by Maetens, Hettiarachchy, Dewettinck, Horax, and Moens (2018). Chips prepared from the UV- and IR-treated soybeans had significantly lower antinutritional factors due to the high reduction in lipoxygenases and trypsin inhibitor activities. However, these chips had lower scores in sensory analysis, particularly for their flavor, aroma, and overall impression. The authors suggested that the use of flavoring and spices may have the potential to increase the consumer acceptability of soybean chips.

Lu et al. (2010) tested UV (up to 16.1 mJ/cm^2) for its potential to inactivate bacteria in beer. They reported that UV significantly reduced inoculated *Escherichia coli* and *Lactobacillus brevis* and naturally occurring lactic acid bacteria and *Enterobacteriaceae*, while the reduction of *S. cerevisiae* was not as efficient. In another study, beer samples were tested with a triangle tasting 24 h after UV treatment at doses between 100 and 400 J/L. For all the UV dose levels tasted, panelists distinguished the differences in the beer samples. Typical off-flavors, including lightstruck, burnt rubber, catty, and oxidation were detected in UV-treated samples (Mezui & Swart, 2010).

11.6 PULSED ELECTRIC FIELDS

In pulsed electric field (PEF) treatment, the food product is placed between two electrodes and exposed to high voltage with short pulses. The strength of the electric fields generally ranges between 10–80 kV/cm (Barbosa-Canovas & Altunakar, 2006). A PEF system mainly includes a high-voltage pulse generator, a treatment chamber, a fluid-handling system, a control, and a monitoring device (Nowosad et al., 2021). PEF treatment can be applied to only fluid food products, and the fluids under treatment need to be homogeneous (Morales-de la Peña, Welti-Chanes, & Martín-Belloso, 2019). Electric field strength, treatment time, pulse shape, pulse width, pulse frequency, pulse polarity, temperature, the characteristics of the microorganism, and the medium are the most important factors for the efficiency of the treatment (Morales-de la Peña et al., 2019).

Microbial inactivation due to PEF depends on the formation of pores in the cell membrane, electroporation, and electric breakdown that causes the leakage of the intracellular matrix. The efficiency of PEF against bacterial spores is limited (Reineke & Mathys, 2020). However, combining PEF with heat treatment and other methods gives promising results (Morales-de la Peña et al., 2019; Reineke & Mathys, 2020; Schweiggert, Carle, & Schieber, 2007).

Even though PEF is considered a non-thermal process, a temperature increase in the product is still possible, thus cooling is required during processing (Lindgren, Aronsson, Galt, & Ohlsson, 2002). The other disadvantage of PEF is the corrosion of the electrodes, as some metal ions can migrate into the food during treatment (Nowosad et al., 2021).

11.6.1 Utilization of PEF during Processing of Grain-Based Foods

PEF can be utilized in a wide range of applications in the food industry, such as drying, freezing, extraction, starch modification, waste valorization, and inactivation of microorganisms, spores, and enzymes (Nowosad et al., 2021). This technology has generally been applied to cereal-based foods in liquid forms, such as beer and rice wine. It is also used for starch and starch-based products.

For instance, Ratanatriwong, Evrendilek, and Zhang (2008) claimed that the combination of PEF with mild heat treatment at 55°C increased the shelf life of a rice pudding product without affecting color and pH values.

Also, changes in the shape of corn starch granules after PEF was reported by Han et al. (2009). They also observed decreases in the peak, breakdown, and final viscosity and the gelatinization temperature of the starch when the electric field strength was increased.

Moreover, Huang et al. (2013) reported that the PEF treatment of Chinese rice wine was effective against *S. cerevisiae*. The treatment led to some alterations in physicochemical properties and improved some quality parameters. The authors also noted that PEF can cause a tailing effect that some of the cells cannot be inactivated.

It was also reported by Zeng et al. (2016) that PEF caused changes in granular packaging, crystalline structure, thermal properties, and digestibility of waxy rice starch. The PEF treatment increased the amounts of rapidly digestible starch and slowly digestible starch while decreasing that of resistant starch in waxy rice starch.

Moreover, Hong et al. (2020) reported that the modification of maize starch by PEF-assisted acetylation improved the texture of noodle samples. They also noted that the digestibility of the noodles prepared with PEF-treated waxy starch was lower compared to that of the samples prepared with native waxy starch.

11.6.2 Effects of PEF on the Organoleptic Quality and Consumer Acceptance of Grain-Based Foods

PEF treatment has been applied for beer pasteurization in several studies. For instance, Evrendilek et al. (2004) evaluated the effects of PEF treatment on the inactivation of microorganisms, sensory properties of beer, and release of metal ions from the electrode. They observed that PEF was efficient for microbial inactivation; however, a significant transfer of Fe, Cr, Zn, and Mn ions to beer from the electrode and losses in flavor and mouth-feel were encountered due to the treatment.

In addition, Elham Alami Milani, Alkhafaji, and Silva (2015) claimed that PEF resulted in a reduction in *S. cerevisiae* ascospores in nine different beers and that the reduction was higher when the PEF treatment was combined with heat treatment. However, PEF treatment caused some changes in aroma, flavor, and lightstruck in some of the beer samples.

It was also reported that the inactivation effect of PEF against *S. cerevisiae* in Chinese rice wine was higher than that of thermosonication (Lyu, Huang, Yang, Wang, & Wang, 2016). The authors also noted that thermosonication and PEF demonstrated an additive effect against *S. cerevisae* when used in combination.

Moreover, Mihindukulasuriya and Jayaram (2020) investigated the effects of PEF on the release of electrode metal ions, organoleptic properties, and shelf life of beer. They reported that there was no Fe ion transfer from the electrode and that the Ti ion transfer was lower than 2 ppm.

The sensory properties of PEF-treated samples were generally better compared to untreated beer samples. The PEF treatment increased the shelf life of beer to 90 days.

11.7 COLD PLASMA

Plasma is defined as the fourth state of the matter. It can also be regarded as an ionized gas that is composed of electrons, positive or negative ions, and neutral molecules (Niemira, 2012). Gases can be ionized to generate plasma by the application of heat, electricity, laser light, radiation, or compression (Niemira, 2012; Sonawane, Marar, & Patil, 2020). It can also be defined depending on how it is generated, such as by dielectric barrier discharge, plasma jet, uniform glow discharge plasma, gliding arc discharge, corona discharge plasma jet, etc. (Niemira, 2012; Sonawane et al., 2020). It is possible to use several types of gases in the generation of plasma, including air, N_2, Ar, He, a mixture of gases, etc. (Niemira, 2012; Sonawane et al., 2020). In addition, plasma can be classified as low pressure, atmospheric pressure, and high pressure depending on the pressure level (Sonawane et al., 2020).

In cold (or non-thermal) plasma technology, thermal degradation is minimized. The microbial inactivation mechanism of the technology depends on the effects of the radicals and reactive and charged species and the UV generated during the treatment (Niemira, 2012).

Apart from microbial decontamination, cold plasma can also be used for inactivating anti-nutrients, decreasing cooking time, improving ultrafiltration and germination, and modifying starch in the food industry (Sonawane et al., 2020).

11.7.1 Applications of Cold Plasma on Grain-Based Foods and Effects on Sensory Properties

Plasma treatments were used for microbial inactivation in several cereal-based food products, such as rice cereal bars (Suhem, Matan, Nisoa, & Matan, 2013) and wheat (Iqbal et al., 2019; Los, Ziuzina, Boehm, & Bourke, 2020; Thomas-Popo et al., 2019). It can also be used for the improvement of germination and functional properties and the fortification of cereal-based foods, as well as starch modification (Akasapu, Ojah, Gupta, Choudhury, & Mishra, 2020; Chaple et al., 2020; Okyere, Bertoft, & Annor, 2019; Thirumdas, Deshmukh, & Annapure, 2016; Thirumdas, Trimukhe, Deshmukh, & Annapure, 2017; Yodpitak et al., 2019).

Plasma treatment of brown rice decreased cooking time and increased softness (Chen, Chen, & Chang, 2012; Thirumdas, Saragapani, Ajinkya, Deshmukh, & Annapure, 2016). In addition, Chen (2014) also observed a reduction in the cooking time and cooking loss of brown rice after low-pressure plasma treatment. Similarly, a decrease in the cooking time and hardness of brown rice was reported in another study (Chen, Hung, Lin, & Liou, 2015). The authors added that the treatment also decreased fat acidity and the activity of α-amylase and lipoxygenase enzymes during storage.

Plasma treatment can also be used for the improvement of the properties of surfaces of polymers, such as food packaging material (Ozdemir, Yurteri, & Sadikoglu, 1999; Pankaj et al., 2014). This can be advantageous in food processing as well. For instance, Misra et al. (2014) utilized non-thermal dielectric barrier discharge plasma treatment for enhancing the oil spreadability of biscuit surfaces. Moreover, when it was used for the modification of starch films, a decrease in oxygen permeability and an increase in tensile strength were observed (Sheikhi et al., 2020).

Campelo et al. (2020) claimed that cold plasma can affect the volatile composition, and thus it can be used for the modification of aroma and flavor. Although the number of studies evaluating the effects of non-thermal plasma on the sensory quality of grain-based foods is quite limited, in a recent study by Park, Puligundla, and Mok (2020) the effects of corona discharge plasma treatment on biochemical and sensory properties of rice grain seedlings and aqueous tea infusions were investigated. They observed that the plasma treatment inactivated the microbial load and did not alter the germination rate but improved the weight and length of the seedlings. The treatment increased α- and β-amylase activities during germination of up to three days. Total phenolic content and antioxidant activity were improved in germinated and ungerminated grains and tea infusions. According to the sensory evaluation, no effects of the cold plasma treatment of rice grains were observed on the turbidity, color, and flavor of the aqueous tea infusions.

11.8 OZONATION

Ozone gas is found in the troposphere, and it has a strong oxidizing and disinfecting ability (Guzel-Seydim, Greene, & Seydim, 2004). The microbial inactivation mechanism of ozone can be explained mainly by the oxidation of proteins, enzymes, and lipids of the cell envelope (Guzel-Seydim et al., 2004; Pandiselvam, Sunoj, Manikantan, Kothakota, & Hebbar, 2017; Victorin, 1992). Ozone can also cause cell death by affecting the nucleic acids (Guzel-Seydim et al., 2004; Pandiselvam et al., 2017).

It has been used for the decontamination of water, fruits and vegetables, spices, and equipment (Guzel-Seydim et al., 2004). It can also be used for reducing the mycotoxin levels in food products (Afsah-Hejri, Hajeb, & Ehsani, 2020; Diao, Hou, & Dong, 2013).

11.8.1 Applications of Ozonation on Grain-Based Foods and Effects on Sensory Properties

Ozone treatment can be utilized in the milling of grains and also can demonstrate positive effects on dough strength, bread volume, shelf life, and modification of starch and proteins (Zhu, 2018). Moreover, ozonation can be helpful in decreasing mycotoxin levels and microbial load in cereals (Alexandre et al., 2019; Luo et al., 2014; Qi et al., 2016; Wang et al., 2016). In addition, ozone has been applied in grain storage for controlling insects (Pandiselvam et al., 2017).

The results of several promising studies indicate that ozone treatments have the potential to improve the sensory quality of grain-based products. For instance, Li et al. (2013) reported that the ozonation of wheat flour inhibited polyphenol oxidase and increased L* value, and the darkening rate of noodles made from the treated flour was decreased while shelf life was increased.

It was reported that the quality of wheat was not altered by ozone treatment (10–60 mg/L for 2–5 h) (Trombete et al., 2016). The treatment did not negatively affect the technological properties (hectoliter weight, extraction rate, wet and dry gluten content, gluten index, dough strength, and extensibility) or chemical and mineral compositions. No changes in the aroma or overall appearance of the flour were detected according to the sensory tests. In addition, ozonation treatment of wheat grains did not cause any color changes in the flour. Furthermore, Mei, Liu, Huang, and Ding (2016) reported that ozonation of wheat flour for 1 h increased the sensory quality scores of steamed bread.

On the contrary, Chittrakorn, Earls, and MacRitchie (2014) reported that ozonation of wheat flour in place of chlorination for up to 36 min increased the cake volume, but it also caused an unpleasant odor to develop.

11.9 CONCLUSIONS

Non-thermal processing techniques are lethal to most microorganisms and can be applied to ensure food safety. Moreover, these techniques can be also used to improve product quality. Their potential for each food type should be taken into account in order to optimize their effects on pathogens and spoilage. Many studies have focused on their potential for microbial decontamination and pathogen reduction. But in many of these studies, the impact of the technology on organoleptic quality and consumer acceptability is not addressed.

For grain-based food products, there has been limited consideration given to organoleptic acceptability of the product and consumer acceptability of the non-thermal technologies. According to reviewed articles, HPP, ultrasound, irradiation (gamma and electron beam), and cold plasma treatments generally do not adversely affect sensory properties depending on the level of treatment dose, intensity, and duration. UV irradiation can extend the shelf life of products such as flour, cake, bread, and chips without affecting the sensory properties of the product, but its use in brewing has been reported to cause off-flavor. Similarly, PEF and ozonation may cause some alterations in the sensory properties of some products.

For most non-thermal technologies, there is a limited number of studies about the effect of these technologies on the sensorial properties and organoleptic characteristics of grain-based food products. Further studies are needed to better understand their impact on organoleptic characteristics and consumer acceptability.

REFERENCES

Afsah-Hejri, L., Hajeb, P., & Ehsani, R. J. (2020). Application of ozone for degradation of mycotoxins in food: A review. *Comprehensive Reviews in Food Science and Food Safety, 19*(4), 1777–1808. doi:10.1111/1541-4337.12594

Agúndez-Arvizu, Z., Fernández-Ramírez, M. V., Arce-Corrales, M. E., Cruz-Zaragoza, E., Meléndrez, R., Chernov, V., & Barboza-Flores, M. (2006). Gamma radiation effects on commercial Mexican bread making wheat flour. *Nuclear Instruments and Methods in Physics Research, Section B: Beam Interactions with Materials and Atoms, 245*(2), 455–458. doi:10.1016/j.nimb.2005.11.141

Akasapu, K., Ojah, N., Gupta, A. K., Choudhury, A. J., & Mishra, P. (2020). An innovative approach for iron fortification of rice using cold plasma. *Food Research International, 136,* 109599. doi:10.1016/j.foodres.2020.109599

Alexandre, A. P. S., Castanha, N., Costa, N. S., Santos, A. S., Badiale-Furlong, E., Augusto, P. E. D., & Calori-Domingues, M. A. (2019). Ozone technology to reduce zearalenone contamination in whole maize flour: Degradation kinetics and impact on quality. *Journal of the Science of Food and Agriculture, 99*(15), 6814–6821. doi:10.1002/jsfa.9966

Andrews, L. S., Ahmedna, M., Grodner, R. M., Liuzzo, J. A., Murano, P. S., Murano, E. A., ... Wilson, P. W. (1998). Food preservation using ionizing radiation. In: G. W. Ware (Ed.), *Reviews of Environmental Contamination and Toxicology Vol. 154.* New York: Springer. doi:10.1007/978-1-4612-2208-8_1

Angioloni, A., & Collar, C. (2012a). Effects of pressure treatment of hydrated oat, finger millet and sorghum flours on the quality and nutritional properties of composite wheat breads. *Journal of Cereal Science, 56*(3), 713–719. doi:10.1016/j.jcs.2012.08.001

Angioloni, A., & Collar, C. (2012b). Promoting dough viscoelastic structure in composite cereal matrices by high hydrostatic pressure. *Journal of Food Engineering, 111*(4), 598–605. doi:10.1016/j.jfoodeng.2012.03.010

Anwar, M. M., Asael, M. A., & El-Adly, N. A. (2015). Gamma radiation influence on rheological and technological characteristics of wheat flour (misr-1) and sensory properties of pan bread. *Journal of Nuclear Technology in Applied Science, 3*(2), 89–97.

Awad, T. S., Moharram, H. A., Shaltout, O. E., Asker, D., & Youssef, M. M. (2012). Applications of ultrasound in analysis, processing and quality control of food: A review. *Food Research International, 48*(2), 410–427. doi:10.1016/j.foodres.2012.05.004

Aziz, N. H., Souzan, R. M., & Shahin Azza, A. (2006). Effect of γ-irradiation on the occurrence of pathogenic microorganisms and nutritive value of four principal cereal grains. *Applied Radiation and Isotopes, 64*(12), 1555–1562. doi:10.1016/j.apradiso.2005.10.006

Azzeh, F. S., & Amr, A. S. (2009). Effect of gamma irradiation on physical characteristics of Jordanian durum wheat and quality of semolina and lasagna products. *Radiation Physics and Chemistry, 78*(9), 818–822. doi:10.1016/j.radphyschem.2009.04.003

Azzeh, F. S., Amr, A. S., & Alazzeh, A. Y. (2011). Disinfestation of semolina and wheat grains using gamma irradiation and its effect before and after milling of Jordanian durum wheat on semolina and lasagna quality. *Journal of Food Processing and Preservation, 35*(5), 656–664. doi:10.1111/j.1745-4549.2011.00515.x

Bao, J., Shu, Q., Xia, Y., Bergman, C., & McClung, A. (2001). Effects of gamma irradiation on aspects of milled rice (*Oryza sativa*) end-use quality. *Journal of Food Quality, 24*(4), 327–336. doi:10.1111/j.1745-4557.2001.tb00612.x

Barbosa-Canovas, G. V., & Altunakar, B. (2006). Pulsed electric fields processing of foods: An overview. In: J. Raso & V. Heinz (Eds.), *Pulsed Electric Fields Technology for the Food Industry: Fundamentals and Applications* (pp. 3–26). New York: Springer. doi:10.1007/978-0-387-31122-7

Barbosa-Cánovas, G. V., Góngora-Nieto, M. M., Rodriguez, J. J., & Swanson, B. G. (2005). Nonthermal processing of foods and emerging technologies. In: G. V. Barbosa-Cánovas (Ed.), *Food Engineering Vol. 3* (pp. 281–310). Paris: EOLSS Publishers/UNESCO.

Bárcenas, M. E., Altamirano-Fortoul, R., & Rosell, C. M. (2010). Effect of high pressure processing on wheat dough and bread characteristics. *LWT – Food Science and Technology, 43*(1), 12–19. doi:10.1016/j.lwt.2009.06.019

Bashir, K., & Aggarwal, M. (2019). Physicochemical, structural and functional properties of native and irradiated starch: A review. *Journal of Food Science and Technology-Mysore, 56*(2), 513–523.

Bermudez-Aguirre, D., Mobbs, T., & Barbosa-Canovas, G. V. (2011). Ultrasound applications in food processing. In: H. Feng, G. V. Barbosa-Canovas, & J. Weiss (Eds.), *Ultrasound Technologies for Food and Bioprocessing* (pp. 65–105). New York: Springer. doi:10.1007/978-1-4419-7472-3_3

Blandino, A., Al-Aseeri, M. E., Pandiella, S. S., Cantero, D., & Webb, C. (2003). Cereal-based fermented foods and beverages. *Food Research International, 36*(6), 527–543. doi:10.1016/S0963-9969(03)00009-7

Boluda-Aguilar, M., Taboada-Rodríguez, A., López-Gómez, A., Marín-Iniesta, F., & Barbosa-Cánovas, G. V. (2013). Quick cooking rice by high hydrostatic pressure processing. *LWT – Food Science and Technology, 51*(1), 196–204. doi:10.1016/j.lwt.2012.09.021

Bonto, A. P., Jearanaikoon, N., Sreenivasulu, N., & Camacho, D. H. (2020). High uptake and inward diffusion of iron fortificant in ultrasonicated milled rice. *LWT, 128*. doi:10.1016/j.lwt.2020.109459

Buzrul, S. (2012). High hydrostatic pressure treatment of beer and wine: A review. *Innovative Food Science and Emerging Technologies, 13*, 1–12. doi:10.1016/j.ifset.2011.10.001

Buzrul, S., Alpas, H., & Bozoglu, F. (2005). Effect of high hydrostatic pressure on quality parameters of lager beer. *Journal of the Science of Food and Agriculture, 85*(10), 1672–1676. doi:10.1002/jsfa.2166

Campelo, P. H., Alves Filho, E. G., Silva, L. M. A., de Brito, E. S., Rodrigues, S., & Fernandes, F. A. N. (2020). Modulation of aroma and flavor using glow discharge plasma technology. *Innovative Food Science and Emerging Technologies, 62*, 102363. doi:10.1016/j.ifset.2020.102363

Canto, A. C. V. da C. S., Monteiro, M. L. G., Costa-Lima, B. R. C. da, Lázaro, C. A., Marsico, E. T., Silva, T. J. P. da, & Conte-Junior, C. A. (2019). Effect of UV-C radiation on *Salmonella* spp. reduction and oxidative stability of caiman (*Caiman crocodilus yacare*) meat. *Journal of Food Safety, 39*(2), e12604. doi:10.1111/jfs.12604

Carcea, M. (2020). Nutritional value of grain-based foods. *Foods, 9*(4), 504. doi:10.3390/foods9040504

Chang, A. C. (2003). The effects of gamma irradiation on rice wine maturation. *Food Chemistry, 83*(3), 323–327. doi:10.1016/S0308-8146(03)00050-5

Chang, A. C. (2005). Study of ultrasonic wave treatments for accelerating the aging process in a rice alcoholic beverage. *Food Chemistry, 92*(2), 337–342. doi:10.1016/j.foodchem.2004.07.027

Chaple, S., Sarangapani, C., Jones, J., Carey, E., Causeret, L., Genson, A., ... Bourke, P. (2020). Effect of atmospheric cold plasma on the functional properties of whole wheat (*Triticum aestivum* L.) grain and wheat flour. *Innovative Food Science and Emerging Technologies, 66*, 102529. doi:10.1016/j.ifset.2020.102529.

Chemat, F., Zill-e-Huma, & Khan, M. K. (2011). Applications of ultrasound in food technology: Processing, preservation and extraction. *Ultrasonics Sonochemistry, 18*(4)(4, SI), 813–835. doi:10.1016/j.ultsonch.2010.11.023

Chen, H. H. (2014). Investigation of properties of long-grain brown rice treated by low-pressure plasma. *Food and Bioprocess Technology, 7*(9), 2484–2491. doi:10.1007/s11947-013-1217-2

Chen, H. H., Chen, Y. K., & Chang, H. C. (2012). Evaluation of physicochemical properties of plasma treated brown rice. *Food Chemistry, 135*(1), 74–79. doi:10.1016/j.foodchem.2012.04.092

Chen, H. H., Hung, C. L., Lin, S. Y., & Liou, G. J. (2015). Effect of low-pressure plasma exposure on the storage characteristics of brown rice. *Food and Bioprocess Technology, 8*(2), 471–477. doi:10.1007/s11947-014-1415-6

Chen, Y., Jiang, W., Jiang, Z., Chen, X., Cao, J., Dong, W., & Dai, B. (2015). Changes in physicochemical, structural, and sensory properties of irradiated brown japonica rice during storage. *Journal of Agricultural and Food Chemistry, 63*(17), 4361–4369. doi:10.1021/jf5047514

Chittrakorn, S., Earls, D., & MacRitchie, F. (2014). Ozonation as an alternative to chlorination for soft wheat flours. *Journal of Cereal Science, 60*(1), 217–221. doi:10.1016/j.jcs.2014.02.013

Collar, C., & Angioloni, A. (2017). High-legume wheat-based matrices: Impact of high pressure on starch hydrolysis and firming kinetics of composite breads. *Food and Bioprocess Technology, 10*(6), 1103–1112. doi:10.1007/s11947-017-1883-6

Condón-Abanto, S., Condón, S., Raso, J., Lyng, J. G., & Álvarez, I. (2016). Inactivation of *Salmonella* typhimurium and *Lactobacillus plantarum* by UV-C light in flour powder. *Innovative Food Science and Emerging Technologies, 35*, 1–8. doi:10.1016/j.ifset.2016.03.008

Crehan, C. M., Troy, D. J., & Buckley, D. J. (2000). Effects of salt level and high hydrostatic pressure processing on frankfurters formulated with 1.5 and 2.5% salt. *Meat Science, 55*(1), 123–130. doi:10.1016/S0309-1740(99)00134-5

de Souza, P. M., & Fernández, A. (2012). Consumer acceptance of UV-C treated liquid egg products and preparations with UV-C treated eggs. *Innovative Food Science and Emerging Technologies, 14*, 107–114. doi:10.1016/j.ifset.2011.12.005

Deng, Y., Zhong, Y., Yu, W., Yue, J., Liu, Z., Zheng, Y., & Zhao, Y. (2013). Effect of hydrostatic high pressure pretreatment on flavor volatile profile of cooked rice. *Journal of Cereal Science, 58*(3), 479–487. doi:10.1016/j.jcs.2013.09.010

Diao, E., Hou, H., & Dong, H. (2013). Ozonolysis mechanism and influencing factors of aflatoxin B1: A review. *Trends in Food Science and Technology, 33*(1), 21–26. doi:10.1016/j.tifs.2013.06.002

Ding, J., & Feng, H. (2018). Controlled germination for enhancing the nutritional value of sprouted grains. In: H. Feng, B. Nemzer, & J. W. DeVries (Eds.), *Sprouted Grains: Nutritional Value, Production, and Applications* (pp. 91–112). Washington: AACC International Press. doi:10.1016/B978-0-12-811525-1.00005-1

Dogu-Baykut, E., & Gunes, G. (2019). Ultraviolet (UV-C) radiation as a practical alternative to decontaminate thyme (*Thymus vulgaris* L.). *Journal of Food Processing and Preservation, 43*(6), e13842. doi:10.1111/jfpp.13842

Du, Z., Xing, J., Luo, X., Wang, L., Pan, L., Li, Y., … Chen, Z. (2019). Characterization of the physical properties of electron-beam-irradiated white rice and starch during short-term storage. *PLOS ONE*, *14*(12), e0226633. doi:10.1371/journal.pone.0226633

Esmaeilzadeh Kenari, R., & Nemati, A. (2020). The effectiveness of ultrasound bath and probe treatments on the quality of baking and shelf life of cupcakes. *Food Science and Nutrition*, *8*(6), 2929–2939. doi:10.1002/fsn3.1595

Evrendilek, G. A., Li, S., Dantzer, W. R., & Zhang, Q. H. (2004). Pulsed electric field processing of beer: Microbial, sensory, and quality analyses. *Journal of Food Science*, *69*(8), M228–M232. doi:10.1111/j.1365-2621.2004.tb09892.x

Fan, X., Huang, R., & Chen, H. (2017). Application of ultraviolet C technology for surface decontamination of fresh produce. *Trends in Food Science and Technology*, *70*, 9–19. doi:10.1016/j.tifs.2017.10.004

Fan, X., & Niemira, B. A. (2020). Gamma ray, electron beam, and X-ray irradiation. In: A. Demirci, H. Feng, & K. Krishnamurthy (Eds.), *Food Safety Engineering* (pp. 471–492). New York: Springer. doi:10.1007/978-3-030-42660-6_18

Farkas, D. F., & Hoover, D. G. (2000). High pressure processing. *Journal of Food Science*, *65*(s8), 47–64. doi:10.1111/j.1750-3841.2000.tb00618.x

FDA. (2000). 21 CFR part 179: Irradiation in the production, processing and handling of food. *Federal Register*, *65*, 71056–71058.

Fischer, S., Russ, W., & Meyer-Pittroff, R. (2002). High pressure advantages for brewery processes. In: R. Hayashi (Ed.), *Trends in High Pressure Bioscience and Biotechnology* (pp. 397–404). Amsterdam: Elsevier. doi:10.1016/S0921-0423(02)80130-7

Gao, S., Lewis, G. D., Ashokkumar, M., & Hemar, Y. (2014). Inactivation of microorganisms by low-frequency high-power ultrasound: 2. A simple model for the inactivation mechanism. *Ultrasonics Sonochemistry*, *21*(1), 454–460. doi:10.1016/j.ultsonch.2013.06.007

Gayán, E., Condón, S., & Álvarez, I. (2014). Biological aspects in food preservation by ultraviolet light: A review. *Food and Bioprocess Technology*, *7*(1), 1–20. doi:10.1007/s11947-013-1168-7

Gómez-López, V. M., Koutchma, T., & Linden, K. (2012). Ultraviolet and pulsed light processing of fluid foods. In: P. C. Cullen, B. K. Tiwari, & V. P. Valdramidis (Eds.), *Novel Thermal and Non-Thermal Technologies for Fluid Foods* (pp. 185–223). New York: Elsevier. doi:10.1016/B978-0-12-381470-8.00008-6

Guerrero-Beltrán, J. A., & Barbosa-Cánovas, G. V. (2004). Advantages and limitations on processing foods by UV light. *Food Science and Technology International*, *10*(3), 137–147. doi:10.1177/1082013204044359

Guzel-Seydim, Z. B., Greene, A. K., & Seydim, A. C. (2004). Use of ozone in the food industry. *LWT – Food Science and Technology*, *37*(4), 453–460. doi:10.1016/j.lwt.2003.10.014

Han, Z., Zeng, X., Zhang, B., & Yu, S. (2009). Effects of pulsed electric fields (PEF) treatment on the properties of corn starch. *Journal of Food Engineering*, *93*(3), 318–323. doi:10.1016/j.jfoodeng.2009.01.040

Heinz, V., & Buckow, R. (2010). Food preservation by high pressure. *Journal für Verbraucherschutz und Lebensmittelsicherheit*, *5*(1), 73–81. doi:10.1007/s00003-009-0311-x

Hong, J., An, D., Liu, C., Li, L., Han, Z., Guan, E., … Bian, K. (2020). Rheological, textural, and digestible properties of fresh noodles: Influence of starch esterified by conventional and pulsed electric field-assisted dual technique with full range of amylose content. *Journal of Food Processing and Preservation*, *44*(8), e14567. doi:10.1111/jfpp.14567

Huang, K., Yu, L., Liu, D., Gai, L., & Wang, J. (2013). Modeling of yeast inactivation of PEF-treated Chinese rice wine: Effects of electric field intensity, treatment time and initial temperature. *Food Research International*, *54*(1), 456–467. doi:10.1016/j.foodres.2013.07.046

Hüttner, E. K., Bello, F. D., & Arendt, E. K. (2010). Fundamental study on the effect of hydrostatic pressure treatment on the bread-making performance of oat flour. *European Food Research and Technology*, *230*(6), 827–835. doi:10.1007/s00217-010-1228-4

Iqbal, T., Farooq, M., Afsheen, S., Abrar, M., Yousaf, M., & Ijaz, M. (2019). Cold plasma treatment and laser irradiation of *Triticum* spp. seeds for sterilization and germination. *Journal of Laser Applications*, *31*(4), 42013. doi:10.2351/1.5109764

Jalali, M., Sheikholeslami, Z., Elhamirad, A. H., Haddad Khodaparast, M. H., & Karimi, M. (2020). The effect of the ultrasound process and pre-gelatinization of the corn flour on the textural, visual, and sensory properties in gluten-free pan bread. *Journal of Food Science and Technology*, *57*(3), 993–1002. doi:10.1007/s13197-019-04132-7

Janve, B., Yang, W., Kozman, A., Sims, C., Teixeira, A., Gunderson, M. A., & Rababah, T. M. (2013). Enhancement of corn nixtamalization by power ultrasound. *Food and Bioprocess Technology, 6*(5), 1269–1280. doi:10.1007/s11947-012-0816-7

Janve, B., Yang, W., & Sims, C. (2015). Sensory and quality evaluation of traditional compared with power ultrasound processed corn (*Zea mays*) tortilla chips. *Journal of Food Science, 80*(6), S1368–S1376. doi:10.1111/1750-3841.12892

Kahdre, M. A., & Yousef, A. E. (2002). Susceptibility of human rotavirus to ozone, high pressure, and pulsed electric field. *Journal of Food Protection, 65*(9), 1441–1446. doi:10.4315/0362-028X-65.9.1441

Kim, H.-S., Kim, B.-Y., & Baik, M.-Y. (2012). Application of ultra high pressure (UHP) in starch chemistry. *Critical Reviews in Food Science and Nutrition, 52*(2), 123–141. doi:10.1080/10408398.2010.498065

Knorr, D., Heinz, V., & Buckow, R. (2006). High pressure application for food biopolymers. *Biochimica et Biophysica Acta (BBA) – Proteins and Proteomics, 3*(3), 619–631. doi:10.1016/j.bbapap.2006.01.017

Knorr, D., Zenker, M., Heinz, V., & Lee, D. U. (2004). Applications and potential of ultrasonics in food processing. *Trends in Food Science and Technology, 15*(5), 261–266. doi:10.1016/j.tifs.2003.12.001

Konak, Ü. I., Certel, M., & Karakaş, B. (2017). Effects of ultraviolet irradiation of batter and different baking methods on cake quality. *Journal of Microbiology, Biotechnology and Food Sciences, 7*(2), 113–117. doi:10.15414/jmbfs.2017.7.2.113-117

Koutchma, T. N. (2019). *Ultraviolet Light in Food Technology: Principles and Applications.* Boca Raton: CRC Press.

Kumar, A., Nayak, R., Purohit, S. R., & Rao, P. S. (2021). Impact of UV-C irradiation on solubility of Osborne protein fractions in wheat flour. *Food Hydrocolloids, 110*, 105845. doi:10.1016/j.foodhyd.2020.105845.

Lacroix, M., & Ouattara, B. (2000). Combined industrial processes with irradiation to assure innocuity and preservation of food products — A review. *Food Research International, 33*(9), 719–724. doi:10.1016/S0963-9969(00)00085-5

Li, M., Peng, J., Zhu, K.-X., Guo, X.-N., Zhang, M., Peng, W., & Zhou, H.-M. (2013). Delineating the microbial and physical–chemical changes during storage of ozone treated wheat flour. *Innovative Food Science and Emerging Technologies, 20*, 223–229. doi:10.1016/j.ifset.2013.06.004

Li, M., Zhu, K., Guo, X., Peng, W., & Zhou, H. (2011). Effect of water activity (a_w) and irradiation on the shelf-life of fresh noodles. *Innovative Food Science and Emerging Technologies, 12*(4), 526–530. doi:10.1016/j.ifset.2011.06.005

Lindgren, M., Aronsson, K., Galt, S., & Ohlsson, T. (2002). Simulation of the temperature increase in pulsed electric field (PEF) continuous flow treatment chambers. *Innovative Food Science and Emerging Technologies, 3*(3), 233–245. doi:10.1016/S1466-8564(02)00044-9

Liu, C., Gong, S., Shao, Y., & Qu, L. (2014). Effects of ultraviolet irradiation treatment on the quality and shelf life of steamed bread. *Journal of Food Safety and Quality, 5*(8), 2416–2421.

Liu, H., Fan, H., Cao, R., Blanchard, C., & Wang, M. (2016). Physicochemical properties and *in vitro* digestibility of sorghum starch altered by high hydrostatic pressure. *International Journal of Biological Macromolecules, 92*, 753–760. doi:10.1016/j.ijbiomac.2016.07.088

Los, A., Ziuzina, D., Boehm, D., & Bourke, P. (2020). Effects of cold plasma on wheat grain microbiome and antimicrobial efficacy against challenge pathogens and their resistance. *International Journal of Food Microbiology, 335*, 108889. doi:10.1016/j.ijfoodmicro.2020.108889

Lu, G., Li, C., Liu, P., Cui, H., Yao, Y., & Zhang, Q. (2010). UV inactivation of microorganisms in beer by a novel thin-film apparatus. *Food Control, 21*(10), 1312–1317. doi:10.1016/j.foodcont.2010.03.007

Luo, D., Wu, R., Zhang, J., Zhang, K., Xu, B., Li, P., ... Li, X. (2018). Effects of ultrasound assisted dough fermentation on the quality of steamed bread. *Journal of Cereal Science, 83*, 147–152. doi:10.1016/j.jcs.2018.07.016

Luo, X., Wang, R., Wang, L., Li, Y., Bian, Y., & Chen, Z. (2014). Effect of ozone treatment on aflatoxin B1 and safety evaluation of ozonized corn. *Food Control, 37*, 171–176. doi:10.1016/j.foodcont.2013.09.043

Lyu, C., Huang, K., Yang, N., Wang, H., & Wang, J. (2016). Combination of thermosonication and pulsed electric fields treatments for controlling *Saccharomyces cerevisiae* in Chinese rice wine. *Food and Bioprocess Technology, 9*(11), 1854–1864. doi:10.1007/s11947-016-1769-z

Maetens, E., Hettiarachchy, N., Dewettinck, K., Horax, R., & Moens, K. (2018). Reductions of anti-nutritional factors of germinated soybeans by ultraviolet and infrared treatments for snack chips preparation. *LWT – Food Science and Technology, 90*, 513–518. doi:10.1016/j.lwt.2018.01.001

Mañas, P., & Pagán, R. (2005). Microbial inactivation by new technologies of food preservation. *Journal of Applied Microbiology, 98*(6), 1387–1399. doi:10.1111/j.1365-2672.2005.02561.x

Marathe, S., Deshpande, R., Tripathy, J., & Jamdar, S. N. (2017). Development of shelf stable ready-to-eat vegetable *pulav* using radiation technology. *Journal of Food Processing and Preservation, 41*(5), e13104. doi:10.1111/jfpp.13104

Marathe, S. A., Machaiah, J. P., Rao, B. Y. K., Pednekar, M. D., & Rao, V. S. (2002). Extension of shelf-life of whole-wheat flour by gamma radiation. *International Journal of Food Science and Technology, 37*(2), 163–168. doi:10.1046/j.1365-2621.2002.00553.x

McCann, T. H., Leder, A., Buckow, R., & Day, L. (2013). Modification of structure and mixing properties of wheat flour through high-pressure processing. *Food Research International, 53*(1), 352–361. doi:10.1016/j.foodres.2013.04.030

Mei, J., Liu, G., Huang, X., & Ding, W. (2016). Effects of ozone treatment on medium hard wheat (*Triticum aestivum* L.) flour quality and performance in steamed bread making. *CyTA – Journal of Food, 14*(3), 449–456. doi:10.1080/19476337.2015.1133714

Mezui, A. M., & Swart, P. (2010). Effect of UV-C disinfection of beer – Sensory analyses and consumer ranking. *Journal of the Institute of Brewing, 116*(4), 348–353. doi:10.1002/j.2050-0416.2010.tb00785.x

Miano, A. C., Ibarz, A., & Augusto, P. E. D. (2017). Ultrasound technology enhances the hydration of corn kernels without affecting their starch properties. *Journal of Food Engineering, 197*, 34–43. doi:10.1016/j.jfoodeng.2016.10.024

Mihindukulasuriya, S. D. F., & Jayaram, S. H. (2020). Release of electrode materials and changes in organoleptic profiles during the processing of liquid foods using pulse electric field treatment. *IEEE Transactions on Industry Applications, 56*(1), 711–717. doi:10.1109/TIA.2019.2955659

Milani, E. A., Alkhafaji, S., & Silva, F. V. M. (2015). Pulsed electric field continuous pasteurization of different types of beers. *Food Control, 50*, 223–229. doi:10.1016/j.foodcont.2014.08.033

Milani, E. A., Ramsey, J. G., & Silva, F. V. M. (2016). High pressure processing and thermosonication of beer: Comparing the energy requirements and *Saccharomyces cerevisiae* ascospores inactivation with thermal processing and modeling. *Journal of Food Engineering, 181*, 35–41. doi:10.1016/j.jfoodeng.2016.02.023

Milani, E. A., & Silva, F. V. M. (2016). Nonthermal pasteurization of beer by high pressure processing: Modelling the inactivation of *Saccharomyces cerevisiae* ascospores in different alcohol beers. *High Pressure Research, 36*(4), 595–609. doi:10.1080/08957959.2016.1190354

Misra, N. N., Sullivan, C., Pankaj, S. K., Alvarez-Jubete, L., Cama, R., Jacoby, F., & Cullen, P. J. (2014). Enhancement of oil spreadability of biscuit surface by nonthermal barrier discharge plasma. *Innovative Food Science and Emerging Technologies, 26*, 456–461. doi:10.1016/j.ifset.2014.10.001

Morales-de la Peña, M., Welti-Chanes, J., & Martín-Belloso, O. (2019). Novel technologies to improve food safety and quality. *Current Opinion in Food Science, 30*, 1–7. doi:10.1016/j.cofs.2018.10.009

Niemira, B. A. (2012). Cold plasma decontamination of foods. *Annual Review of Food Science and Technology, 3*(1), 125–142. doi:10.1146/annurev-food-022811-101132

Nowosad, K., Sujka, M., Pankiewicz, U., & Kowalski, R. (2021). The application of PEF technology in food processing and human nutrition. *Journal of Food Science and Technology, 58*(2), 397–411. doi:10.1007/s13197-020-04512-4

Ochoa-Velasco, C. E., Ávila-Sosa, R., Hernández-Carranza, P., Ruíz-Espinosa, H., Ruiz-López, I. I., & Guerrero-Beltrán, J. Á. (2020). Mathematical modeling used to evaluate the effect of UV-C light treatment on microorganisms in liquid foods. *Food Engineering Reviews, 12*(3), 290–308. doi:10.1007/s12393-020-09219-y

Okyere, A. Y., Bertoft, E., & Annor, G. A. (2019). Modification of cereal and tuber waxy starches with radio frequency cold plasma and its effects on waxy starch properties. *Carbohydrate Polymers, 223*, 115075. doi:10.1016/j.carbpol.2019.115075

Ozdemir, M., Yurteri, C. U., & Sadikoglu, H. (1999). Physical polymer surface modification methods and applications in food packaging polymers. *Critical Reviews in Food Science and Nutrition, 39*(5), 457–477. doi:10.1080/10408699991279240

Pa, N. F. C., Chin, N. L., Yusof, Y. A., & Aziz, N. A. (2014). Power ultrasound assisted mixing effects on bread physical properties. In: N. L. Chin, H. C. Man, & R. A. Talib (Ed.), *2nd International Conference on Agricultural and Food Engineering, CAFEi2014 – New Trends Forward* (pp. 60–66). Amsterdam: Elsevier. doi:10.1016/j.aaspro.2014.11.009

Pandiselvam, R., Sunoj, S., Manikantan, M. R., Kothakota, A., & Hebbar, K. B. (2017). Application and kinetics of ozone in food preservation. *Ozone: Science and Engineering, 39*(2), 115–126. doi:10.1080/019195 12.2016.1268947

Pankaj, S. K., Bueno-Ferrer, C., Misra, N. N., Milosavljević, V., O'Donnell, C. P., Bourke, P., ... Cullen, P. J. (2014). Applications of cold plasma technology in food packaging. *Trends in Food Science and Technology, 35*(1), 5–17. doi:10.1016/j.tifs.2013.10.009

Park, H., Puligundla, P., & Mok, C. (2020). Cold plasma decontamination of brown rice grains: Impact on biochemical and sensory qualities of their corresponding seedlings and aqueous tea infusions. *LWT, 131*, 109508. doi:10.1016/j.lwt.2020.109508

Park, J. N., Song, B. S., Kim, J. H., Choi, J., Sung, N. Y., Han, I. J., & Lee, J. W. (2012). Sterilization of ready-to-cook *Bibimbap* by combined treatment with gamma irradiation for space food. *Radiation Physics and Chemistry, 81*(8), 1125–1127. doi:10.1016/j.radphyschem.2012.02.042

Pasqualone, A., Gambacorta, G., Summo, C., Caponio, F., Di Miceli, G., Flagella, Z., ... Lenucci, M. S. (2016). Functional, textural and sensory properties of dry pasta supplemented with lyophilized tomato matrix or with durum wheat bran extracts produced by supercritical carbon dioxide or ultrasound. *Food Chemistry, 213*, 545–553. doi:10.1016/j.foodchem.2016.07.006

Piyasena, P., Mohareb, E., & McKellar, R. C. (2003). Inactivation of microbes using ultrasound: A review. *International Journal of Food Microbiology, 87*(3), 207–216. doi:10.1016/S0168-1605(03)00075-8

Polat, H., Dursun Capar, T., Inanir, C., Ekici, L., & Yalcin, H. (2020). Formulation of functional crackers enriched with germinated lentil extract: A response surface methodology Box-Behnken design. *LWT, 123*, 109065. doi:10.1016/j.lwt.2020.109065

Polesi, L. F., Piedade, J., Amistá, M. J. de M., Milagres, R. C. R. M., Spoto, M. H. F., Sarmento, S. B. S., & Canniatti-Brazaca, S. G. (2012). Irradiação de arroz e alterações nas suas propriedades físico-químicas e sensoriais. *Boletim centro de pesquisa de processamento de alimentos, 30*(2), 255–262.

Priyadarshini, A., Rajauria, G., O'Donnell, C. P., & Tiwari, B. K. (2019). Emerging food processing technologies and factors impacting their industrial adoption. *Critical Reviews in Food Science and Nutrition, 59*(19), 3082–3101. doi:10.1080/10408398.2018.1483890

Qi, L., Li, Y., Luo, X., Wang, R., Zheng, R., Wang, L., ... Chen, Z. (2016). Detoxification of zearalenone and ochratoxin A by ozone and quality evaluation of ozonised corn. *Food Additives and Contaminants. Part A, 33*(11), 1700–1710. doi:10.1080/19440049.2016.1232863

Rao, V. S., Srirangarajan, A. N., Kamat, A. S., Adhikari, H. R., & Nair, P. M. (1994). Studies on extension of shelf-life of rawa by gamma irradiation. *Journal of Food Science and Technology, 31*(4), 311–315.

Rastogi, N. K., Raghavarao, K. S. M. S., Balasubramaniam, V. M., Niranjan, K., & Knorr, D. (2007). Opportunities and challenges in high pressure processing of foods. *Critical Reviews in Food Science and Nutrition, 47*(1), 69–112. doi:10.1080/10408390600626420

Ratanatriwong, P., Evrendilek, G. A., & Zhang, Q. H. (2008). Pulsed electric field processing of formulated rice pudding. *Journal of Food Safety, 28*(1), 126–141. doi:10.1111/j.1745-4565.2007.00100.x

Reineke, K., & Mathys, A. (2020). Endospore inactivation by emerging technologies: A review of target structures and inactivation mechanisms. *Annual Review of Food Science and Technology, 11*(1), 255–274. doi:10.1146/annurev-food-032519-051632

Reineke, K., Mathys, A., Heinz, V., & Knorr, D. (2013). Mechanisms of endospore inactivation under high pressure. *Trends in Microbiology, 21*(6), 296–304. doi:10.1016/j.tim.2013.03.001

Ros-Polski, V., Koutchma, T., Xue, J., Defelice, C., & Balamurugan, S. (2015). Effects of high hydrostatic pressure processing parameters and NaCl concentration on the physical properties, texture and quality of white chicken meat. *Innovative Food Science and Emerging Technologies, 30*, 31–42. doi:10.1016/j. ifset.2015.04.003

Roy, M., Ghosh, S., & Chatterjee, S. (1991). Gamma irradiation of rice grains. *Journal of Food Science and Technology, 28*(6), 337–340.

Santos, L. M. R., Oliveira, F. A., Ferreira, E. H. R., & Rosenthal, A. (2017). Application and possible benefits of high hydrostatic pressure or high-pressure homogenization on beer processing: A review. *Food Science and Technology International = Ciencia y Tecnologia de los Alimentos Internacional, 23*(7), 561–581. doi:10.1177/1082013217714670

Schweiggert, U., Carle, R., & Schieber, A. (2007). Conventional and alternative processes for spice production – A review. *Trends in Food Science and Technology, 18*(5), 260–268. doi:10.1016/j.tifs. 2007.01.005

Shao, Y., Tang, F., Xu, F., Wang, Y., & Bao, J. (2013). Effects of γ-irradiation on phenolics content, antioxidant activity and physicochemical properties of whole grainrice. *Radiation Physics and Chemistry, 85*, 227–233. doi:10.1016/j.radphyschem.2013.01.022

Sheikhi, Z., Hosseini, S. M., Khani, M. R., Farhoodi, M., Abdolmaleki, K., Shokri, B., … Mirmoghtadaie, L. (2020). Treatment of starch films with a glow discharge plasma in air and O_2 at low pressure. *Food Science and Technology International, 27*(3), 276–285. doi:10.1177/1082013220948641

Shi, F., Zhao, H., Song, H., Guo, W., Wang, L., Cui, X., … Li, S. (2020). Effects of electron-beam irradiation on inoculated *Listeria innocua*, microbiological and physicochemical quality of fresh noodles during refrigerated storage. *Food Science and Nutrition, 8*(1), 114–123. doi:10.1002/fsn3.1277

Sirisoontaralak, P., & Noomhorm, A. (2006). Changes to physicochemical properties and aroma of irradiated rice. *Journal of Stored Products Research, 42*(3), 264–276. doi:10.1016/j.jspr.2005.04.001

Smelt, J. P. P. M., Rijke, A. G. F., & Hayhurst, A. (1994). Possible mechanism of high pressure inactivation microorganisms. *High Pressure Research, 12*(4–6), 199–203.

Sonawane, S. K., Marar, T., & Patil, S. (2020). Non-thermal plasma: An advanced technology for food industry. *Food Science and Technology International = Ciencia y Tecnologia de los Alimentos Internacional, 26*(8), 727–740. doi:10.1177/1082013220929474

Štulíková, K., Bulíř, T., Nešpor, J., Jelínek, L., Karabín, M., & Dostálek, P. (2020). Application of high-pressure processing to assure the storage stability of unfiltered lager beer. *Molecules, 25*(10), 2414. doi:10.3390/molecules25102414

Suhem, K., Matan, N., Nisoa, M., & Matan, N. (2013). Inhibition of *Aspergillus flavus* on agar media and brown rice cereal bars using cold atmospheric plasma treatment. *International Journal of Food Microbiology, 161*(2), 107–111. doi:10.1016/j.ijfoodmicro.2012.12.002

Sung, W.-C., Hong, M.-C., & Chang, T.-S. (2008). Effects of storage and gamma irradiation on (japonica) waxy rice. *Radiation Physics and Chemistry, 77*(1), 92–97. doi:10.1016/j.radphyschem.2007.03.002

Taghinia, P., Ataye-Salehi, E., & Sheikholeslami, Z. (2015). Impact of pretreated rice bran on wheat dough performance and Barbari bread quality. *Journal of Agricultural Science and Technology, 17*, 135–144.

Tan, M. C., Chin, N. L., & Yusof, Y. A. (2011). Power ultrasound aided batter mixing for sponge cake batter. *Journal of Food Engineering, 104*(3), 430–437. doi:10.1016/j.jfoodeng.2011.01.006

Thirumdas, R., Deshmukh, R. R., & Annapure, U. S. (2016). Effect of low temperature plasma on the functional properties of basmati rice flour. *Journal of Food Science and Technology, 53*(6), 2742–2751. doi:10.1007/s13197-016-2246-4

Thirumdas, R., Saragapani, C., Ajinkya, M. T., Deshmukh, R. R., & Annapure, U. S. (2016). Influence of low pressure cold plasma on cooking and textural properties of brown rice. *Innovative Food Science and Emerging Technologies, 37*, 53–60. doi:10.1016/j.ifset.2016.08.009

Thirumdas, R., Trimukhe, A., Deshmukh, R. R., & Annapure, U. S. (2017). Functional and rheological properties of cold plasma treated rice starch. *Carbohydrate Polymers, 157*, 1723–1731. doi:10.1016/j. carbpol.2016.11.050

Thomas-Popo, E., Mendonça, A., Misra, N. N., Little, A., Wan, Z., Moutiq, R., ... Keener, K. (2019). Inactivation of Shiga-toxin-producing *Escherichia coli, Salmonella enterica* and natural microflora on tempered wheat grains by atmospheric cold plasma. *Food Control, 104,* 231–239. doi:10.1016/j.foodcont.2019.04.025

Tian, Yaoqi, Li, D., Zhao, J., Xu, X., & Jin, Z. (2014). Effect of high hydrostatic pressure (HHP) on slowly digestible properties of rice starches. *Food Chemistry, 152,* 225–229. doi:10.1016/j.foodchem.2013.11.162

Tian, Yuting, Huang, J., Xie, T., Huang, L., Zhuang, W., Zheng, Y., & Zheng, B. (2016). Oenological characteristics, amino acids and volatile profiles of Hongqu rice wines during pottery storage: Effects of high hydrostatic pressure processing. *Food Chemistry, 203,* 456–464. doi:10.1016/j.foodchem.2016.02.116

Tiozon Jr., R. N., Camacho, D. H., Bonto, A. P., Oyong, G. G., & Sreenivasulu, N. (2021). Efficient fortification of folic acid in rice through ultrasonic treatment and absorption. *Food Chemistry, 335,* 127629. doi:10.1016/j.foodchem.2020.127629

Toepfl, S., Mathys, A., Heinz, V., & Knorr, D. (2006). Review: Potential of high hydrostatic pressure and pulsed electric fields for energy efficient and environmentally friendly food processing. *Food Reviews International, 22*(4), 405–423. doi:10.1080/87559120600865164

Trombete, F., Minguita, A., Porto, Y., Freitas-Silva, O., Freitas-Sá, D., Freitas, S., ... Fraga, M. (2016). Chemical, technological, and sensory properties of wheat grains (*Triticum aestivum* L) as affected by gaseous ozonation. *International Journal of Food Properties, 19*(12), 2739–2749. doi:10.1080/10942912.2016.1144067

Vallons, K. J. R., & Arendt, E. K. (2009). Effects of high pressure and temperature on the structural and rheological properties of sorghum starch. *Innovative Food Science and Emerging Technologies, 10*(4), 449–456. doi:10.1016/j.ifset.2009.06.008

Vallons, K. J. R., Ryan, L. A. M., & Arendt, E. K. (2011). Promoting structure formation by high pressure in gluten-free flours. *LWT – Food Science and Technology, 44*(7), 1672–1680. doi:10.1016/j.lwt.2010.11.024

Vallons, K. J. R., Ryan, L. A. M., Koehler, P., & Arendt, E. K. (2010). High pressure-treated sorghum flour as a functional ingredient in the production of sorghum bread. *European Food Research and Technology, 231*(5), 711–717. doi:10.1007/s00217-010-1316-5

Victorin, K. (1992). Review of the genotoxicity of ozone. *Mutation Research/Reviews in Genetic Toxicology, 277*(3), 221–238. doi:10.1016/0165-1110(92)90045-B

Villavicencio, A. L. C. H., Araújo, M. M., Fanaro, G. B., Rela, P. R., & Mancini-Filho, J. (2007). Sensorial analysis evaluation in cereal bars preserved by ionizing radiation processing. *Radiation Physics and Chemistry, 76*(11), 1875–1877. doi:10.1016/j.radphyschem.2007.04.012

Wang, L., Shao, H., Luo, X., Wang, R., Li, Y., Li, Y., ... Chen, Z. (2016). Effect of ozone treatment on deoxynivalenol and wheat quality. *PLOS ONE, 11*(1), e0147613. doi:10.1371/journal.pone.0147613

Warchalewski, J. R., Gralik, J., Zawirska-Wojtasiak, R., Zabielski, J., & Kusnierz, R. (1998). The evaluation of wheat grain odour and colour after gamma and microwave irradiation. *Electronic Journal of Polish Agricultural Universities, 1*(1).

Welti-Chanes, J., López-Malo, A., Palou, E., Bermúdez, D., Guerrero-Beltrán, J. A., & Barbosa-Cánovas, G. V. (2004). Fundamentals and applications of high pressure processing to foods. In: G. V. Barbosa-Cánovas, M. S. Tapia, M. P. Cano, O. Martín-Belloso, & A. Martínez (Eds.), *Novel Food Processing Technologies* (pp. 157–182). Boca Raton: CRC Press. doi:10.1201/9780203997277.ch8

Yang, H., Han, M., Wang, X., Han, Y., Wu, J., Xu, X., & Zhou, G. (2015). Effect of high pressure on cooking losses and functional properties of reduced-fat and reduced-salt pork sausage emulsions. *Innovative Food Science and Emerging Technologies, 29,* 125–133. doi:10.1016/j.ifset.2015.02.013

Yang, Y., Xia, Y., Wang, G., Tao, L., Yu, J., & Ai, L. (2019). Effects of boiling, ultra-high temperature and high hydrostatic pressure on free amino acids, flavor characteristics and sensory profiles in Chinese rice wine. *Food Chemistry, 275,* 407–416. doi:10.1016/j.foodchem.2018.09.128

Yin, H., Dong, J., Yu, J., Chang, Z., Qian, Z., Liu, M., ... Wang, D. (2016). A preliminary study about the influence of high hydrostatic pressure processing on the physicochemical and sensorial properties of a cloudy wheat beer. *Journal of the Institute of Brewing, 122*(3), 462–467. doi:10.1002/jib.344

Yodpitak, S., Mahatheeranont, S., Boonyawan, D., Sookwong, P., Roytrakul, S., & Norkaew, O. (2019). Cold plasma treatment to improve germination and enhance the bioactive phytochemical content of germinated brown rice. *Food Chemistry*, *289*, 328–339. doi:10.1016/j.foodchem.2019.03.061

Zahn, S., Schneider, Y., Zucker, G., & Rohm, H. (2005). Impact of excitation and material parameters on the efficiency of ultrasonic cutting of bakery products. *Journal of Food Science*, *70*(9), E510–E513.

Zeng, F., Gao, Q., Han, Z., Zeng, X., & Yu, S. (2016). Structural properties and digestibility of pulsed electric field treated waxy rice starch. *Food Chemistry*, *194*, 1313–1319. doi:10.1016/j.foodchem.2015.08.104

Zhang, X., Wang, L., Cheng, M., Wang, R., Luo, X., Li, Y., & Chen, Z. (2015). Influence of ultrasonic enzyme treatment on the cooking and eating quality of brown rice. *Journal of Cereal Science*, *63*, 140–146. doi:10.1007/978-3-030-42660-6

Zhang, Z. H., Wang, L. H., Zeng, X. A., Han, Z., & Brennan, C. S. (2019). Non-thermal technologies and its current and future application in the food industry: A review. *International Journal of Food Science and Technology*, *54*(1), 1–13. doi:10.1111/ijfs.13903

Zhao, X., Sun, H., Zhu, H., Liu, H., Zhang, X., & Feng, Z. (2019). Effect of packaging methods and storage conditions on quality characteristics of flour product naan. *Journal of Food Science and Technology*, *56*(12), 5362–5373. doi:10.1007/s13197-019-04007-x

Zhu, F. (2016). Impact of γ-irradiation on structure, physicochemical properties, and applications of starch. *Food Hydrocolloids*, *52*, 201–212. doi:10.1016/j.foodhyd.2015.05.035

Zhu, F. (2018). Effect of ozone treatment on the quality of grain products. *Food Chemistry*, *264*, 358–366. doi:10.1016/j.foodchem.2018.05.047

Zhu, K.-X., Su, C.-Y., Guo, X.-N., Peng, W., & Zhou, H.-M. (2011). Influence of ultrasound during wheat gluten hydrolysis on the antioxidant activities of the resulting hydrolysate. *International Journal of Food Science and Technology*, *46*(5), 1053–1059. doi:10.1111/j.1365-2621.2011.02585.x

Chapter 12

Safety Issues for Non-Thermal Food Processing Applications in Grain Industries

Aysegul Mutlu-Ingok, Dilara Devecioglu, Dilara Nur Dikmetas
and Funda Karbancioglu-Guler

CONTENTS

12.1 INTRODUCTION

Thermal processes are among the most common methods with practical and economic properties to ensure food safety, but the applications of these processes on temperature-sensitive products are limited (Wuytack et al., 2002). The use of chemical preservatives is the other common way to eliminate microorganisms. However, only a few chemicals are available to apply to grains due to health, safety, environmental and economic perspectives (Collins, 2006). Additionally, the usage of these chemicals has been forbidden in many countries due to their adverse effects. Moreover, nowadays, consumers' interest in foods that are more natural with fewer or no chemical additives is rising (Schmidt et al., 2018). Non-thermal processes are emerging and promising techniques for the decontamination of both microorganisms and toxins in food matrices (Kalagatur et al., 2018). These processes, including high pressure, pulsed electric field, ultrasound, cold plasma, and ultraviolet light applications, may be used for food preservation on a large scale (Woldemariam et al., 2019). In addition to the individual use of non-thermal applications, combining them with mild heat treatments improves the process's efficiency. Moreover, these non-thermal technologies can be used in combination with each other.

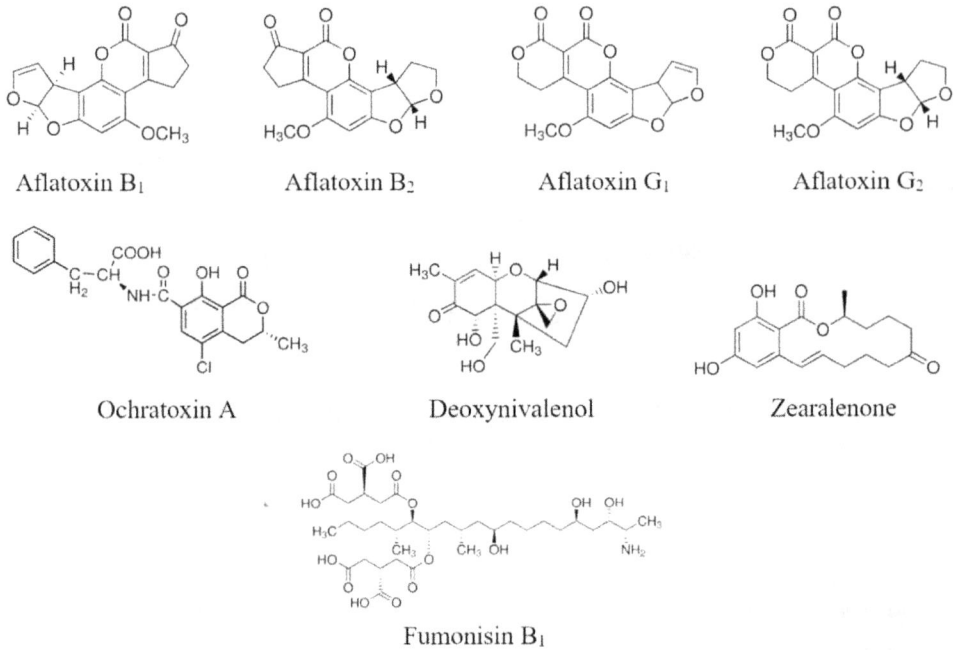

Figure 12.1   Some commonly studied mycotoxins in grain and grain products.

Grains and grain products are among the product groups where non-thermal applications are most commonly used. Contamination of cereals with microorganisms and pests causes significant losses during storage, transportation and marketing (Mostafavi et al., 2012). The losses of grains are commonly due to the activities of several insects and microorganisms including bacteria and fungus. The presence of insects in cereals leads to quantitative and qualitative results (Yadav et al., 2014). Non-thermal applications can contribute to food safety by eliminating mainly biological and chemical hazards. In terms of chemical hazards, non-thermal processes have advantages such as reducing allergens, free radicals, and pesticide residues in grains. Among the contaminants, mycotoxins are also crucial from both economic and health perspectives. Some of the commonly studied mycotoxins in grain and grain products are shown in Figure 12.1. One of the most common advantages of non-thermal processes is that, in addition to its potential to ensure food safety, there is minimal or no change in the food's sensory and nutritional characteristics.

The present chapter focuses on the studies about applications of several non-thermal processes to ensure the safety of grains and grain products, mainly on microbial safety. In addition to microbial safety, several advantages of these processes in chemical safety aspects such as reducing allergens, free radicals and pesticide residues are also discussed under each non-thermal technology heading.

12.2 APPLICATION OF COLD PLASMA IN GRAIN INDUSTRIES

Plasma is an ionized gas that is considered the fourth state of matter. Cold plasma provides food safety by eliminating hazards, especially biological ones, and is preferred as an alternative to thermal processes. Its key advantage is establishing the inactivation of microorganisms at low temperatures without any of the adverse effects of high temperatures on food quality (Shankar et al., 2014; Schlüter and Fröhling, 2014). There are different categories of food products including milk and dairy products, fruits, vegetables and processed foods to which cold plasma has been applied. Grains and food products derived from grains have also been subjected to different cold plasma treatment applications in some studies (Table 12.1); however, available studies are limited.

The effectiveness of cold plasma technology significantly depends on several parameters including instrument setup, treatment conditions, gas composition, food model and hazard type (Dasan, Boyacı and Mutlu, 2016; Misra et al., 2011). Dasan, Boyacı and Mutlu (2016) studied the atmospheric pressure fluidized bed plasma system to inactivate *Aspergillus flavus* and *Aspergillus parasiticus* in maize. The experiment was carried out at 100% voltage, 25 kHz frequency and 655 W by two reactors (diameters: 49 mm and 65 mm). At the end of the 5 min treatment, remarkable reductions in *Aspergillus flavus* and *Aspergillus parasiticus* were observed and the treatment using nitrogen instead of air as the plasma-forming gas was found less effective. The application of a dielectric barrier discharge atmospheric cold plasma was studied in a model media prepared with two grains (wheat and barley) against biofilm formed by different bacterial cultures (*Bacillus subtilis, Lactobacillus plantarum, Lactobacillus brevis, Escherichia coli* and *Bacillus atrophaeus*). Depending on the model media and culture type, the efficiency of the treatment showed variation in reduction. While direct treatment was more effective than indirect treatment, the biofilm formed by *Bacillus atrophaeus* was highly resistant among the selected cultures (Los et al., 2017).

In addition to cold plasma studies performed with crops, different grain-based products including Korean rice cake (Han et al., 2020), brown cereal rice bars (Suhem et al., 2013) and wheat flour (Ramanan et al., 2018) have become a field of interest. Han et al. (2020) showed the reduction of pathogens inoculated in Korean rice cake in the case of washing with water treated with surface dielectric barrier discharge cold plasma. Similarly, Suhem et al. (2013) has reported that cold plasma may be used against *Aspergillus flavus* in rice cereal bars to ensure food safety. They achieved a fungal reduction by 40 W cold plasma application for 20 min and suggested that method as a cheap and useful alternative for providing food safety. The effect of treatment on the pests, one of the major microbiological safety problems, was shown by Ramanan et al. (2018) in refined wheat flour. They tested the treatment efficiency to control *Tribolium castaneum* and suggested an optimum condition as 2,500 V for 15 min at 3.7 cm electrode distance. The mortality rate for three stages of pest (egg, larva and adult) were found as 93%, 93% and 100%, respectively. With the increasing number of studies suggesting the usage of cold plasma, the concerns about its effect on food quality have increased. While Misra et al. (2015) showed changes in both the structural and functional properties of wheat flour, Mahendran et al. (2016) demonstrated that maida flour has no color change after cold plasma treatment against *Tribolium castaneum*. Brown rice was also studied as a type of grain treated by this technology. Park et al. (2020) achieved a reduction in the number of microbial contaminants such as aerobic bacteria, yeasts and molds using corona discharge plasma jet under atmospheric pressure. The study of Lee et al. (2016) tested the effectiveness of cold plasma on brown rice inoculated with *Bacillus cereus, Bacillus subtilis* and *Escherichia coli* O157:H7 for microbial safety and also quality. They indicated a positive impact on textural properties, while Lee et al. (2019) showed an increase in shelf life and changes to several physicochemical parameters regarding treatment conditions.

Additionally, Thomas-Popo et al. (2019) subjected tempered wheat grains to atmospheric cold plasma to inactivate shiga toxin–producing *Escherichia coli* strains and *Salmonella enterica*

TABLE 12.1 STUDIES CONDUCTED WITH COLD PLASMA TREATMENT

Grain or Product	Hazard	System	Feed Gas	Reduction	References
		Microorganism			
Barley	*Escherichia coli, Bacillus atrophaeus, Penicillium verrucosum*	Atmospheric cold plasma	Air	2.1–2.4 log	Los et al. (2018)
Brown rice	Aerobic bacteria, yeasts and molds	Corona discharge plasma jet	Air	1.53–1.74 log for 1–10 min	Park et al. (2020)
Brown rice cereal bar	*Aspergillus flavus*	Cold atmospheric plasma	Argon	4 log for 20 min	Suhem et al. (2013)
Maize	*Aspergillus flavus* and *Aspergillus parasiticus* spores	Atmospheric pressure fluidized bed plasma	Air Nitrogen	4.99–5.48 log 4.54–4.68 log	Dasan, Boyacı and Mutlu (2016)
Maize seeds	Epiphytic filamentous fungi	Atmospheric pressure coplanar surface barrier discharge	Air	1 log	Zahoranová et al. (2014)
Wheat	*Bacillus amyloliquefaciens, Bacillus atrophaeus* endospores	Low-pressure inductively coupled plasma integrated into circulating fluidized bed	Air, gas mixture	1.5 log	Butscher (2016)
Wheat	*Bacillus amyloliquefaciens* endospores	Low-pressure inductively coupled plasma integrated into circulating fluidized bed	Argon and oxygen	2.6 log for 74 sec	Butscher et al. (2015)
Wheat	*Bacillus atrophaeus, Aspergillus niger, Aspergillus flavus, Aspergillus candidus, Penicillium citrinum, Penicillium verrucosum, Penicillum chrysogenum*	Atmospheric cold plasma dielectric barrier discharge	Air	0.41–3.88 log (cfu/g) for 5–20 min	Los et al. (2020)
Wheat	*Escherichia coli, Bacillus atrophaeus, Penicillium verrucosum*	Atmospheric cold plasma	Air	1.5–2.5 log	Los et al. (2018)
Wheat	*Geobacillus stearothermophilus* endospores	Atmospheric pressure dielectric barrier discharge	Argon	3 log for 60 min	Butscher et al. (2016)

(Continued)

TABLE 12.1 (CONTINUED)   STUDIES CONDUCTED WITH COLD PLASMA TREATMENT

Grain or Product	Hazard	System	Feed Gas	Reduction	References
Wheat and barley food media	*Bacillus subtilis, Lactobacillus plantarum, Lactobacillus brevis, Escherichia coli, Bacillus atrophaeus* biofilms	Dielectric barrier discharge atmospheric cold plasma	Air, direct treatment	1.72–4.01 log	Los et al. (2017)
Wheat, barley, oat, lentil, rye, corn, chickpea	*Aspergillus parasiticus* and *Penicillium* spores	Low-pressure inductively coupled plasma	Air, SF6	3 log	Selcuk et al. (2008)
Tempered wheat	*Escherichia coli, Salmonella enterica*	Atmospheric cold plasma	Air	3.09–4.84 log for 20 min	Thomas-Popo et al. (2019)
Mycotoxin					
Corn	Aflatoxin B_1	High-voltage atmospheric cold plasma	Air, modified atmosphere gas (MA65)	62–82% for 10 min	Shi et al. (2017)
Maize	Aflatoxin B_1 Fumonisin B_1	Pulsed dielectric barrier discharge cold atmospheric pressure plasma	Helium Oxygen	64–65%	Wielogorska et al. (2019)

serotypes (Enteritidis, Heidelberg, Typhimurium, Gaminara and Oranienburg). Though these microbial cultures were inactivated, yeasts and molds present as natural microflora on wheat grains were more sensitive to treatment. *Escherichia coli*, *Bacillus atrophaeus* and *Penicillium verrucosum* were selected to test the microbiological efficiency of atmospheric cold plasma on both barley and wheat. *Bacillus atrophaeus* was the most resistant culture and the highest sensitivity was observed in *Escherichia coli* without any undesired effect on quality.

Different studies have focused on the biological hazards of wheat including *Geobacillus stearothermophilus* (Butscher et al., 2016), *Bacillus atrophaeus, Aspergillus niger, Aspergillus. flavus, Aspergillus candidus, Penicillium citrinum, Penicillium verrucosum* and *Penicillum chrysogenum* (Los et al., 2020). While Butscher et al. (2016) indicated the importance of treatment time in accordance with variable microbial reduction, Los et al. (2020) observed a decrease in the number of various microbial cultures with 20 min exposure time and *Bacillus atrophaeus* was the most resistant microorganism, similar to the results of Los et al. (2018).

Besides biological hazards, toxins are one of the major threats for the grain industry and may also be subject to cold plasma treatment. Pulsed dielectric barrier discharge cold atmospheric pressure plasma was assessed by Wielogorska et al. (2019). They treated maize samples contaminated by aflatoxin B_1 (50 µg/ml) and fumonisin B_1 (200 µg/ml) for 10 min and reported above 50% reduction in the exposed mycotoxin concentration. In addition, significant detoxification of other mycotoxins such as ochratoxin A, deoxynivalenol, zearalenone and enniatin B has also been shown by using this technique. Furthermore, Shi et al. (2017) showed observable degradation of aflatoxin B_1 with high-voltage atmospheric cold plasma under changing treatment parameters including relative humidity, direct or indirect treatment and treatment time for contaminated corn samples. As a result of degradation kinetics, the highest degradation rate was observed for a 10-min treatment time in 40% relative humidity without any change in direct and indirect treatment.

Either the degradation or inhibition mechanism of the process is related not only to the type of hazard but also to treatment conditions. Until now, several mechanisms have been demonstrated including photodesorption, erosion, etching, cell membrane perforation, electrostatic disruption, oxidation, the formation of reactive species and damage in DNA structure (Schlüter and Fröhling, 2014). Among them, the formation of several reactive oxygen species based upon selected gas for treatment has been linked with the antimicrobial effectiveness of the process in several studies. While Dasan, Boyacı and Mutlu (2016) focused on the many oxygen and nitrogen species related to the antimicrobial impact of the process, Los et al. (2020) mentioned the DNA/RNA damage because of physical processes during treatment.

Agrochemicals including pesticides are chemical substances used to control food quality and optimize food production, but their improper usage raises many problems. The cold plasma process has been shown as effective for the degradation of pesticides present in mango (Phan et al., 2018), blueberries (Sarangapani et al., 2017) and water (Sarangapani et al., 2016). Although the presence of pesticide residues in maize grains and flour (Mahugija et al., 2017) and maize and cowpea (Akoto et al., 2013) were found, the usage of cold plasma to contend with agrochemical residues on grains is an unfocused research area.

Cold plasma technology still has unexplained effects on the quality of food products and its mechanism is not fully understood. Hence, more multidisciplinary research and a better understanding are required in order to focus on the applicability of the process and scale up to both pilot and commercial scale.

12.3 APPLICATION OF ELECTROLYZED WATER IN GRAIN INDUSTRIES

Electrolyzed water has been used to decontaminate food products, equipment and non-food-contact surfaces. Additionally, electrolyzed oxidizing water has been recognized as an efficient antimicrobial agent by the Japan Ministry of Health, Labour and Welfare (MHLW) and the United States Food and Drug Administration (FDA) (Chen et al., 2020a; AlZain, 2020). It has been categorized as acidic, neutral or alkaline according to its pH value (AlZain, 2020). The active chlorine content on slightly acidic electrolyzed water, in particular, results in a significant germicidal impact (Chen et al., 2020a). However, there has been limited research on the use of electrolyzed water in grain processing (Table 12.2). Slightly acidic electrolyzed water (SAEW) has gained more attention for use in the food industry due to its neutral pH (5–6) and lower available chlorine concentrations (Li et al., 2015).

Wheat tempering has been conducted at different available chlorine concentrations with SAEW. It was reported that the total plate count (TPC) and yeast/mold count (YMC) were significantly reduced when the SAEW concentration increased. However, mesophilic aerobic spores (MAS) changed only slightly. That bactericidal effect has been connected to oxidation-reduction potential (ORP-reactions) and a high concentration of hypochlorous acid (HClO) (Chen et al., 2020a).

TABLE 12.2   STUDIES CONDUCTED WITH ELECTROLYZED WATER TREATMENT

Grain or Product	Hazard	Method	Treatment Conditions	Reduction	References
		Microorganism			
Barley rice	*Bacillus cereus* vegetative	Acidic electrolyzed water	3 h	1.98–2.76 log	Park et al. (2009)
	Bacillus cereus spore			1.31–2.49 log	
Barley rice	*Bacillus cereus* vegetative	Alkali electrolyzed water	3 h	1.87–2.31 log	Park et al. (2009)
	Bacillus cereus spore			1.29–2.04 log	
Brown rice	Total plate count	Strong acidic electrolyzed water	*	2.24–2.36 log	Liu et al. (2013)
Brown rice	Total plate count	Slightly acidic electrolyzed water	*	1.57–2.54 log	Liu et al. (2013)
Millet	Total bacteria count	Slightly acidic electrolyzed water	15–30 mg/L, 2 h	0.76–1.07 log	Li et al. (2015)
Wheat grains Straight flour Clear flour bran	Total plate count Total yeast/mold counts	Slightly acidic electrolyzed water	70 mg/L, 15 min	0.65–0.93 log 0.49–0.73 log	Chen et al. (2020a)
		Mycotoxin			
Wheat grains	Deoxynivalenol	Acidic electrolyzed water	20 g/120 ml, 45 min	30.32–57.48%	Lyu et al. (2018)
Wheat grains	Deoxynivalenol	Alkali electrolyzed water	20 g/120 ml, 45 min	28.9–61.60%	Lyu et al. (2018)

* Not indicated.

Liu et al. (2013) investigated the effect of pre-soaking and ultrasonic treatment in brown rice with alkali and acidic electrolyzed water treatment. Strong acidic electrolyzed treatments have caused reduction of 2.24 and 2.36 log cfu/g in the total plate count, while slightly acidic electrolyzed treatment results in 1.57 and 2.4 log cfu/g reduction. Antimicrobial activity was improved with strong and slightly acidic electrolyzed water when used with presoaking and ultrasonic treatment, with a reduction of 2.89 and 3.75 log cfu/g. The microbial counts of brown millet have been reduced to 6.57, 6.30 and 6.61 log cfu/g from 7.72 log cfu/g after 2 h SAEW. Additionally, SAEW treatment could not improve the germination of sprouts when compared with the control group (Li et al., 2015).

Audenaert et al. (2012) reported that neutralized electrolyzed water treatment of wheat grains for a short time (4 h) resulted in a reduction of the germination capacity of *Fusarium graminearum*. In a long-time treatment (14 days), *Fusarium graminearum* was inhibited by 50% compared to the control sample. In addition to *in vivo* studies, *in vitro* results showed an increase in deoxynivalenol biosynthesis at sublethal concentrations. However, another study determined the optimum pH values for acid and alkali electrolyzed water where the maximum reduction of deoxynivalenol and *Fusarium graminearum* content in wheat grains was achieved. In that study, deoxynivalenol elimination was also investigated at different pH levels. When the sample pH value was 2.5, 3.5, 4.5, 5.5 and 6.5, the deoxynivalenol level after acid electrolyzed water treatment was reduced by 0.32%, 46.19%, 46.98%, 57.48% and 48.29%, respectively. Thus, an efficient pH was found as 5.5 for acid electrolyzed water treatment. Moreover, for alkali electrolyzed water treatment, the highest deoxynivalenol reduction was observed at pH 9.5 with 61.6% (Lyu et al., 2018).

Park et al. (2009) studied the combination of electrolyzed water and citric acid's efficiency for eliminating *Bacillus cereus* vegetative cells and spores on different grains such as brown rice, glutinous rice, barley rice and Job's tear rice. At all tested temperatures, acidic and alkali electrolyzed water treatment combined with citric acid treatment significantly reduced (more than 5 log) the *Bacillus cereus* cells and spores in brown rice. In another study, the maximum inactivation of *Bacillus cereus* spores in brown rice was achieved with 150 mg/L acidic electrolyzed water treatment for 30 minutes with no adverse impact (Zhang et al., 2018).

12.4 APPLICATION OF HIGH PRESSURE IN GRAIN INDUSTRIES

High-pressure processing (HPP) is another type of non-thermal method used to inactivate both pathogens and spoilage microorganisms in foods while preserving their sensorial and nutritional characteristics. HPP can be applied to both liquid and solid (with high moisture content) foods (Yordanov and Angelova, 2010). This technology is also advantageous as it has a much smaller effect on low-molecular-weight components such as flavor compounds, vitamins and pigments, etc. than thermal treatments. In this respect, foods that were treated with high-pressure applications are very similar to fresh food products (Muntean et al., 2016). This process not only contributed to food safety but also increased the shelf life of food products. This process covers the pressure ranges from 50 to 1000 MPa (Considine et al., 2008). The use of HPP has important potential in terms of the minimally processed, additive-free and safe food products demanded by consumers. This technology was first applied in fruit juices and jams and later in many products, including rice products (Murchie et al., 2005).

One of the most important purposes of high-pressure applications is microbial inactivation (Yordanov and Angelova, 2010). Microorganisms have the potential to show their reactions under stress conditions against high-pressure applications. Depending on the intensity of the application, some changes occur in the cell and microorganisms are destroyed. After pressure applications above 100 MPa, pathogens in food can be killed at room temperature, thus avoiding the energy consumption required by thermal processes for heating and cooling (Huang et al., 2014). Sensitivities of microorganisms against pressure applications vary depending on the species of microorganisms, and moreover, strains of the same species (Yordanov and Angelova, 2010). As well, Gram-negative bacteria are more sensitive when compared to Gram-positive bacteria to high-pressure homogenization (Wuytack et al., 2002).

Moreover, yeast and molds were also more sensitive to high-pressure applications than Gram-positive bacteria. Bacterial spores were one of the most resistant to HPP. The resistance of microorganisms to high pressure can be related to their growth phase (Yordanov and Angelova, 2010). Differences in the sensitivities of microorganisms to high-pressure applications could also be attributed to the differences in their cell membranes. Such processes affect the membrane and result in leakage within the membrane (Murchie et al., 2005). In addition to providing microbial inactivation, the high-pressure application also contributes to the texture of the product. Another advantage of this process is that the processing time is not affected by the size of the product because the pressure is distributed uniformly to the product (Boluda-Aguilar et al., 2013).

As with other non-thermal techniques, the high-pressure application can be combined with other technologies such as mild heat treatment. Daryaei et al. (2013) found that a combination of 600 MPa pressure application and heat treatment at 60°C for 15 min can effectively inactivate *Bacillus cereus* spores in rice. In another study in which heat treatment was applied after HPP, it was reported that the application of pressure at 400 MPa and heat treatment at 100°C extended the shelf life of rice by up to 56 days (Yu et al., 2018). High-pressure applications can also be combined with antimicrobial packaging systems. The use of high-pressure applications in combination with antimicrobial packaging systems can be advantageous in terms of less processed food production. In this context, pasteurization-like high-pressure treatment (up to 800 MPa at ambient temperature) combined with packaging systems was successfully achieved to inhibit the growth of *Botrytis cinerea* (Raouche et al., 2011).

The effect of high-pressure applications on molds and mycotoxins has also been studied. The complete reduction of *Fusarium graminearum*, deoxynivalenol and zearalenone in maize grains were reported at 550 MPa of pressure, 45°C of temperature and 20 min of pressure-holding time. Decreasing the load of fungi by HPP was correlated with the dropping levels of deoxynivalenol and zearalenone. Additionally, it has been reported that spore inactivation was more effective in peptone water (0.99 a_w) at lower pressure levels compared to maize grains (0.85 a_w) due to its relatively high water activity (Kalagatur et al., 2018).

In addition to the contribution of high-pressure applications to ensure microbial food safety, high hydrostatic pressure applications may result in the release of allergenic proteins. In this way, the allergenicity of rice can be reduced (Kato et al., 2000). As long as it does not adversely affect the quality characteristics of the food, using it in combination with different techniques may have a positive effect on reducing allergenicity (Huang et al., 2014). In a study investigating protein release after high-pressure application, the release of major rice allergens (16 kDa albumin, α-globulin and 33 kDa globulin) from polished rice grains was detected after immersion in distilled water and pressure applications at 100–400 MPa. It has been explained by the occurrence of partial disruption in the endosperm cell, the permeation of the surrounding solution into the endosperm cell, the dissolution of proteins, and then the diffusion of proteins from the endosperm cell into the surroundings (Kato et al., 2000).

Moreover, it has been reported that allergenic proteins have almost disappeared in the presence of proteolytic enzymes (Kato et al., 2000). It is also known that protease treatment can improve the release of allergenic proteins by high-pressure application. Since the treatment of pressure promotes the penetration of the protease solution into the endosperm cell, the use of protease enzymes with pressure applications increases the efficiency in terms of the release of allergenic proteins (Estrada-Girón et al., 2005). Further, it should also be noted that the solubilization of proteins is also affected by the surrounding solution that grains are immersed in for pressurization (Estrada-Girón et al., 2005).

As mentioned above, high-pressure applications can ensure food safety with no or minimal changes in the sensorial and nutritional characteristics. However, several factors should be noticed before applications that affect the microbial resistance such as species and strain differences of microorganisms, type of mycotoxin, growth phase of microorganisms, intensity and duration of applications, etc. Another advantage of this process is that it is also possible to increase this method's effectiveness by using it in combination with other methods.

12.5 APPLICATION OF IRRADIATION IN GRAIN INDUSTRIES

Irradiation has been recognized as an environmentally friendly non-thermal technique to improve food quality and food safety. Reducing microbial load and improving shelf life are among the most important advantages (Ito et al., 2018). Based on these advantages, studies on irradiation are increasing day by day. In addition to contributing to microbial safety, a reduction in viscosity has been reported in both grain and grain products by irradiation (Ito et al., 2018). As of 2005, the disinfection of grains and fruits among irradiated foods worldwide was in a share of 20% with 82,000 tons (Kume et al., 2009). Considering that grains are often contaminated by various microorganisms, mainly at the time of harvest, it is certainly an important function for the food industry to ensure food safety and improve food stability and quality.

Radiation application has many contributions in terms of ensuring food safety, especially microbial safety of grains. Insect control in these products is also important, as it is a major problem. Besides food safety, the contribution of radiation to the product's quality is not to be underestimated. To take advantage of these benefits, radiation technology can be used alone or in combination with other technologies. Since the irradiation of foods results in little temperature increase, the nutrient loss is minimal (Manupriya et al., 2020). Even though food irradiation has been accepted as a reliable method in more than 40 countries, consumer concerns on this subject continue (Kilonzo-Nthenge, 2012). While it can be used individually, irradiation can be combined with other technologies such as mild heat treatment. There are different radiation sources for food applications based on their penetrating abilities: X-rays, gamma and electron beam (Cleland, 2006; Khaneghah et al., 2020). Among them, gamma irradiation was the most common one (Begum et al., 2020).

Gamma irradiation can be classified into three groups: the low dose (up to 1 kGy), the medium dose (1 to 10 kGy) and the high dose (10 to 50 kGy) (Urbain, 1986). Gamma irradiation was suggested for a wide range of traditional rice products. This process can reduce microbial growth. For cereal grains, gamma irradiation (10 kGy) has been useful for reducing the microbial load (Khaneghah et al., 2020). A dose of 10 kGy was sufficient for decontaminating microorganisms (Aziz et al., 2006). Several studies proved that phenomenon. One of them treated Korean whole rice porridge, decreasing the total bacteria even at 1 kGy (Yang et al., 2007). Gamma irradiation was also suggested for treating harmful pathogens in ready-to-eat *Kimbab* from Korea (Chung et al., 2007) and for the disinfection of sorghum flours (Mukisa et al., 2012). In that product, while 10 kGy has a negligible effect on the total aerobic count of unmalted flours, the same dose has resulted in 3 log cycles of reduction in malted flour. Moreover, re-irradiation with 25 kGy caused complete decontamination. It was also stated that the effectiveness of this process is related to the initial microbial load as well as the diversity of microorganisms (Mukisa et al., 2012).

Comparing gamma and X-ray irradiation in the case of D_{10} values, X-ray-irradiated *Escherichia coli*, *Salmonella* Typhimurium and *Listeria monocytogenes* had higher D_{10} values than gamma-irradiated ones. Meanwhile, the dose rate did not significantly affect the D_{10} values of the rice pathogens (Begum et al., 2020). The log reduction for *Listeria monocytogenes* was higher than for *Escherichia coli* and *Salmonella* Typhimurium at all tested radiation doses independent of the radiation source and rate. In the same study, irradiation was combined with essential oil applications. The treatment of the combination of irradiation and oregano/thyme EO increased the radiosensitization of both *Escherichia coli*, *Salmonella* Typhimurium and *Listeria monocytogenes* (Begum et al., 2020).

Electron beam irradiation (EBI) is another type of irradiation. Electron beam application is different from the others as it is carried out without using any radioactive source. Although its penetration ability is low, this mechanical application seems more temperate to consumers. It also has the potential to reduce the microbial load. In a study investigating the effect of electron-beam irradiation on husked and unhusked rice, it was concluded that a higher dose of radiation was required to inactivate *Escherichia coli*, *Bacillus cereus*, sulfite-reducing clostridia and fungi in unhusked rice (7.5 kGy) when compared with husked rice (1.1 kGy) (Sarrías et al., 2003). The positive effect of EBI on increasing the shelf life of rice by reducing significantly the total bacterial count has also been reported. It also contributes to quality throughout its shelf life (Luo et al., 2019).

Furthermore, since the presence of fungi in grain and grain products may result in mycotoxin production, it is an important risk factor for both economic losses and health risks (Khaneghah et al., 2020). Although many factors affect mycotoxin production, some of these factors are not controllable by humans (Luo et al., 2017). When evaluated from this perspective, the detoxification of toxins becomes an important issue. In the case of microbial food safety, EBI is recommended for eliminating these risks. Two different mechanisms can realize this detoxification effect. The detoxification ability of EBI was mainly expressed as two mechanisms: the first one is direct contact of produced electrons with mycotoxins and the second one is indirect degradation caused by reactive species such as H and OH radicals (Khaneghah et al., 2020). With this non-thermal technique, not only can a number of microorganisms be inactivated but also several toxins can be detoxified. Several factors affect the success of the process, such as the irradiation dose, initial microbial load, food composition, type of microorganism or toxin, etc. (Khaneghah et al., 2020). Other studies conducted with irradiation are summarized in Table 12.3.

In addition to microbial inactivation, radiation is also suggested as a non-thermal technique for insect control. Insects in grains can be resistant to some chemical pesticides. Moreover, chemical residues of these insecticides can also pose serious problems (Lorenz and Miller, 1975). Methyl bromide, one of the most widely used chemicals in pest control in stored products because of its broad-spectrum effectiveness, has strict restrictions in many countries due to its adverse effects on the environment (Ahmed, 2001). When compared with bacteria, insects have approximately the same resistance levels to radiation as vegetative forms. Although this effect can be observed in the later stages, radiation usually does not cause rapid death in insects. In insects that survive in this way, deformations in organs and tissues can be observed.

In most cases, increasing radiation doses can cause a shortening of the life span of the insect, while in some cases, the life processes of the insect slow down and this may result in a longer life length (Lorenz and Miller, 1975). Generally the irradiation dose used for disinfection of grains did not have a negative effect on the quality and functional properties of the grain. Eggs are generally the most sensitive to irradiation applications. The susceptibility of insects to the radiation process is also affected by physical factors. For example, the sensitivity of insects to radiation increases with the increase in temperature during the process (Ahmed, 2001). In a study investigating the effect of gamma and microwave irradiation on wheat grain, it was stated that gamma-irradiated wheat grain (at 0.1 kGy), which was used as a food, extended the development time of the larvae of *Tribolium confusum* Duv. by seven days (Warchalewski et al., 2000). Relatively low doses of gamma irradiation were suggested for the disinfection of the Mediterranean flour moth, *Ephestia kuehniella* Zeller. Gamma irradiation has been reported to delay development periods in both eggs and larvae. A radiation level of 300 Gy was required to sterilize pupae and adults, while 250 Gy was sufficient for complete larvae death (Ayvaz and Tunçbilek, 2006). In another research, the sufficient level of radiation was reported as 300 Gy for preventing the reproduction of Mexican rice borers, *Eoreuma loftini* (Dyar) (Lepidoptera: Pyralidae) irradiated as late pupae (Darmawi et al., 1998). For controlling rice weevil, *Sitophilus oryzae* (L.) (Coleoptera:

TABLE 12.3   STUDIES CONDUCTED WITH IRRADIATION TREATMENT

Grain or Product	Hazard	Type of Irradiation	Dose (kGy)	Results	References
Microorganism					
Adlay	Total aerobic bacteria, molds and yeasts	Gamma	4	Enterobacteria inactivation	Wen et al. (2008)
			6	Yeast or fungi inactivation	
Barley	*Fusarium* infection	Electron beam	6–8	21–62% decrease	Kottapalli et al. (2006)
			10	50–98% decrease	
Brown rice, Milled rice	Total viable bacteria	Electron beam	1–5	73.3–100% inactivation	Luo et al. (2019)
Fresh gnocchi	Mesophilic bacteria, physicrophilic bacteria, molds and yeasts	Gamma	5	2–6.70 log reduction	Cassares et al. (2020)
Husked rice	*Bacillus cereus* spores	Electron Beam	1.1–7.5	0.33–2.96 log reduction	Sarrías et al. (2003)
Kimbab	*Salmonella* Typhimurium, *Escherichia coli*, *Listeria ivanovii*, *Staphylococcus aureus*	Gamma	1	2–3 log reduction	Chung et al. (2007)
			3	Not detected	
Maize seeds	Total fungi	Electron beam	0.1–6.2	Reduced but not inactivate completely	Nemţanu et al. (2014)
Red sorghum flour	Total aerobic count and aerobic spore count	Gamma	10	0.1–3.5 log reduction	Mukisa et al. (2012)
			10+25	Complete decontamination	
Rice	*Escherichia coli*, *Salmonella* Typhimurium and *Listeria monocytogenes*	Gamma	1.5	3.6–5.7 log reduction	Begum et al. (2020)
Rice	*Escherichia coli*, *Salmonella* Typhimurium and *Listeria monocytogenes*	X-ray	1.5	3.3–4.4 log reduction	Begum et al. (2020)
Waxy rice	Aerobic plate count	Gamma	0.5–2	No significant difference	Sung et al. (2008)
Wheat, barley, maize, sorghum	Coliforms and "coagulase- positive" staphylococci	Gamma	1	Inhibited	Aziz et al. (2006)
Wheat, barley, maize, sorghum	Total aerobic bacteria	Gamma	5	2 log reduction	Aziz et al. (2006)
			10	More than 3 log reduction	

(Continued)

TABLE 12.3 (CONTINUED)   STUDIES CONDUCTED WITH IRRADIATION TREATMENT

Grain or Product	Hazard	Type of Irradiation	Dose (kGy)	Results	References
Wheat, barley, maize, sorghum	Fungi	Gamma	5	Inhibited	Aziz et al. (2006)
		Mycotoxin			
Corn	Ochratoxin A Zearalenone	Electron beam	50	67.9% degradation 71.1% degradation	Luo et al. (2017)
Corn	Aflatoxin B$_1$	Gamma	6–10	89.59–95.47% reduction	Serra et al. (2018)

Curculionidae) in rice, a dose of 120 Gy was enough for adult mortality and sterility. This dose also resulted in no adult emerging within the immature life stage consisting of a mixture of eggs, larvae and pupae. In brief, a dose of 120 Gy radiation was suggested for controlling *Sitophilus oryzae* (L.) (Coleoptera: Curculionidae) (Follett et al., 2013).

Studies on the effects of radiation applications on microorganisms and insects as well as their effects on allergens are also available in the literature. Lectin, one of the most common agents in food intolerance, has been investigated in this respect. While low doses of gamma radiation (0.1 kGy) stimulated lectin, administration of high doses (above 1 kGy) caused changes in the lectin structure and significant losses in its activity (Vaz et al., 2011). Although the efficiency of ionizing radiation for controlling the allergenicity in food is known, several factors such as the dose rate, pH, temperature, humidity, etc. that can affect the characteristics of foods should also be considered (Harder et al., 2017).

12.6 APPLICATION OF OZONE IN GRAIN INDUSTRIES

Ozone, an unstable strong oxidizing agent, is a chemical disinfectant for microorganisms containing bacteria, fungi and viruses (Wang et al., 2016a; Piacentini et al., 2017; Young et al.,2006). The initial microbial load is one of the most important factors affecting the antimicrobial activity of ozone; if the microbial population is large, the antimicrobial activity of ozone will decrease (Kumar and Sabikhi, 2019). After ozone application, the microbial cells are burst due to the formation of pores in the cell wall and water movement from outside to inside. These bursts cause microbial cell death. Additionally, microorganisms' cell membranes, sulfhydryl groups and the amino acids of enzymes as well as peptides and proteins are affected by the oxidizing power of ozone (Kumar and Sabikhi, 2019).

Ozone can be used as a fungicide or insecticide (Raila et al., 2006). It also has some advantages over fumigants. These advantages are that it rapidly decomposes into molecular oxygen without leaving any residue, it does not produce toxic compounds and it can be produced on-site (Wang et al., 2016a; Kumar and Sabikhi, 2019). Moreover, the gas form of ozone is effective on large volumes with high penetration capacity (Silva et al., 2020). Due to its advantages over conventional food processing technologies, ozone has numerous applications such as disinfectant in water treatment, food processing and preservation (Tiwari et al., 2010). Additionally, it has been used to control pest development in stored grains, disinfect flours and degrade toxic molecules

(Lullien-Pellerin; 2012). In cereal grains, ozone has been used as an antifungal agent and for the destruction of mycotoxins (Tiwari et al., 2010).

In the food industry, ozone application is gaining a lot of attention day by day and it has been confirmed as generally recognized as safe for the usage of both aqueous or gaseous ozone in food processing (Tiwari et al., 2010). Moreover, it has potent disinfectant characteristics because of its triatomic oxygen (O_3) content (Piacentini et al., 2017). Many researchers studied the effect of the ozone process on insects and microorganisms, especially during grain storage. Moreover, the reduction of toxic chemical levels is another important topic of literature study.

Grains are usually stored in bulk silos at ambient temperatures for 36 months. Insects have caused severe damage to stored grains and could develop resistance to insecticides commonly used (Lullien-Pellerin, 2012). Storage grains are very susceptible to several insects, such as *Tribolium castaneum* (McDonough et al., 2011a), *Tribolium confusum*, *Ephestia kuehniella* (Işikber & Öztekin, 2009), *Oryzaephilus surinamensis* and *Rhyzopertha dominica* (Sousa et al., 2008; Subramanyam et al., 2017). However, pesticide usage has been restricted due to increasing environmental concerns and newly adopted legislation in Europe and the USA. So ozone could be produced as an alternative to fungicide (Lullien-Pellerin, 2012). The type of insect species, and in particular its life stages, affects the efficiency of ozone in stored products. McDonough et al. (2011a) evaluated high ozone concentrations applied in a short time to kill all life stages – eggs, larvae, pupae and adults – of six different insects on stored grains. *Plodia interpunctella* adults were 100% killed when treated with 500 ppm ozone at 60 min; however, at the same concentration and exposure time, *Tribolium castaneum* mortality was 12%.

Moreover, a higher concentration is needed to obtain 100% mortality for eggs and pupae than for larvae. It can be understood from this study that the sensitivity levels of different insect species and stages are different. Therefore, larval and pupal stages are more sensitive to ozone than the adult stage. Some studies conducted with ozone treatment against insects are summarized in Table 12.4.

Similar results were obtained by Işikber and Öztekin (2009), who found a significant difference in susceptibility to gaseous ozone treatment between *Ephestia kuehniella* Zeller and *Tribolium confusum* and their different life stages in hard winter wheat. At a lower ozone concentration of 5 ppm, the mortality rate of *Ephestia kuehniella* Zeller was higher than in untreated stored grain. When the ozone concentration increased from 10 ppm to 100 ppm, there was only a little change in mortality rate (Moraglio et al., 2018). In the literature, researchers found contradictory results about the most efficient temperatures for pest control with ozone. For example, Silva et al. (2020) reported that when O_3 concentration reached 0.1 mg/L, stored rice showed 100% mortality of *Sitophilus zeamais* insects at six days. The insect mortality rate in grains had not been affected by low and high temperatures. Hansen et al. (2013) observed stored crushed wheat, raisins and mixed birdseed treated with a high dose of ozone (approximately 131 ppm) for eight days at 7.9°C, 20°C and 31.6°C, all of which led to full mortality in all stages of *Plodia interpunctella* except eggs at 31.6°C. However, stored crushed wheat treated with a high dose of ozone for eight days led to full mortality at all three temperatures and all stages of *Sitophilus granarius*. Based on these studies, there was no relation between the temperature and the insect mortality rate in stored grain. The rate of ozone degradation was not found to be primarily temperature-dependent in wheat and maize. However, the half-life of ozone changed according to the wheat grade. For example, the half-life of ozone is 122 s in grade 4 wheat and is 242 s in grade 2 wheat (Hardin et al., 2010). Unlike these studies, the average *Sitophilus zeamais* mortality with 50 ppm ozone for 24 h was 12.7%, 3.95%, 73.21% and 100% at 20, 30, 35 and 40°C, respectively (Sousa et al., 2006).

TABLE 12.4   STUDIES CONDUCTED WITH OZONE TREATMENT AGAINST INSECTS

Grain or Product	Insect Species	Life Stage	Concentration and Exposure Time	Mortality Rate (%)	References
Corn	*Sitophilus zeamais*	Adult	50 ppm, 48 h	100	Faroni, Pereira and Sousa (2007)
Durum wheat flour	*Ephestia kuehniella* Zeller	Eggs Larvae Pupae	100 ppm, 24 h	62 68 85	Moraglio et al. (2018)
Maize	*Sitophilus zeamais*	Adult	47,800 ppm, 1.5 min	100	McDonough et al. (2011b)
Maize	*Tribolium castaneum* *Sitophilus zeamais* *Plodia interpunctella*	Adult	25 ppm, 5 days	77 99 100	Kells et al. (2001)
Popcorn kernel	*Sitophilus zeamais*	Adult	500 µg/L, 10 h 2200 µg/L, 3.12 h 4500 µg/L, 2.1 h	100	Silva et al. (2019)
Red flour	*Tribolium castaneum*	Adult	47,800 ppm, 1.5 min	100	McDonough et al. (2011b)
Rice	*Sitophilus zeamais*	Adult	0.1 mg/L, 6 days	100	Silva et al. (2020)
Wheat	*Ephestia kuehniella*	Adult Pupae Larvae Eggs	13.9 mg/L, 2 h	100	Işikber and Öztekin (2009)
Wheat	*Rhyzopertha dominica*	Adult	0.5 g/m³, 60 min 1 g/m³, 30 min	64 5	Subramanyam et al. (2017)
Wheat	*Rhyzopertha dominica*	Adult	0.21 g/m³, 4 h	96	Xinyi et al. (2019)
Wheat	*Sitophilus granarius* *Plodia interpunctella*	Adult Larvae Eggs	135 ppm, 8 days	100	Hansen et al. (2013)
Wheat	*Tribolium confusum*	Adult Pupae Eggs Larvae	13.9 mg/L, 2 h	4.2–14.1 4.2–14.1 4.2–14.1 74	Işikber and Öztekin (2009)

Ozone in aqueous or gaseous form has been used to inactivate microbial decontamination in several food grains (Table 12.5). Several log reductions in cereal grains can be achieved according to the literature, depending on temperature, pH, relative humidity and ozone concentration (Tiwari et al., 2010). Fungal contamination is the major issue determining the quality of grain. Kottapalli et al. (2005) reported that treating malting barley with ozone caused a 38% to 54% reduction in *Fusarium* without any adverse effect. Gaseous ozone treatment of 100–200 ppm for

TABLE 12.5   STUDIES CONDUCTED WITH OZONE TREATMENT AGAINST MICROORGANISMS AND MYCOTOXINS

Grain or Product	Hazard	Treatment Conditions	Reduction (%)	References
Microorganism				
Barley	*Fusarium* spp.	11 mg/g, 15 min 26 mg/g, 15 min	24 36	Kottapalli et al. (2005)
Malting barley grains	*Fusarium oxysporum, Fusarium graminearum, Fusarium poae*	60 mg/kg, 80 min	66.3–93.3	Piacentini et al. (2017)
Maize	*Fusarium verticilloides*	100 ppm, 30 min	100	Mylona et al. (2014)
Maize	*Aspergillus* spp. *Fusarium* spp. *Mucor* spp.	500 ppm, 1h	50–76	White et al. (2013)
Rice	*Aspergillus* spp. *Penicillium* spp.	0.393 kg/m^3	90	Savi et al. (2020)
Wheat	*Fusarium graminearum*	60 μmol/mol, 180 min	100	Savi et al. (2014)
Wheat seeds	Total fungal count	45 min	92.86	Granella et al. 2018
Mycotoxin				
Bran	Deoxynivalenol	80 mg/L, 10 min	76.21	Sun et al. (2016)
Corn	Aflatoxin	10–12 wt, 96 h	92	Prudente and King (2002)
Corn	Deoxynivalenol	80 mg/L, 10 min	70.65	Sun et al. (2016)
Corn	Zearalenone Ochratoxin A	100 mg/L, 180 min	70.7–90.7	Qi et al. (2016)
Corn flour	AflatoxinB_1 AflatoxinB_2 Aflatoxin G_1	75 mg/L, 60 min	70.73–78.76	Luo et al. (2014)
Durum wheat	Deoxynivalenol Deoxynivalenol-3-Glc	55 g O_3/h, 6 h	29–44	Piemontese et al. (2018)
Pasta	Deoxynivalenol Deoxynivalenol-3-Glc	55 g O_3/h, 6 h	11–42	Piemontese et al. (2018)
Wheat	Deoxynivalenol	60 μmol/mol, 120 min	100	Savi et al. (2014)
Wheat	Deoxynivalenol	75 μg/ml, 30 min 75 μg/ml, 60 min 75 μg/ml, 90 min	26.40 39.16 53.48	Wang et al. (2016a)
Wheat	Deoxynivalenol	80 mg/L, 10 min	74.86	Sun et al. (2016)

30–60 min on maize showed a significant reduction in *Fusarium verticillioides*. However, less reduction in mycelial growth may occur. Fumonisins have not been produced at 0.88 a_w after ozone treatment, while fumonisin production was completely inhibited at 0.92 a_w for 30 days of storage (Mylona et al., 2014). In dry paddy rice stored in silos at three different concentrations (10, 20 and 40 mg/L), total fungi load was highly reduced. From this study, different fungal strains show different gas susceptibility in the order *Acremonium > Alternaria > Aureobasidium > Aspergillus > Penicillium*. Furthermore, after ozone treatment, a_w was reduced (Beber-Rodrigues et al., 2015). Rice grain's total fungi count was also reduced with ozone application. It was concluded that the increased exposure times were not sufficient to inhibit the total fungal count completely. This result shows that increasing the ozone concentration is crucial to maximizing fungal decontamination (Savi et al., 2020).

Ozone can be used as an antifungal agent in silo-stored rice, but strategies to homogenize ozone spread need to be adopted for effective use (Savi et al., 2020). Granella et al. (2018) reported that the fungal count reduced from 1.87 cfu/g to 0.13 cfu/g when ozone-treated wheat seeds were dried at 50°C for 45 minutes. They concluded that ozone exposure time and drying air temperature did not significantly affect physiological qualities such as germination, vigor and electrical conductivity. Ozone treatment at 60 μmol/mol after 30, 60 and 120 min in pericarp wheat grain significantly decreased the amount of deoxynivalenol (205.9, 123.7 mg/kg and no detection) compared to the control group (1065, 923 and 981 mg/kg), while endosperm had the same impact. Depend on these findings, ozone treatment has been more effective in controlling deoxynivalenol's external part than its endosperm (Savi et al., 2014). Wang et al. (2016a) studied ozone's efficiency on wheat kernels and different flour fractions against deoxynivalenol. Ozone treatment at 75 mg/L O_3 of wheat kernels after 30, 60 or 90 min reduced deoxynivalenol by 26.4%, 89.16% and 53.48%, respectively. Additionally, the same impact was observed in bran and shorts. However, the deoxynivalenol degradation rate was not significantly changed among different flour fractions at different exposure times. Piemontese et al. (2018) studied gaseous ozone's efficiency on durum wheat grain in reducing deoxynivalenol, deoxynivalenol-3-Glc, bacteria, fungi and yeasts. These studies were conducted at an ozone concentration of 55 g/h flushed for 6, 8 and 10 h. The bacterial count (in all treatments) and fungal count at 6 and 8 h of treatment were remarkably reduced. Additionally, yeasts (in all treatments) and fungi after 10 h of treatment were completely inhibited. Feoxynivalenol and deoxynivalenol-3-Glc counts were reduced in both durum wheat and pasta that was made with ozonated grains. Moreover, Sun et al. (2016) found that deoxynivalenol content was also significantly reduced in wheat, corn and bran treated with saturated aqueous ozone at 80 mg/L for 15 minutes. For maize, White et al. (2013) reported that 500 and 1,000 ppm ozone were significantly effective in controlling *Aspergillus*, *Fusarium* and *Mucor*, while 15,000 ppm ozone was effective for *Penicillium*. It can be understood from this research that *Penicillium* was more resistant to ozone treatment than *Aspergillus* or *Fusarium*. Furthermore, Agriopoulou et al. (2016) reported that the order of sensitivity of aflatoxins to ozone treatment was aflatoxin G_1 > aflatoxin B_1 > aflatoxin G_2 > aflatoxin B_2. Torlak et al. (2016) observed a reduction in aflatoxin B, aerobic plate and total yeast and mold counts in poultry feed by applying 2.8 mg/L and 5.3 mg/L gaseous ozone for 240 min. However, the ozonation processes at those levels have increased thiobarbituric acid reactive substances in feed samples to 4.4 and 5.3 mg/kg from 2.4 mg/kg, respectively. As a result, ozonation was effective in controlling aflatoxin B_1 and microorganisms, but optimization was required to protect food quality. Besides, the impact of ozone propagation and distribution in storage silos on soybean efficiency has been evaluated. Three different portions, which were inferior, central and superior, were selected and a higher fungal reduction was observed in the inferior partition. Total elimination in inferior portions with 180 min exposure was achieved, while elimination at central and superior partitions was 96% and 80%, respectively. Additionally, at shorter exposure times, *Aspergillus* and *Penicillium* were more resistant to ozone treatment than *Fusarium* (Gomes et al., 2020).

Bifenthrin and pirimiphos-methyl pesticides, used against pests, fungi and weeds, are widely used in wheat storage units. The maximum residue limits for bifenthrin and pirimiphos-methyl pesticides are approximately 0.5 mg/kg and 5.0 mg/kg, according to EU regulations (2009). Savi et al. (2016) reported that the levels of bifenthrin and pirimiphos-methyl were reduced by 37.5% and 71.6% with the ozone application, respectively (Savi et al., 2016). Additionally, some studies have been conducted to reduce pesticide residue in grain foods by ozone application (Savi et al., 2015; de Freitas et al., 2017). Besides these beneficial impacts of ozone treatment, the number of radical species could be changed depending on the cereal type and their parts. Łabanowska et al. (2016) found that the number of radicals in whole grains of oat, barley and wheat was increased more than two times. Ozone treatment reduced the radical amount in oat seed coats while increasing it in wheat and barley seed coats. The radical formation differs depending on the cereal genotype.

Ozone is an effective method for disinfecting grains; it is also a greener alternative for preventing or inactivating insects, microorganisms and mycotoxins. However, the efficacy of ozone treatment depends on certain factors, such as application period, concentration and environmental factors. It should be considered that prolonged exposure might cause a significant change in grains.

12.7 APPLICATION OF PULSED LIGHT IN GRAIN INDUSTRIES

Pulsed light (PL) technology involves producing high-power pulses at a short time with inert gas-flash lamps. Among other relevant non-thermal technologies, pulsed light is an alternative to traditional thermal treatment with FDA approval (John and Ramaswamy, 2018; Wang et al., 2016b). High power pulses of radiation have different spectrums including sun (200–1,100 nm), infrared (IR), visible light (VL) and ultraviolet (UV) (Mahendran et al., 2019). In the food industry, PL is used to decontaminate food contact surfaces, equipment, media and food packaging materials. Additionally, it effectively reduces microbial contamination in foods (Mahendran et al., 2019). Pulsed light treatment can be used to kill many pathogenic and spoilage microorganisms in foods including bacteria, yeasts, molds and viruses (John and Ramaswamy, 2018). The microbial inactivation mechanism of PL has been explained with photophysical (structural cell damage), chemical (cell death due to DNA lesions) and thermal (cell death due to disruption and structural changes) mechanisms (Heinrich et al., 2015). The main advantages of using pulsed light are lower energy cost and environmental impact, the lack of residual compound and the significant microbial reduction in short treatment times (Zenklusen et al., 2018). However, the potential usage of PL in the decontamination of grains was limited. Some examples of the studies conducted with pulsed light treatment against different microorganisms and mycotoxins are shown in Table 12.6.

Chen et al. (2019) used intense pulsed light alternatively to conventional pasteurization methods in wheat flour. Intense pulsed light's (IPL) ability to inactivate *Cronobacter sakazakii* and *Enterococcus faecium* was assessed to optimize processing parameters including water activity, particle size and lamp height. Approximately a 5 log cfu/g reduction was observed for both *Cronobacter sakazakii* and *Enterococcus faecium* in wheat flour after four passes. The *Cronobacter sakazakii* inactivation rate was higher than that of *Enterococcus faecium*, which might result

TABLE 12.6   STUDIES CONDUCTED WITH PULSED LIGHT TREATMENT

Grain or Product	Hazard	Method	Treatment Conditions	Reduction	References
Microorganism					
Corn meal	*Aspergillus niger*	Pulsed UV light	5.6 J/cm², 20–100 s	0.176–1.352 log	Jun et al. (2003)
Malting barley	*Aspergillus carbonarius Aspergillus flavus*	Pulsed light	6–18 J/cm², 15 s	1.2–1.7 log	Zenklusen et al. (2018)
Wheat seed	*Mold*	Pulsed light	51.2 J/g	3–4 log	Aron Maftei et al. (2014)
Wheat flour	*Cronobacter sakazakii Enterococcus faecium*	Intense pulsed light	29.36 J/cm², 28 s	2.79–4.92 log	Chen et al. (2019)
Wheat flour	*Cronobacter sakazakii Salmonella* spp. *Bacillus cereus*	Catalytic intense pulsed light	17.46 J/cm², 60 s	2.8–5.42 log	Chen et al. (2020b)
Wheat flour	*Salmonella enterica*	Pulsed light emitting diode	1199 J/cm², 60 s	1.61–2.22 log	Subedi et al. (2020)
Wheat flour	*Salmonella* spp.	Intense pulsed light–emitting diode	270–1620 J/cm², 10–60 min	0.89–2.91 log	Du et al. (2020)
Wheat flour	*Saccharomyces cerevisiae*	Pulsed UV light	24.31 J/cm²	1 log	Fine and Graviş (2004)
Wheat flour	*Cronobacter sakazakii Salmonella* spp. *Bacillus cereus*	Intense pulsed light	17.46 J/cm², 60 s	1.53–4.15 log	Chen et al. (2020b)
Mycotoxin					
Rice fran	Aflatoxin B₁ Aflatoxin B₂	Pulsed light	0.52 J/cm²/pulse, 15 s	86.7–90.3%	Wang et al. (2016b)
Rough rice	Aflatoxin B₁ Aflatoxin B₂	Pulsed light	0.52 J/cm²/pulse, 80 s	39.2–75%	Wang et al. (2016b)

from the antibiotic resistance of the gram-positive pathogen. Additionally, they found that when the thickness of the sample is increased, the inactivation rate is decreased. A prototype of IPL with TiO_2 photocatalysis has been developed to disinfect powdered food products made with wheat flour. The results showed that *Cronobacter sakazakii*, *Salmonella* spp. and *Bacillus cereus* were significantly inactivated by 5.15, 4.95 and 2.8 log cfu/g reductions, respectively, after catalytic intense pulsed light treatment. Catalytic intense pulsed light enables approximately one more log reduction than intense pulsed light against bacteria. The *Bacillus cereus* spore is found to be the most resistant bacteria to IPL treatment. It was concluded that a high-energy intensity of each pulse, a high peak intensity and a short duration result in a higher inactivation rate (Chen et al., 2020b). Additionally, *Salmonella* reduction in wheat flour was investigated with pulsed light–emitting diode (LED) treatments. Increasing the treatment time results in higher energy dosage and higher microbial inactivation. Due to wheat flour's opaque surfaces, inactivation of microorganisms was obtained with higher photothermal energy. The temperature of wheat flour increased during LED treatment, and the reactive oxygen species produced contribute to microbial inactivation with photothermal effects (Subedi et al., 2020). Du et al. (2020) investigated the inactivation ability of *Salmonella* cocktail in wheat flour with a pulsed light–emitting diode treatment at 365 nm. In a semi-closed system with 365 nm pulsed light treatment, the maximum reduction was observed as 2.91 log in wheat flour. However, prolonged treatment times were required to obtain higher inactivation rates. Commercial application of this technology requires a reduction of processing time and control of water activity. The bactericidal effect of intense pulsed light treatment was also achieved for cassia seeds and glutinous millet with cyclone type and belt type. When using the belt-type intense pulsed light, the maximum microbial reduction in cassia seeds and glutinous millet was observed to be 2.63 and 0.55 log/g, respectively, while it was 0.63 and 0.55 log/g with the cyclone type. The study also concluded that cassia seeds were more vulnerable to PL treatment due to their large particle size than glutinous millet (Hwang et al., 2018). One study by Zenklusen et al. (2018) evaluated the pulsed light inactivation efficiency in malting barley against *Aspergillus carbonarius* and *Aspergillus flavus*. These authors found 1.2–1.7 log reductions in *Aspergillus carbonarius* and *Aspergillus flavus* counts after 15 s of PL treatment, respectively. However, when the treatment times were higher than 15 s, some slight modification was shown of the structure of grains, including altering the pericarp, cell walls and amyloplasts of the endosperm after 75 s of treatment. Pulsed UV light technology has also been studied on grain products including wheat flour and corn flour. Jun et al. (2003) studied the inactivation effects of pulsed light on *Aspergillus niger* spores in cornmeal. They found that when the electric power and treatment time increased, inactivation was also increased by optimizing process conditions. The maximum log reduction (4.93) was observed at 3.8 kV for 100 s with a 3-cm distance from the strobe. Fine and Gravis (2004) applied pulsed UV light to wheat flour to inactivate *Saccharomyces cerevisiae*. Efficient inactivation was observed after four to six rounds of treatment with UV light. Also, they found a synergistic thermal effect on microorganisms with pulsed UV light.

There are also some studies that investigate pulsed light technology's ability to degrade and detoxify aflatoxin in grain products. The mycotoxin degradation rate of PL treatment is affected by exposure time, product moisture content, temperature, mycotoxin structure and product structure (Deng et al., 2020; Abuagela et al., 2018). Wang et al. (2016b) studied the aflatoxin detoxification ability of pulsed light in rough rice and rice bran. Rough rice was pulsed light–treated at a distance of 9 cm for 80 s while rice bran was treated for 15 s. After PL treatment, rough rice showed a 75% (132 to 33 ppb) reduction in aflatoxin B_1 and 39% (45 to 27 ppb) reduction in aflatoxin B_2. Additionally, in rice bran, pulsed light treatment reduced aflatoxin B_1 and aflatoxin B_2 by 90.3% and 86.7%, respectively. It was also concluded that aflatoxin B_1 is more resistant than aflatoxin B_2 to pulsed light degradation. Aflatoxin B_2 is more stable under PL treatment because it does not contain a double bond; however, aflatoxin B_1 is more vulnerable because of its C8–C9 double bond.

12.8 APPLICATION OF ULTRAVIOLET LIGHT IN GRAIN INDUSTRIES

Ultraviolet (UV) light is applied as an alternative method for reducing microbial load in food products and sterilizing food contact surfaces like food packages and food processing environments (Taze et al., 2019). The processing wavelength changing from 200 nm to 400 nm is subdivided into three categories as UV-C (200–280 nm), UV-B (280–315 nm) and UV-A (315–400 nm) (Sastry et al., 2000). This technique affects the genetic material of microorganisms and inhibits their reproduction mechanisms by that effect, depending on parameters including type of strain, culture and growth phase (Falguera et al., 2011). Although different studies have reported the reduction of microbial load and the inhibition of both the microorganisms and their metabolites in food products through the UV method (Corrêa et al., 2020; Tremarin et al., 2017; Mikš-Krajnik et al., 2017; Menezes et al., 2020; Feliciano and Gabriel, 2019), there are few studies that deal with the application of UV treatment to grain and grain products. Some studies conducted with ultraviolet light treatment against microorganisms and mycotoxins are shown in Table 12.7.

Fungal cultures and their metabolites, especially fungal toxins, are major risks for consumers and are commonly observed in plant-based food products. Ghanghro et al. (2016) addressed the loss of grains during storage due to fungal growth and mycotoxin production. Aflatoxins have gained great attention for being controlled and eliminated by UV treatment (Patras et al., 2017). Ferreira et al. (2020) reported that the treatment of brown, black and red rice grains with UV light for a certain time reduced mycotoxins and controlled fungal growth. While the fungal colonies in the untreated grains increased during six months, UV-C application for 3 h decreased colonies in grains of brown, black and red rice by 21%, 56% and 88%, respectively. The treatment conditions are important parameters for sufficient inactivation. In that study, the effect of treatment time was also studied and showed that an increase in exposure time increased UV-C's efficiency in reducing both mycotoxins and fungal colonies. Ataila et al. (2004) determined the mycotoxin production by *Aspergillus parasiticus*, *Fusarium verticillioides*, *Scopulariopsis fusca* and *Verticillium lecanii* on wheat grain stored at different relative humidities (50%, 74% and 80%) and treated with two different wavelengths of UV (254 nm and 362 nm) for various exposure times (30, 60 and 120 min). Depending on the exposure time and wavelength, while none of some mycotoxins was detected, other mycotoxins were reduced at different levels. When the grains were exposed to UV light at both wavelengths for 30 and 60 min, aflatoxin G_1 and ochratoxin B were eliminated. The degradation of other mycotoxins (aflatoxins B_1, B_2 and G_2; ochratoxin A, sterigmatocystin, deoxynivalenol) was observed under certain treatment conditions. This study also showed the importance of exposure conditions because some specific treatment parameters caused an increase in the toxin-producing ability of cultures, which might be related to the chemical changes observed after UV exposure.

The study by Popović et al. (2018) focused on the effectiveness of UV-C treatment to reduce the growth of *Fusarium graminearum* and *Penicillium verrucosum*, as well as mycotoxins produced by these cultures. The maize and wheat kernels were exposed to 3D UV-C due to being a potential risky grain type for studying cultures. There was a noticeable reduction in *Penicillium verrucosum* culture on maize and wheat kernels, contrary to the *Fusarium graminearum* spores. However, reduction in the same cultures that were naturally occurring on maize and wheat kernels was lower. In mycotoxins, the highest to lowest reductions by using 3D UV-C were zearalenone A, deoxynivalenol and ochratoxin A. The reduction of overall mycotoxins on maize kernels (52–17%) was higher than on wheat kernels (48–46%). That difference was correlated with the surface characteristics of kernels. Ghanghro et al. (2016) tried to develop an effective UV treatment system to reduce aflatoxin in wheat grain that is open or in sacks. The reduction rate of aflatoxin showed a correlation with exposure time. The effectiveness was lower for wheat grain in sacks, but aflatoxin's reduction reached even 66% after 160 min of treatment. In the other study, the deoxynivalenol level in corn silage was reduced by UV-C at 254 nm after both 30 min and 60 min of exposure time (Murata et al., 2011)

Hidaka and Kubota (2006) studied the impact of UV at 254 nm on bacterial and fungal cultures in wheat samples and found that a treatment time of 6.3 h for the bacterial culture and 5.6 h for the fungal culture was enough to provide 90% sterilization. Jajić et al. (2016) showed a significant reduction in deoxynivalenol levels in maize samples. They applied both UV-A

TABLE 12.7   STUDIES CONDUCTED WITH ULTRAVIOLET LIGHT TREATMENT

Grain or Products	Hazard	Treatment Conditions	Result	References
Microorganism				
Black rice	Fungal colonies	1–3 h (2.06–6.18 kJ/cm²), 254 nm	22–56% decrease	Ferreira et al. (2020)
Brown rice	Fungal colonies	1–3 h (2.06–6.18 kJ/cm²), 254 nm	26% increase 21% decrease	Ferreira et al. (2020)
Maize kernel	*Penicillium verrucosum* spores (inoculated)	500 mJ/cm²	94.5% reduction	Popović et al. (2018)
Red rice	Fungal colonies	1–3 h (2.06–6.18 kJ/cm²), 254 nm	79–88% decrease	Ferreira et al. (2020)
Wheat	*Aspergillus* and *Penicillium*	5.6 h, 254 nm	90% sterilization	Hidaka and Kubota (2006)
Wheat	*Bacillus* and *Pseudomonas*	6.3 h, 254 nm	90% sterilization	Hidaka and Kubota (2006)
Wheat flour	*Salmonella typhimurium* and *Lactobacillus plantarum*	2 m vertical tunnel with twelve 480 W UV-C lamps, 4 s	4 log, 10 cycles inactivation	Condón-Abanto et al. (2016)
Wheat kernel	*Penicillium verrucosum* spores (inoculated)	500 mJ/cm²	94% reduction	Popović et al. (2018)
Mycotoxin				
Black rice	Mycotoxins (423.7 µg/kg)	1–3 h (2.06–6.18 kJ/cm²), 254 nm	297.1–172.8 µg/kg	Ferreira et al. (2020)
Brown rice	Mycotoxins (211.8 µg/kg)	1–3 h (2.06–6.18 kJ/cm²), 254 nm	200.8–166.7 µg/kg	Ferreira et al. (2020)
Corn silage	Deoxynivalenol	30–60 min, 254 nm	22–21% reduction	Murata et al. (2011)
Maize	Deoxynivalenol (1.902–5.334 mg/kg)	30–120 min, 254–368 nm, 15–30 cm	17–148% remaining content	Jajić et al. (2016)
Maize kernel	Zearalenone, deoxynivalenol, ochratoxin A	10000 mJ/cm²	52%, 30%, 17% reduction	Popović et al. (2018)
Red rice	Mycotoxins (262.4 µg/kg)	1–3 h (2.06–6.18 kJ/cm²), 254 nm	144.8–133.9 µg/kg	Ferreira et al. (2020)
Wheat grain (in sacks)	Aflatoxin	5–160 min, 254 nm	9–73% reduction	Ghanghro et al. (2016)
Wheat grain (open)	Aflatoxin	5–160 min, 254 nm	18–90% reduction	Ghanghro et al. (2016)
Wheat kernel	Zearalenone, deoxynivalenol, ochratoxin A	10000 mJ/cm²	48%, 15%, 6% reduction	Popović et al. (2018)

at 368 nm and UV-C at 254 nm, then compared results with parameters of UV wavelength (254 nm and 368nm), treatment time (30, 60 and 120 min) and treatment distance (15 and 30 cm). Apart from three conditions (first: 30 cm, 30 min, 254 nm, 1.902 mg/kg deoxynivalenol control sample; second: 15 cm, 30 min, 254 nm, 5.334 mg/kg deoxynivalenol control sample; third: 30 cm, 60 min, 254 nm, 5.334 mg/kg deoxynivalenol control sample), all of the other conditions resulted in an important reduction in the deoxynivalenol level of the samples. In the case of non-reducing conditions, the deoxynivalenol level in the control sample increased by 23%, 6% and 48% in order of condition ranking. The highest reductions were obtained at 368 nm for 30 min with a 30-cm treatment distance from 1.902 to 0.308 mg/kg and at 254 nm for 120 min with the same distance from 5.334 to 0.891 mg/kg. In addition to grains, their processed products are important for ensuring food safety from biological and chemical hazards. In this respect, Condón-Abanto et al. (2016) paid attention to UV-C's ability to inactivate *Salmonella* Typhimurium and *Lactobacillus plantarum* in wheat flour and showed inactivation in the studied cultures.

Alongside studies concerning the impact of the process directly on grains, the impact of UV on hazards isolated from grains might be considered. Hussein et al. (2015) isolated an aflatoxin B_1 producer *Aspergillus flavus* from corn seeds and investigated the effect of UV light. UV application for 30, 60 and 90 min inhibited the growth of *A. flavus* with 84.73–88.50–92.53% at 220 nm and 81.46–94.40–87.37% at 365 nm, respectively. The inhibition rate of UV changed based on the wavelength of the UV light treatment. While the highest inhibition was observed for 120 min at 220 nm, 60 min resulted in the highest inhibition at 365 nm. An *in vitro* study of the inhibition of mycotoxins produced by *Aspergillus flavus* including aflatoxin B_1, aflatoxin B_2, aflatoxin G_1, aflatoxin G_2 and the aflatoxin total with UV was conducted by Sharareh et al. (2015). That study also provided insight into understanding the changes in the structure of mycotoxins. The treatment at 366 nm for 10 min showed reductions of 98%, 99.5%, 99.8%, 100% and 99.1% in the amounts of aflatoxin B_1, aflatoxin B_2, aflatoxin G_1, aflatoxin G_2 and total aflatoxins, respectively. Therefore, the possible success of the method in the model system was mentioned.

The quality parameters of studied food products are critical for both consumers and producers. In some studies, the effect of UV treatment on food quality was also discussed in several aspects. Ferreira et al. (2020) carried out analyses to check quality parameters after UV treatment. While the bound phenolics were released in both black and red rice grains after 1 h of treatment, cooking and color parameters showed no change. However, the increasing treatment duration from 1 to 3 h decreased the total phenolics in black (695.4 to 661.7 µg/g) and red rice (724.9 to 623.7 µg/g). In the study by Popović et al. (2018), the moisture, crude protein and germination properties of maize and wheat kernels exposed to 5,000 mJ/cm^2 UV-C were compared with those of untreated kernels. Except in reducing the germination of wheat from 91.7% to 75%, there were no noteworthy changes in other quality characteristics. Ghanghro et al. (2016) also reported that UV-C treatment had no impact on the crude protein of wheat. In the study of Murata et al. (2011), the content of α-tocopherol and β-carotene in corn silage was preserved after UV at 254 nm for 15 min, 30 min and 60 min. Although Hidaka and Kubota (2006) aimed to reach 90% sterilization by increasing exposure time, they observed slight and acceptable germination changes as well as quality changes.

In addition to the single effect of UV treatment, the synergistic effect of combined processes for microbial inactivation has been shown (Jung et al., 2008), but the number of reported studies for grain and grain products should be increased. Future research should also look into the impact of UV treatment on food quality and other relevant properties.

12.9 CONCLUSION

Current non-thermal strategies to achieve the food safety of grain and grain products are comparable to other chemical, biological and physical approaches. By using non-thermal technologies, food products can be produced by maintaining sensorial and nutritional characteristics. In the case of the food safety of grains, biological and chemical hazards are the main topics. Non-thermal technologies can combine with other non-thermal or thermal treatments to eliminate the hazards in grain and grain products. Several factors should be considered when using these technologies, such as dose, concentration, treatment time, etc. However, understanding the complexity and parameters of these methods and their mechanisms is the main challenge. Although general comprehension of these methods has been obtained by the studies conducted to date, there is still a necessity to increase knowledge.

REFERENCES

Abuagela, M. O., Iqdiam, B. M., Mostafa, H., Gu, L., Smith, M. E., & Sarnoski, P. J. (2018). Assessing pulsed light treatment on the reduction of aflatoxins in peanuts with and without skin. *International Journal of Food Science and Technology*, *53*(11), 2567–2575.

Agriopoulou, S., Koliadima, A., Karaiskakis, G., & Kapolos, J. (2016). Kinetic study of aflatoxins' degradation in the presence of ozone. *Food Control*, *61*, 221–226.

Ahmed, M. (2001). Disinfestation of stored grains, pulses, dried fruits and nuts, and other dried foods. In: R. Molins (Ed.), *Food Irradiation: Principles and Applications* (pp. 77–112). New York: John Wiley & Sons.

Akoto, O., Andoh, H., Darko, G., Eshun, K., & Osei-Fosu, P. (2013). Health risk assessment of pesticides residue in maize and cowpea from Ejura, Ghana. *Chemosphere*, *92*(1), 67–73.

AlZain, S. (2020). Effect of chemical, microwave irradiation, steam autoclave, ultraviolet light radiation, ozone and electrolyzed oxidizing water disinfection on properties of impression materials: A systematic review and meta-analysis study. *The Saudi Dental Journal*, *32*(4), 161–170.

Aron Maftei, N., Ramos-Villarroel, A. Y., Nicolau, A. I., Martín-Belloso, O., & Soliva-Fortuny, R. (2014). Pulsed light inactivation of naturally occurring moulds on wheat grain. *Journal of the Science of Food and Agriculture*, *94*(4), 721–726.

Ataila, M., Hassanein, N. M., El-Beih, A. A., & Youssef, Y. A. (2004). Effect of fluorescent and UV light on mycotoxin production under different relative humidities in wheat grains. *Acta Pharmaceutica Sciencia*, *46*(3), 1006–1012.

Audenaert, K., Monbaliu, S., Deschuyffeleer, N., Maene, P., Vekeman, F., Haesaert, G., De Saeger, S., & Eeckhout, M. (2012). Neutralized electrolyzed water efficiently reduces Fusarium spp. in vitro and on wheat kernels but can trigger deoxynivalenol (DON) biosynthesis. *Food Control*, *23*(2), 515–521.

Ayvaz, A., & Tunçbilek, A. Ş. (2006). Effects of gamma radiation on life stages of the Mediterranean flour moth, Ephestia kuehniella Zeller (Lepidoptera: Pyralidae). *Journal of Pest Science*, *79*(4), 215.

Aziz, N. H., Souzan, R. M., & Azza, A. S. (2006). Effect of γ-irradiation on the occurrence of pathogenic microorganisms and nutritive value of four principal cereal grains. *Applied Radiation and Isotopes*, *64*(12), 1555–1562.

Beber-Rodrigues, M., Savi, G. D., & Scussel, V. M. (2015). Ozone effect on fungi proliferation and genera susceptibility of treated stored dry paddy rice (*Oryza sativa* L.). *Journal of Food Safety*, *35*(1), 59–65.

Begum, T., Follett, P. A., Hossain, F., Christopher, L., Salmieri, S., & Lacroix, M. (2020). Microbicidal effectiveness of irradiation from gamma and X-ray sources at different dose rates against the foodborne illness pathogens *Escherichia coli*, *Salmonella* Typhimurium and *Listeria monocytogenes* in rice. *LWT*, *132*. http://www.ncbi.nlm.nih.gov/pubmed/109841.

Boluda-Aguilar, M., Taboada-Rodríguez, A., López-Gómez, A., Marín-Iniesta, F., & Barbosa-Cánovas, G. V. (2013). Quick cooking rice by high hydrostatic pressure processing. *LWT-Food Science and Technology*, *51*(1), 196–204.

Butscher, D. (2016). *Nonthermal Plasma Inactivation of Microorganisms on Granular Food Products* (Doctoral dissertation, ETH Zurich).

Butscher, D., Schlup, T., Roth, C., Müller-Fischer, N., Gantenbein-Demarchi, C., & von Rohr, P. R. (2015). Inactivation of microorganisms on granular materials: Reduction of *Bacillus amyloliquefaciens* endospores on wheat grains in a low pressure plasma circulating fluidized bed reactor. *Journal of Food Engineering, 159*, 48–56.

Butscher, D., Zimmermann, D., Schuppler, M., & von Rohr, P. R. (2016). Plasma inactivation of bacterial endospores on wheat grains and polymeric model substrates in a dielectric barrier discharge. *Food Control, 60*, 636–645.

Cassares, M., Sakotani, N. L., Kunigk, L., Vasquez, P., & Jurkiewicz, C. (2020). Effect of gamma irradiation on shelf life extension of fresh pasta. *Radiation Physics and Chemistry.* http://www.ncbi.nlm.nih.gov/pubmed/108940.

Chen, D., Cheng, Y., Peng, P., Liu, J., Wang, Y., Ma, Y., Anderson, E., Chen, C., Chen, P., & Ruan, R. (2019). Effects of intense pulsed light on *Cronobacter sakazakii* and *Salmonella* surrogate *Enterococcus faecium* inoculated in different powdered foods. *Food Chemistry, 296*, 23–28.

Chen, Y. X., Guo, X. N., Xing, J. J., Sun, X. H., & Zhu, K. X. (2020a). Effects of wheat tempering with slightly acidic electrolyzed water on the microbial, biological, and chemical characteristics of different flour streams. *LWT, 118.* http://www.ncbi.nlm.nih.gov/pubmed/108790.

Chen, D., Wiertzema, J. R., Peng, P., Cheng, Y., Wang, Y., Liu, J., Ma, Y., Mosher, W., Kang, M., Min, M., Chen, P., Baumler, D. J., Chen, C., Lee, L., Vickers, Z., Feirtag, J., & Chen, P. (2020b). Catalytic intense pulse light inactivation of *Cronobacter sakazakii* and other pathogens in non-fat dry milk and wheat flour. *Food Chemistry, 332*, 127420.

Chung, H. J., Lee, N. Y., Jo, C., Shin, D. H., & Byun, M. W. (2007). Use of gamma irradiation for inactivation of pathogens inoculated into *Kimbab*, steamed rice rolled by dried laver. *Food Control, 18*(2), 108–112.

Cleland, M. R. (2006). Advances in gamma ray, electron beam, and X-ray technologies for food irradiation. In: C. H. Sommers and X. Fan (Eds.), *Food Irradiation Research and Technology* (pp. 11–35). Ames, IA: IFT Press, Blackwell Publishing.

Collins, P. J. (2006, October). Resistance to chemical treatments in insect pests of stored grain and its management. In: *Proceedings of the 9th International Working Conference on Stored Product Protection* (Vol. 15), Sao Paulo, Brazil.

Condón-Abanto, S., Condón, S., Raso, J., Lyng, J. G., & Álvarez, I. (2016). Inactivation of *Salmonella typhimurium* and *Lactobacillus plantarum* by UV-C light in flour powder. *Innovative Food Science and Emerging Technologies, 35*, 1–8.

Considine, K. M., Kelly, A. L., Fitzgerald, G. F., Hill, C., & Sleator, R. D. (2008). High-pressure processing–effects on microbial food safety and food quality. *FEMS Microbiology Letters, 281*(1), 1–9.

Corrêa, T. Q., Blanco, K. C., Garcia, É. B., Perez, S. M. L., Chianfrone, D. J., Morais, V. S., & Bagnato, V. S. (2020). Effects of ultraviolet light and curcumin-mediated photodynamic inactivation on microbiological food safety: A study in meat and fruit. *Photodiagnosis and Photodynamic Therapy, 30*, 101678.

Darmawi, Legaspi, J., & Hallman, G. J. (1998). Radiation doses for quarantine security against Mexican rice borer and sugarcane borer. In *Proceedings of the 1998 Annual International Research Conference on Methyl Bromide Alternatives and Emissions Reductions (Orlando, Florida)*, Fresno, p. 117, 1–2.

Daryaei, H., Balasubramaniam, V. M., & Legan, J. D. (2013). Kinetics of Bacillus cereus spore inactivation in cooked rice by combined pressure–heat treatment. *Journal of Food Protection, 76*(4), 616–623.

Dasan, B. G., Boyaci, I. H., & Mutlu, M. (2016). Inactivation of aflatoxigenic fungi (*Aspergillus* spp.) on granular food model, maize, in an atmospheric pressure fluidized bed plasma system. *Food Control, 70*, 1–8.

de Freitas, R. D. S., Faroni, L. R. D. A., de Queiroz, M. E. L. R., Heleno, F. F., & Prates, L. H. F. (2017). Degradation kinetics of pirimiphos-methyl residues in maize grains exposed to ozone gas. *Journal of Stored Products Research, 74*, 1–5.

Deng, L. Z., Tao, Y., Mujumdar, A. S., Pan, Z., Chen, C., Yang, X. H., Liu, Z., Wang, H., & Xiao, H. W. (2020). Recent advances in nonthermal decontamination technologies for microorganisms and mycotoxins in low-moisture foods. *Trends in Food Science and Technology.*

Du, L., Prasad, A. J., Gänzle, M., & Roopesh, M. S. (2020). Inactivation of Salmonella spp. in wheat flour by 395 nm pulsed light emitting diode (LED) treatment and the related functional and structural changes of gluten. *Food Research International*, *127*/ http://www.ncbi.nlm.nih.gov/pubmed/108716.

Estrada-Girón, Y., Swanson, B. G., & Barbosa-Cánovas, G. V. (2005). Advances in the use of high hydrostatic pressure for processing cereal grains and legumes. *Trends in Food Science and Technology*, *16*(5), 194–203.

European Union Commission, E. U. (2009). Pesticide EU – Maximum residue limit database. Regulation No. 396/2005. Updated on 18/02/2009. Available from http://ec.europa.eu/food/plant/protection/pesticides/index_en.htm (accessed 01.11.2020).

Falguera, V., Pagán, J., Garza, S., Garvín, A., & Ibarz, A. (2011). Ultraviolet processing of liquid food: A review: Part 2: Effects on microorganisms and on food components and properties. *Food Research International*, *44*(6), 1580–1588.

Faroni, L. R. D., Pereira, A. M., Sousa, A. H., Silva, M. T. C., & Urrichi, W. I. (2007). Influence of corn grain mass temperature on ozone toxicity to Sitophilus zeamais (Coleoptera: Curculionidae) and quality of oil extracted from ozonized grains. In *IOA Conference and Exhibition* (Vol. 1, pp. 1–6). València: IOA.

Feliciano, R. J., & Gabriel, A. A. (2019). Juice composition, physicochemistry, and efficacy of ultraviolet radiation against *Cryptococcus albidus*. *Journal of Food Composition and Analysis*, *84*. http://www.ncbi.nlm.nih.gov/pubmed/103313.

Ferreira, C. D., Lang, G. H., da Silva Lindemann, I., da Silva Timm, N., Hoffmann, J. F., Ziegler, V., & de Oliveira, M. (2020). Postharvest UV-C irradiation for fungal control and reduction of mycotoxins in brown, black, and red rice during long-term storage. *Food Chemistry*, *339*, 127810.

Fine, F., & Gervais, P. (2004). Efficiency of pulsed UV light for microbial decontamination of food powders. *Journal of Food Protection*, *67*(4), 787–792.

Follett, P. A., Snook, K., Janson, A., Antonio, B., Haruki, A., Okamura, M., & Bisel, J. (2013). Irradiation quarantine treatment for control of *Sitophilus oryzae* (Coleoptera: Curculionidae) in rice. *Journal of Stored Products Research*, *52*, 63–67.

Ghanghro, A. B., Channa, M. J., Sheikh, A. S., Nizamani, S. M., & Ghanghro, I. H. (2016). Assessment of aflatoxin level in stored wheat of godowns of Hyderabad division and decontamination by UV radiation. *International Journal of Biological Science*, *8*(1), 8–16.

Gomes, T., Canever, S. B., Savi, G. D., Piacentini, K. C., Cargnin, M., Furtado, B. G., Feltrin, A. C., Quadri, M. B., & Angioletto, E. (2020). Modeling and experimental of mould disinfestation of soybean silos with ozone. *Ozone: Science and Engineering*, *42*(2), 183–193.

Granella, S. J., Christ, D., Werncke, I., Bechlin, T. R., & Coelho, S. R. M. (2018). Effect of drying and ozonation process on naturally contaminated wheat seeds. *Journal of Cereal Science*, *80*, 205–211.

Han, J. Y., Song, W. J., Kang, J. H., Min, S. C., Eom, S., Hong, E. J., Ryu, S., Kim, Sb, Cho, S., & Kang, D. H. (2020). Effect of cold atmospheric pressure plasma-activated water on the microbial safety of Korean rice cake. *LWT*, *120*. http://www.ncbi.nlm.nih.gov/pubmed/108918.

Hansen, L. S., Hansen, P., & Jensen, K. M. V. (2013). Effect of gaseous ozone for control of stored product pests at low and high temperature. *Journal of Stored Products Research*, *54*, 59–63.

Harder, M. N. C., Arthur, V., & Harder, L. N. C. (2017). Gamma radiation effect on allergenic food. In: S. S. Athari (Ed.), Allergen (pp. 139–150). London: IntechOpen.

Hardin, J. A., Jones, C. L., Bonjour, E. L., Noyes, R. T., Beeby, R. L., Eltiste, D. A., & Decker, S. (2010). Ozone fumigation of stored grain; closed-loop recirculation and rate of ozone consumption. *Journal of Stored Products Research*, *46*(3), 149–154.

Heinrich, V., Zunabovic, M., Bergmair, J., Kneifel, W., & Jäger, H. (2015). Post-packaging application of pulsed light for microbial decontamination of solid foods: A review. *Innovative Food Science and Emerging Technologies*, *30*, 145–156.

Hidaka, Y., & Kubota, K. (2006). Study on the sterilization of grain surface using UV radiation. *Japan Agricultural Research Quarterly: JARQ*, *40*(2), 157–161.

Huang, H. W., Hsu, C. P., Yang, B. B., & Wang, C. Y. (2014). Potential utility of high-pressure processing to address the risk of food allergen concerns. *Comprehensive Reviews in Food Science and Food Safety*, *13*(1), 78–90.

Hussein, H. Z., Tuama, R. H., & Ali, A. M. (2015). Study the effect of ozone gas and ultraviolet radiation and microwave on the degradation of aflatoxin b1 produce by *Aspergillus flavus* on stored Maize grains. *IOSR Journal of Agricultural Veterinary Science, 8*(5), 5–12.

Hwang, H. J., Cheigh, C. I., & Chung, M. S. (2018). Comparison of bactericidal effects of two types of pilot-scale intense-pulsed-light devices on cassia seeds and glutinous millet. *Innovative Food Science and Emerging Technologies, 49*, 170–175.

Işikber, A. A., & Öztekin, S. (2009). Comparison of susceptibility of two stored-product insects, *Ephestia kuehniella* Zeller and *Tribolium confusum* du Val to gaseous ozone. *Journal of Stored Products Research, 45*(3), 159–164.

Ito, V. C., Bet, C. D., Wojeicchowski, J. P., Demiate, I. M., Spoto, M. H. F., Schnitzler, E., & Lacerda, L. G. (2018). Effects of gamma radiation on the thermoanalytical, structural and pasting properties of black rice (*Oryza sativa* L.) flour. *Journal of Thermal Analysis and Calorimetry, 133*(1), 529–537.

Jajić, I., Jakšić, S., Krstović, S., & Abramović, B. (2016). Preliminary results on deoxynivalenol degradation in maize by UVA and UVC irradiation. *Contemporary Agriculture, 65*(3–4), 7–12.

John, D., & Ramaswamy, H. S. (2018). Pulsed light technology to enhance food safety and quality: A mini-review. *Current Opinion in Food Science, 23*, 70–79.

Jun, S., Irudayaraj, J., Demirci, A., & Geiser, D. (2003). Pulsed UV-light treatment of corn meal for inactivation of *Aspergillus niger* spores. *International Journal of Food Science and Technology, 38*(8), 883–888.

Jung, Y. J., Oh, B. S., & Kang, J. W. (2008). Synergistic effect of sequential or combined use of ozone and UV radiation for the disinfection of *Bacillus subtilis* spores. *Water Research, 42*(6–7), 1613–1621.

Kalagatur, N. K., Kamasani, J. R., Mudili, V., Krishna, K., Chauhan, O. P., & Sreepathi, M. H. (2018). Effect of high pressure processing on growth and mycotoxin production of Fusarium graminearum in maize. *Food Bioscience, 21*, 53–59.

Kato, T., Katayama, E., Matsubara, S., Omi, Y., & Matsuda, T. (2000). Release of allergenic proteins from rice grains induced by high hydrostatic pressure. *Journal of Agricultural and Food Chemistry, 48*(8), 3124–3129.

Kells, S. A., Mason, L. J., Maier, D. E., & Woloshuk, C. P. (2001). Efficacy and fumigation characteristics of ozone in stored maize. *Journal of Stored Products Research, 37*(4), 371–382.

Khaneghah, A. M., Moosavi, M. H., Oliveira, C. A. F., Vanin, F., & Sant'Ana, A. S. (2020). Electron beam irradiation to reduce the mycotoxin and microbial contaminations of cereal-based products: An overview. *Food and Chemical Toxicology, 143*. http://www.ncbi.nlm.nih.gov/pubmed/111557.

Kilonzo-Nthenge, A. K. (2012). Gamma irradiation for fresh produce. In: Feriz Andovic (Ed.), *Gamma Irradiation* (pp. 251–262). InTech. ISBN: 978-953-51-0316-5.

Kottapalli, B., Wolf-Hall, C. E., & Schwarz, P. (2005). Evaluation of gaseous ozone and hydrogen peroxide treatments for reducing Fusarium survival in malting barley. *Journal of Food Protection, 68*(6), 1236–1240.

Kottapalli, B., Wolf-Hall, C. E., & Schwarz, P. (2006). Effect of electron-beam irradiation on the safety and quality of Fusarium-infected malting barley. *International Journal of Food Microbiology, 110*(3), 224–231.

Kumar, C. T. M., & Sabikhi, L. (2019). Ozone application in food processing. In: O. Chauhan (Ed.), *Non-Thermal Processing of Foods* (pp. 189–243). Florida: CRC Press.

Kume, T., Furuta, M., Todoriki, S., Uenoyama, N., & Kobayashi, Y. (2009). Status of food irradiation in the world. *Radiation Physics and Chemistry, 78*(3), 222–226.

Łabanowska, M., Kurdziel, M., & Filek, M. (2016). Changes of paramagnetic species in cereal grains upon short-term ozone action as a marker of oxidative stress tolerance. *Journal of Plant Physiology, 190*, 54–66.

Lee, J. H., Woo, K. S., Yong, H. I., Jo, C., Lee, S. K., Lee, B. W., Lee, Y., Lee, B., & Kim, H. J. (2019). Physicochemical properties of brown rice according to the characteristics of cultivars treated with atmospheric pressure plasma. *Journal of Cereal Science, 87*, 138–142.

Lee, K. H., Kim, H. J., Woo, K. S., Jo, C., Kim, J. K., Kim, S. H., Park, H. Y., Oh, S., & Kim, W. H. (2016). Evaluation of cold plasma treatments for improved microbial and physicochemical qualities of brown rice. *LWT, 73*, 442–447.

Li, X., Hao, J., Liu, X., Liu, H., Ning, Y., Cheng, R., Tan, B., & Jia, Y. (2015). Effect of the treatment by slightly acidic electrolyzed water on the accumulation of γ-aminobutyric acid in germinated brown millet. *Food Chemistry*, *186*, 249–255.

Liu, R., He, X., Shi, J., Nirasawa, S., Tatsumi, E., Li, L., & Liu, H. (2013). The effect of electrolyzed water on decontamination, germination and γ-aminobutyric acid accumulation of brown rice. *Food Control*, *33*(1), 1–5.

Lorenz, K., & Miller, B. S. (1975). Irradiation of cereal grains and cereal grain products. *C R C Critical Reviews in Food Science and Nutrition*, *6*(4), 317–382.

Los, A., Ziuzina, D., Akkermans, S., Boehm, D., Cullen, P. J., Van Impe, J., & Bourke, P. (2018). Improving microbiological safety and quality characteristics of wheat and barley by high voltage atmospheric cold plasma closed processing. *Food Research International*, *106*, 509–521.

Los, A., Ziuzina, D., Boehm, D., & Bourke, P. (2020). Effects of cold plasma on wheat grain microbiome and antimicrobial efficacy against challenge pathogens and their resistance. *International Journal of Food Microbiology*, *355*, 108889.

Los, A., Ziuzina, D., Boehm, D., Cullen, P. J., & Bourke, P. (2017). The potential of atmospheric air cold plasma for control of bacterial contaminants relevant to cereal grain production. *Innovative Food Science and Emerging Technologies*, *44*, 36–45.

Lullien-Pellerin, V. (2012). Ozone in grain processing. In: C. O'Donnell, B. Tiwari, P. J. Cullen, & R. G. Rice (Eds.), *Ozone in Food Processing* (pp. 81–96). Chicester, UK: Wiley-Blackwell.

Luo, X., Li, Y., Yang, D., Xing, J., Li, K., Yang, M., Wang, R., Wang, L., Zhang, Y., & Chen, Z. (2019). Effects of electron beam irradiation on storability of brown and milled rice. *Journal of Stored Products Research*, *81*, 22–30.

Luo, X., Qi, L., Liu, Y., Wang, R., Yang, D., Li, K., Wang, L., Li, Y., Zhang, Y., & Chen, Z. (2017). Effects of electron beam irradiation on zearalenone and ochratoxin A in naturally contaminated corn and corn quality parameters. *Toxins*, *9*(3), 84.

Luo, X., Wang, R., Wang, L., Li, Y., Wang, Y., & Chen, Z. (2014). Detoxification of aflatoxin in corn flour by ozone. *Journal of the Science of Food and Agriculture*, *94*(11), 2253–2258.

Lyu, F., Gao, F., Zhou, X., Zhang, J., & Ding, Y. (2018). Using acid and alkaline electrolyzed water to reduce deoxynivalenol and mycological contaminations in wheat grains. *Food Control*, *88*, 98–104.

Mahendran, R., Ramanan, K. R., Barba, F. J., Lorenzo, J. M., López-Fernández, O., Munekata, P. E., Roohinejad, S., Sant'Ana, A. S., & Tiwari, B. K. (2019). Recent advances in the application of pulsed light processing for improving food safety and increasing shelf life. *Trends in Food Science and Technology*, *88*, 67–79.

Mahendran, R., Ratish Ramanan, K., Sargunam, R., & Sarumathi, R. (2016). Effect of cold plasma on mortality of *Tribolium castaneum* on maida flour. *Scientific Journal of Agricultururual Engineering*, *3*, 37–44.

Mahugija, J. A. M., Kayombo, A., & Peter, R. (2017). Pesticide residues in raw and processed maize grains and flour from selected areas in Dar es Salaam and Ruvuma, Tanzania. *Chemosphere*, *185*, 137–144.

Manupriya, B. R., Somashekarappa, H. M., Patil, S. L., Shenoy, K. B., & Shenoy, K. B. (2020). Study of gamma irradiation effects on the physico-chemical properties of wheat flour (*Triticum aestivum*, L.). *Radiation Physics and Chemistry*, *172*. http://www.ncbi.nlm.nih.gov/pubmed/108693.

McDonough, M. X., Mason, L. J., & Woloshuk, C. P. (2011a). Susceptibility of stored product insects to high concentrations of ozone at different exposure intervals. *Journal of Stored Products Research*, *47*(4), 306–310.

McDonough, M. X., Campabadal, C. A., Mason, L. J., Maier, D. E., Denvir, A., & Woloshuk, C. (2011b). Ozone application in a modified screw conveyor to treat grain for insect pests, fungal contaminants, and mycotoxins. *Journal of Stored Products Research*, *47*(3), 249–254.

Menezes, N. M. C., Longhi, D. A., Ortiz, B. O., Junior, A. F., & de Aragão, G. M. F. (2020). Modeling the inactivation of *Aspergillus fischeri* and *Paecilomyces niveus* ascospores in apple juice by different ultraviolet light irradiances. *International Journal of Food Microbiology*, *333*, 108773.

Mikš-Krajnik, M., Feng, L. X. J., Bang, W. S., & Yuk, H. G. (2017). Inactivation of *Listeria monocytogenes* and natural microbiota on raw salmon fillets using acidic electrolyzed water, ultraviolet light or/and ultrasounds. *Food Control*, *74*, 54–60.

Misra, N. N., Kaur, S., Tiwari, B. K., Kaur, A., Singh, N., & Cullen, P. J. (2015). Atmospheric pressure cold plasma (ACP) treatment of wheat flour. *Food Hydrocolloids, 44*, 115–121.

Misra, N. N., Tiwari, B. K., Raghavarao, K. S. M. S., & Cullen, P. J. (2011). Nonthermal plasma inactivation of food-borne pathogens. *Food Engineering Reviews, 3*(3–4), 159–170.

Moraglio, S. T., Bosco, L., Pogolotti, C., & Tavella, L. (2018). Effect of ozone gas against life stages of Ephestia kuehniella Zeller (Lepidoptera: Pyralidae) in laboratory and a storehouse. *Journal of Stored Products Research, 79*, 132–138.

Mostafavi, H. A., Mirmajlessi, S. M., & Fathollahi, H. (2012). The potential of food irradiation: Benefits and limitations. In: A. H. Eissa (Ed.), *Trends in Vital Food and Control Engineering* (pp. 43–68). doi:10.5772/2354.

Mukisa, I. M., Muyanja, C. M., Byaruhanga, Y. B., Schüller, R. B., Langsrud, T., & Narvhus, J. A. (2012). Gamma irradiation of sorghum flour: Effects on microbial inactivation, amylase activity, fermentability, viscosity and starch granule structure. *Radiation Physics and Chemistry, 81*(3), 345–351.

Muntean, M. V., Marian, O., Barbieru, V., Cătunescu, G. M., Ranta, O., Drocas, I., & Terhes, S. (2016). High pressure processing in food industry–characteristics and applications. *Agriculture and Agricultural Science Procedia, 10*, 377–383.

Murata, H., Yamaguchi, D., Nagai, A., & Shimada, N. (2011). Reduction of deoxynivalenol contaminating corn silage by short-term ultraviolet irradiation: A pilot study. *Journal of Veterinary Medical Science*, 1103250473. http://www.ncbi.nlm.nih.gov/pubmed/1103250473.

Murchie, L. W., Cruz-Romero, M., Kerry, J. P., Linton, M., Patterson, M. F., Smiddy, M., & Kelly, A. L. (2005). High pressure processing of shellfish: A review of microbiological and other quality aspects. *Innovative Food Science and Emerging Technologies, 6*(3), 257–270.

Mylona, K., Kogkaki, E., Sulyok, M., & Magan, N. (2014). Efficacy of gaseous ozone treatment on spore germination, growth and fumonisin production by *Fusarium verticillioides* in vitro and in situ in maize. *Journal of Stored Products Research, 59*, 178–184.

Nemţanu, M. R., Braşoveanu, M., Karaca, G., & Erper, I. (2014). Inactivation effect of electron beam irradiation on fungal load of naturally contaminated maize seeds. *Journal of the Science of Food and Agriculture, 94*(13), 2668–2673.

Park, H., Puligundla, P., & Mok, C. (2020). Cold plasma decontamination of brown rice grains: Impact on biochemical and sensory qualities of their corresponding seedlings and aqueous tea infusions. *LWT, 131*. http://www.ncbi.nlm.nih.gov/pubmed/109508.

Park, Y. B., Guo, J. Y., Rahman, S. M. E., Ahn, J., & Oh, D. H. (2009). Synergistic effect of electrolyzed water and citric acid against *Bacillus cereus* cells and spores on cereal grains. *Journal of Food Science, 74*(4), M185–M189.

Patras, A., Julakanti, S., Yannam, S., Bansode, R. R., Burns, M., & Vergne, M. J. (2017). Effect of UV irradiation on aflatoxin reduction: A cytotoxicity evaluation study using human hepatoma cell line. *Mycotoxin Research, 33*(4), 343–350.

Phan, K. T. K., Phan, H. T., Boonyawan, D., Intipunya, P., Brennan, C. S., Regenstein, J. M., & Phimolsiripol, Y. (2018). Non-thermal plasma for elimination of pesticide residues in mango. *Innovative Food Science and Emerging Technologies, 48*, 164–171.

Piacentini, K. C., Savi, G. D., & Scussel, V. M. (2017). The effect of ozone treatment on species of Fusarium growth in malting barley (*Hordeum vulgare* L.) grains. *Quality Assurance and Safety of Crops and Foods, 9*(4), 383–389.

Piemontese, L., Messia, M. C., Marconi, E., Falasca, L., Zivoli, R., Gambacorta, L., Perrone, G., & Solfrizzo, M. (2018). Effect of gaseous ozone treatments on DON, microbial contaminants and technological parameters of wheat and semolina. *Food Additives and Contaminants: Part A, 35*(4), 761–772.

Popović, V., Fairbanks, N., Pierscianowski, J., Biancaniello, M., Zhou, T., & Koutchma, T. (2018). Feasibility of 3D UV-C treatment to reduce fungal growth and mycotoxin loads on maize and wheat kernels. *Mycotoxin Research, 34*(3), 211–221.

Prudente, A. D. & King, J. M. (2002). Efficacy and safety evaluation of ozonation to degrade aflatoxin in corn. *Journal of Food Science, 67*(8), 2866–2872.

Qi, L., Li, Y., Luo, X., Wang, R., Zheng, R., Wang, L., Li, Y., Yang, D., Fang, W., & Chen, Z. (2016). Detoxification of zearalenone and ochratoxin A by ozone and quality evaluation of ozonised corn. *Food Additives and Contaminants: Part A, 33*(11), 1700–1710.

Raila, A., Lugauskas, A., Steponavicius, D., Railiene, M., Steponaviciene, A., & Zvicevicius, E. (2006). Application of ozone for reduction of mycological infection in wheat grain. *Annals of Agricultural and Environmental Medicine, 13*(2), 287–294.

Ramanan, K. R., Sarumathi, R., & Mahendran, R. (2018). Influence of cold plasma on mortality rate of different life stages of Tribolium castaneum on refined wheat flour. *Journal of Stored Products Research, 77*, 126–134.

Raouche, S., Mauricio-Iglesias, M., Peyron, S., Guillard, V., & Gontard, N. (2011). Combined effect of high pressure treatment and anti-microbial bio-sourced materials on microorganisms' growth in model food during storage. *Innovative Food Science and Emerging Technologies, 12*(4), 426–434.

Sarangapani, C., Misra, N. N., Milosavljevic, V., Bourke, P., O'Regan, F., & Cullen, P. J. (2016). Pesticide degradation in water using atmospheric air cold plasma. *Journal of Water Process Engineering, 9*, 225–232.

Sarangapani, C., O'Toole, G., Cullen, P. J., & Bourke, P. (2017). Atmospheric cold plasma dissipation efficiency of agrochemicals on blueberries. *Innovative Food Science and Emerging Technologies, 44*, 235–241.

Sarrías, J. A., Valero, M., & Salmerón, M. C. (2003). Elimination of *Bacillus cereus* contamination in raw rice by electron beam irradiation. *Food Microbiology, 20*(3), 327–332.

Sastry, S. K., Datta, A. K., & Worobo, R. W. (2000). Ultraviolet light. *Journal of Food Science, 65*, 90–92.

Savi, G. D., Gomes, T., Canever, S. B., Feltrin, A. C., Piacentini, K. C., Scussel, R., Oliveira, D., Machado-de-Ávila, R. A., Cargnin, M., & Angioletto, E. (2020). Application of ozone on rice storage: A mathematical modeling of the ozone spread, effects in the decontamination of filamentous fungi and quality attributes. *Journal of Stored Products Research, 87*. http://www.ncbi.nlm.nih.gov/pubmed/101605.

Savi, G. D., Piacentini, K. C., & Scussel, V. M. (2015). Reduction in residues of deltamethrin and fenitrothion on stored wheat grains by ozone gas. *Journal of Stored Products Research, 61*, 65–69.

Savi, G. D., Piacentini, K. C., Bittencourt, K. O., & Scussel, V. M. (2014). Ozone treatment efficiency on Fusarium graminearum and deoxynivalenol degradation and its effects on whole wheat grains (*Triticum aestivum* L.) quality and germination. *Journal of Stored Products Research, 59*, 245–253.

Savi, G. D., Piacentini, K. C., Bortolotto, T., & Scussel, V. M. (2016). Degradation of bifenthrin and pirimiphos-methyl residues in stored wheat grains (*Triticum aestivum* L.) by ozonation. *Food Chemistry, 203*, 246–251.

Schlüter, O., & Fröhling, A. (2014). Cold plasma for bioefficient food processing. In: C. A. Batt & M.-L. Tortorello (Eds.), *Encyclopedia of Food Microbiology* (pp. 948–953), 2nd edition. London: Academic Press.

Schmidt, M., Zannini, E., & Arendt, E. K. (2018). Recent advances in physical post-harvest treatments for shelf-life extension of cereal crops. *Foods, 7*(4), 45.

Selcuk, M., Oksuz, L., & Basaran, P. (2008). Decontamination of grains and legumes infected with Aspergillus spp. and Penicillum spp. by cold plasma treatment. *Bioresource Technology, 99*(11), 5104–5109.

Serra, M. S., Pulles, M. B., Mayanquer, F. T., Vallejo, M. C., Rosero, M. I., Ortega, J. M., & Naranjo, L. N. (2018). Evaluation of the use of gamma radiation for reduction of aflatoxin b1 in corn (Zea mays) used in the production of feed for broiler chickens. *Journal of Agricultural Chemistry and Environment, 7*(1), 21–33.

Shankar, R., Kaushik, U., & Bhat, S. A. (2014). The emerging technology in the sector of food technology-the non-thermal technology. *International Journal of Innovation and Applied Studies, 6*(4), 941.

Sharareh, M., Maryam, T., Azadeh, K., & Ali, M. S. (2015). Effect of UV irradiation on quantity and structure of aflatoxins using HPLC, GC/MS and FT-IR techniques. *Ludus Vitalis, 11*(1), 64–69.

Shi, H., Ileleji, K., Stroshine, R. L., Keener, K., & Jensen, J. L. (2017). Reduction of aflatoxin in corn by high voltage atmospheric cold plasma. *Food and Bioprocess Technology, 10*(6), 1042–1052.

Silva, M. V. A., Faroni, L. R. A., Martins, M. A., Sousa, A. H., & Bustos-Vanegas, J. D. (2020). CFD simulation of ozone gas flow for controlling Sitophilus zeamais in rice grains. *Journal of Stored Products Research, 88*. http://www.ncbi.nlm.nih.gov/pubmed/101675.

Silva, M. V. A., Faroni, L. R. A., Sousa, A. H., Prates, L. H. F., & Abreu, A. O. (2019). Kinetics of the ozone gas reaction in popcorn kernels. *Journal of Stored Products Research*, *83*, 168–175.

Sousa, A. D., Faroni, L. D. A., Guedes, R. N. C., Tótola, M. R., & Urruchi, W. I. (2008). Ozone as a management alternative against phosphine-resistant insect pests of stored products. *Journal of Stored Products Research*, *44*(4), 379–385.

Sousa, A. H., Faroni, L. R. D. A., Pereira, A. D. M., Cardoso, F. D. S., & Heberle, E. (2006, October). Influence of grain mass temperature on ozone toxicity to Sitophilus zeamais (Coleoptera: Curculionidae). In *9th International Working Conference on Stored Product Protection*, Sao Paolo, Brasil, pp. 15–18.

Subedi, S., Du, L., Prasad, A., Yadav, B., & Roopesh, M. S. (2020). Inactivation of Salmonella and quality changes in wheat flour after pulsed light-emitting diode (LED) treatments. *Food and Bioproducts Processing*, *121*, 166–177.

Subramanyam, B., Xinyi, E., Savoldelli, S., & Sehgal, B. (2017). Efficacy of ozone against *Rhyzopertha dominica* adults in wheat. *Journal of Stored Products Research*, *70*, 53–59.

Suhem, K., Matan, N., Nisoa, M., & Matan, N. (2013). Inhibition of *Aspergillus flavus* on agar media and brown rice cereal bars using cold atmospheric plasma treatment. *International Journal of Food Microbiology*, *161*(2), 107–111.

Sun, C., Ji, J., Wu, S., Sun, C., Pi, F., Zhang, Y., Tang, L., & Sun, X. (2016). Saturated aqueous ozone degradation of deoxynivalenol and its application in contaminated grains. *Food Control*, *69*, 185–190.

Sung, W. C., Hong, M. C., & Chang, T. S. (2008). Effects of storage and gamma irradiation on (*japonica*) waxy rice. *Radiation Physics and Chemistry*, *77*(1), 92–97.

Taze, B. H., Akgun, M. P., Yildiz, S., Kaya, Z., & Unluturk, S. (2019). UV processing and storage of liquid and solid foods: Quality, microbial, enzymatic, nutritional, organoleptic, composition and properties effects. *Innovative Food Processing Technologies*, *2*, 277–305. https://doi.org/10.1016/B978-0-08-100596-5.22938-7.

Thomas-Popo, E., Mendonça, A., Misra, N. N., Little, A., Wan, Z., Moutiq, R., Coleman, S., & Keener, K. (2019). Inactivation of Shiga-toxin-producing *Escherichia coli, Salmonella enterica* and natural microflora on tempered wheat grains by atmospheric cold plasma. *Food Control*, *104*, 231–239.

Tiwari, B. K., Brennan, C. S., Curran, T., Gallagher, E., Cullen, P. J., & O'Donnell, C. P. (2010). Application of ozone in grain processing. *Journal of Cereal Science*, *51*(3), 248–255.

Torlak, E., Akata, I., Erci, F., & Uncu, A. T. (2016). Use of gaseous ozone to reduce aflatoxin b1 and microorganisms in poultry feed. *Journal of Stored Products Research*, *68*, 44–49.

Tremarin, A., Brandão, T. R., & Silva, C. L. (2017). Application of ultraviolet radiation and ultrasound treatments for *Alicyclobacillus acidoterrestris* spores inactivation in apple juice. *LWT*, *78*, 138–142.

Urbain, W. M. (1986). General effects of ionizing radiation on foods. In: *Food Irradiation* (118–123). Orlando: Academic Press, Inc.

Vaz, A. F., Costa, R. M., Coelho, L. C., Oliva, M. L., Santana, L. A., Melo, A. M., & Correia, M. T. (2011). Gamma irradiation as an alternative treatment to abolish allergenicity of lectins in food. *Food Chemistry*, *124*(4), 1289–1295.

Wang, B., Mahoney, N. E., Pan, Z., Khir, R., Wu, B., Ma, H., & Zhao, L. (2016b). Effectiveness of pulsed light treatment for degradation and detoxification of aflatoxin b1 and B2 in rough rice and rice bran. *Food Control*, *59*, 461–467.

Wang, L., Shao, H., Luo, X., Wang, R., Li, Y., Li, Y., Luo, Y., & Chen, Z. (2016a). Effect of ozone treatment on deoxynivalenol and wheat quality. *PLOS ONE*, *11*(1), e0147613.

Warchalewski, J. R. Prądzynska, A., Gralik, J, J., & Nawrot., J. (2000). The effect of gamma and microwave irradiation of wheat grain on development parameters of some stored grain pests. *Food/Nahrung*, *44*(6), 411–414.

Wen, H. W., Chung, H. P., Wang, Y. T., Hsieh, P. C., Lin, I. H., & Chou, F. I. (2008). Efficacy of gamma irradiation for protection against postharvest insect damage and microbial contamination of adlay. *Postharvest Biology and Technology*, *50*(2–3), 208–215.

White, S. D., Murphy, P. T., Leandro, L. F., Bern, C. J., Beattie, S. E., & van Leeuwen, J. H. (2013). Mycoflora of high-moisture maize treated with ozone. *Journal of Stored Products Research*, *55*, 84–89.

Wielogorska, E., Ahmed, Y., Meneely, J., Graham, W. G., Elliott, C. T., & Gilmore, B. F. (2019). A holistic study to understand the detoxification of mycotoxins in maize and impact on its molecular integrity using cold atmospheric plasma treatment. *Food Chemistry*, *301*, 125281.

Woldemariam, H. W., & Emire, S. A. (2019). High pressure processing of foods for microbial and mycotoxins control: Current trends and future prospects. *Cogent Food and Agriculture, 5*(1). http://www.ncbi.nlm. nih.gov/pubmed/1622184.

Wuytack, E. Y., Diels, A. M., & Michiels, C. W. (2002). Bacterial inactivation by high-pressure homogenisation and high hydrostatic pressure. *International Journal of Food Microbiology, 77*(3), 205–212.

Xinyi, E., Li, B., & Subramanyam, B. (2019). Efficacy of ozone against adults and immature stages of phosphine susceptible and resistant strains of Rhyzopertha dominica. *Journal of Stored Products Research, 83*, 110–116.

Yadav, D. N., Anand, T., Sharma, M., & Gupta, R. K. (2014). Microwave technology for disinfestation of cereals and pulses: An overview. *Journal of Food Science and Technology, 51*(12), 3568–3576.

Yang, Y. H., Oh, S. H., Kwon, O. Y., Byun, M. W., Lee, J. W., Park, S. C., & Kim, M. R. (2007). Effect of gamma irradiation on the microbial and physicochemical properties of Ong-keun jook (Korean whole rice porridge). *Journal of the East Asian Society of Dietary Life, 17*(1), 130–135.

Yordanov, D. G., & Angelova, G. V. (2010). High pressure processing for foods preserving. *Biotechnology and Biotechnological Equipment, 24*(3), 1940–1945.

Young, J. C., Zhu, H., & Zhou, T. (2006). Degradation of trichothecene mycotoxins by aqueous ozone. *Food and Chemical Toxicology, 44*(3), 417–424.

Yu, L., Muralidharan, S., Lee, N. A., Lo, R., Stokes, J. R., Fitzgerald, M. A., & Turner, M. S. (2018). The impact of variable high pressure treatments and/or cooking of rice on bacterial populations after storage using culture-independent analysis. *Food Control, 92*, 232–239.

Zahoranová, A., Henselová, M., Hudecová, D., Kalináková, B., Kovácik, D., Medvecká, V.,& Cernák, M. (2014). Study of low-temperature plasma treatment of plant seeds. In: *14th International Symposium on High Pressure Low Temperature Plasma Chemistry–- Book of Contributions* (pp. 563–567).

Zenklusen, M. H., Coronel, M. B., Castro, M. Á., Alzamora, S. M., & González, H. H. L. (2018). Inactivation of *Aspergillus carbonarius* and *Aspergillus flavus* in malting barley by pulsed light and impact on germination capacity and microstructure. *Innovative Food Science and Emerging Technologies, 45*, 161–168.

Zhang, C., Xia, X., Li, B., & Hung, Y. C. (2018). Disinfection efficacy of electrolyzed oxidizing water on brown rice soaking and germination. *Food Control, 89*, 38–45.

Index

For Product Safety Concerns and Information please contact our EU
representative GPSR@taylorandfrancis.com
Taylor & Francis Verlag GmbH, Kaufingerstraße 24, 80331 München, Germany